Location-based Services

Location-based Services
Fundamentals and Operation

Axel Küpper
Ludwig Maximilian University Munich, Germany

John Wiley & Sons, Ltd

Copyright © 2005   John Wiley & Sons Ltd, The Atrium, Southern Gate, Chichester,
West Sussex PO19 8SQ, England

Telephone (+44) 1243 779777

Email (for orders and customer service enquiries): cs-books@wiley.co.uk
Visit our Home Page on www.wiley.com

All Rights Reserved. No part of this publication may be reproduced, stored in a retrieval system or transmitted in any form or by any means, electronic, mechanical, photocopying, recording, scanning or otherwise, except under the terms of the Copyright, Designs and Patents Act 1988 or under the terms of a licence issued by the Copyright Licensing Agency Ltd, 90 Tottenham Court Road, London W1T 4LP, UK, without the permission in writing of the Publisher. Requests to the Publisher should be addressed to the Permissions Department, John Wiley & Sons Ltd, The Atrium, Southern Gate, Chichester, West Sussex PO19 8SQ, England, or emailed to permreq@wiley.co.uk, or faxed to (+44) 1243 770620.

Designations used by companies to distinguish their products are often claimed as trademarks. All brand names and product names used in this book are trade names, service marks, trademarks or registered trademarks of their respective owners. The Publisher is not associated with any product or vendor mentioned in this book.

This publication is designed to provide accurate and authoritative information in regard to the subject matter covered. It is sold on the understanding that the Publisher is not engaged in rendering professional services. If professional advice or other expert assistance is required, the services of a competent professional should be sought.

Other Wiley Editorial Offices

John Wiley & Sons Inc., 111 River Street, Hoboken, NJ 07030, USA

Jossey-Bass, 989 Market Street, San Francisco, CA 94103-1741, USA

Wiley-VCH Verlag GmbH, Boschstr. 12, D-69469 Weinheim, Germany

John Wiley & Sons Australia Ltd, 42 McDougall Street, Milton, Queensland 4064, Australia

John Wiley & Sons (Asia) Pte Ltd, 2 Clementi Loop #02-01, Jin Xing Distripark, Singapore 129809

John Wiley & Sons Canada Ltd, 22 Worcester Road, Etobicoke, Ontario, Canada M9W 1L1

Wiley also publishes its books in a variety of electronic formats. Some content that appears in print may not be available in electronic books.

Library of Congress Cataloging-in-Publication Data

Küpper, Axel.
  Location-based services : fundamentals and operation / Axel Küpper.
    p. cm.
  ISBN-13 978-0-470-09231-6 (HB)
  ISBN-10 0-470-09231-9 (HB)
  1. Mobile computing. I. Title.
  QA76.59.K87 2005
  004.165 – dc22
                                              2005018110

British Library Cataloguing in Publication Data

A catalogue record for this book is available from the British Library

ISBN-13 978-0-470-09231-6 (HB)
ISBN-10 0-470-09231-9 (HB)

Typeset in 10/12pt Times by Laserwords Private Limited, Chennai, India.
Printed and bound in Great Britain by Antony Rowe Ltd, Chippenham, Wiltshire.
This book is printed on acid-free paper responsibly manufactured from sustainable forestry
in which at least two trees are planted for each one used for paper production.

Contents

Preface xi

List of Abbreviations xiii

1 Introduction 1
   1.1 What are Location-based Services? 1
   1.2 Application Scenarios 3
       1.2.1 Business Initiatives 3
       1.2.2 Public Initiatives 7
   1.3 LBS Actors 10
   1.4 Standardization 11
   1.5 Structure of this Book 13

Part I Fundamentals 15

2 What is Location? 17
   2.1 Location Categories 17
   2.2 Spatial Location 19
       2.2.1 Coordinate Systems 19
       2.2.2 Datums 23
       2.2.3 Map Projections 27
   2.3 Conclusion 33

3 Spatial Databases and GIS 35
   3.1 What are Spatial Databases and GIS? 36
   3.2 Geographic versus Spatial Data Models 36
   3.3 Representing Spatial Objects 38
       3.3.1 Raster Mode 38
       3.3.2 Vector Mode 39
       3.3.3 Representing Topological Relationships 41
       3.3.4 Database Approaches for Spatial Objects 43
   3.4 Features and Themes 45
       3.4.1 Conceptual Schemes 47
       3.4.2 Operations 48

		3.4.3	Topological Predicates	49
		3.4.4	Queries	50
	3.5	Algorithms of Computational Geometry		53
	3.6	Geography Markup Language		55
	3.7	Conclusion		59

4 Basics of Wireless Communications — 61
4.1 Signals 61
4.1.1 Modulation 63
4.1.2 Representing Signals in the Frequency Domain 64
4.1.3 Signal Spectrum and Bandwidth 66
4.2 Propagation of Radio Signals 68
4.2.1 The Electromagnetic Spectrum 68
4.2.2 Antennas 70
4.2.3 Speed of Electromagnetic Waves 72
4.2.4 Attenuation 75
4.2.5 Multipath Propagation 76
4.2.6 Doppler Effect 78
4.3 Multiplexing and Multiple Access 79
4.3.1 SDM and SDMA 80
4.3.2 FDM and FDMA 80
4.3.3 TDM and TDMA 81
4.3.4 CDM and CDMA 82
4.4 Conclusion 86

5 Cellular Networks and Location Management — 89
5.1 Overview of Cellular Systems 90
5.2 Principles of Cellular Networks 91
5.2.1 GSM Architecture 93
5.2.2 GPRS Architecture 95
5.2.3 UMTS Architecture 96
5.3 Mobility Management 97
5.4 Common Concepts of Location Management 99
5.4.1 Location Update and Paging 99
5.4.2 Database Concepts 102
5.5 Location Management in CS Networks 103
5.5.1 Identifiers and Addresses 104
5.5.2 Localization and Routing 106
5.5.3 Location Updates 107
5.6 Location Management in PS Networks 109
5.6.1 Localization and Routing 109
5.6.2 Characteristics of CS and PS Traffic 112
5.6.3 Location Updates 115
5.7 Conclusion 119

Part II Positioning   121

6 Fundamentals of Positioning   123
- 6.1 Classification of Positioning Infrastructures 126
  - 6.1.1 Integrated and Stand-alone Infrastructures 127
  - 6.1.2 Network and Terminal-based Positioning 128
  - 6.1.3 Satellites, Cellular, and Indoor Infrastructures 128
- 6.2 Basic Positioning Methods 130
  - 6.2.1 Proximity Sensing 130
  - 6.2.2 Lateration 131
  - 6.2.3 Angulation 138
  - 6.2.4 Dead Reckoning 140
  - 6.2.5 Pattern Matching 142
  - 6.2.6 Hybrid Approaches 142
- 6.3 Range Measurements 143
  - 6.3.1 Time Measurements 144
  - 6.3.2 Received Signal Strength 148
- 6.4 Accuracy and Precision 148
- 6.5 Error Sources 151
- 6.6 Conclusion 154

7 Satellite Positioning   155
- 7.1 Historical Background 155
- 7.2 Orbital Motion of Satellite Systems 157
  - 7.2.1 Satellite Orbits 157
  - 7.2.2 Keplerian Elements 160
- 7.3 Global Positioning System 162
  - 7.3.1 GPS Segments 162
  - 7.3.2 Satellite Constellation 164
  - 7.3.3 Pilot Signals and Spreading Codes 165
  - 7.3.4 Navigation Message 168
  - 7.3.5 GPS Services 170
  - 7.3.6 GPS Positioning 171
  - 7.3.7 GPS Error Budget 174
- 7.4 Differential GPS 177
- 7.5 Galileo 179
- 7.6 Conclusion 183

8 Cellular Positioning   185
- 8.1 Positioning in GSM Networks 185
  - 8.1.1 GSM Air Interface 186
  - 8.1.2 GSM Positioning Components 190
  - 8.1.3 Cell-Id Combined with Timing Advance 192
  - 8.1.4 E-OTD 194
  - 8.1.5 U-TDoA 208

         8.2 Positioning in UMTS Networks 211
              8.2.1 UMTS Air Interfaces 211
              8.2.2 UMTS Positioning Components 217
              8.2.3 Cell-based Methods 218
              8.2.4 OTDoA-IPDL 220
              8.2.5 RIT Measurements in UMTS 221
         8.3 Assisted GPS in GSM and UMTS 225
         8.4 Positioning in other Cellular Systems 229
         8.5 Conclusion 230

9 **Indoor Positioning**   **233**
         9.1 WLAN Positioning 233
              9.1.1 Principles of WLAN Positioning 234
              9.1.2 WLAN Fingerprinting 236
         9.2 RFID Positioning 239
         9.3 Indoor Positioning with GPS 240
         9.4 Non Radiolocation Systems 241
              9.4.1 Infrared-based Systems 241
              9.4.2 Ultrasound-based Systems 243
         9.5 Conclusion 244

Part III LBS Operation   247

10 **Interorganizational LBS Operation**   **249**
         10.1 LBS Supply Chain 250
         10.2 Scenarios of the LBS Supply Chain 252
         10.3 Supplier/Consumer Patterns for Location Dissemination 254
              10.3.1 Querying 254
              10.3.2 Reporting 255
              10.3.3 Evaluation of Querying and Reporting 256
         10.4 Privacy Protection 257
              10.4.1 Characteristics of Privacy Protection for LBSs 258
              10.4.2 Definition of Privacy 259
              10.4.3 Concepts and Mechanisms for Privacy Protection 261
         10.5 Conclusion 269

11 **Architectures and Protocols for Location Services**   **271**
         11.1 GSM and UMTS Location Services 273
              11.1.1 LCS Network Architecture 275
              11.1.2 LCS Functional Entities 276
              11.1.3 Location Procedures 279
              11.1.4 Privacy Options 283
              11.1.5 Outlook to Future Releases 286
         11.2 Enhanced Emergency Services 288
              11.2.1 Wired Enhanced Emergency Services 289
              11.2.2 Wireless Enhanced Emergency Services 290

11.3 Mobile Location Protocol 294
    11.3.1 MLP Structure and Location Service 294
    11.3.2 Example 297
    11.3.3 Outlook to Future Releases 300
11.4 WAP Location Framework 301
    11.4.1 WAP Overview 301
    11.4.2 WAP Location Services 302
11.5 Parlay/OSA 305
11.6 Geopriv 307
    11.6.1 Geopriv Entities 308
    11.6.2 Location Objects 309
    11.6.3 Geopriv Outlook 311
11.7 Conclusion 312

12 LBS Middleware   315
12.1 Conceptual View of an LBS Middleware 316
12.2 Location API for J2ME 318
    12.2.1 Overview of J2ME 318
    12.2.2 Location API for J2ME 320
12.3 OpenGIS Location Services 322
    12.3.1 Information Model 326
    12.3.2 Core Services 328
12.4 Conclusion 335

13 LBS – The Next Generation   337

Bibliography   343

Index   353

Preface

Location-based Services (LBSs) are IT services for providing information that has been created, compiled, selected, or filtered taking into consideration the current locations of the users or those of other persons or mobile objects. They can also appear in conjunction with conventional services like telephony and related added value features, for example, to realize location-based routing of calls or location-based charging. The attractiveness of LBSs results from the fact that their participants do not have to enter location information manually, but that they are automatically pinpointed and tracked. Therefore, the key technology is positioning, for which various methods exist differing from each other in a number of quality parameters and other circumstances. Once location information is derived, it needs to be processed in several ways, including its transformation into the format of another spatial reference system, its correlation with other location information or geographic content, the generation of maps, or the calculation of navigation instructions, to name only a few. Usually, these tasks are not carried out on a single mobile device or PC but are adopted by many actors involved in the operation of the respective LBS. Thus, the operation of LBSs is an interorganizational matter for which various actors like network operators, service and content providers have to cooperate on a distributed infrastructure. This imposes a number of challenges like the real-time exchange of location information between these actors or the saving of privacy interests of individuals the location information refers to.

This book explains the fundamentals and operation of LBSs and gives a thorough introduction to the key technologies and organizational procedures mentioned earlier. It is dedicated to all those who want to obtain an understanding of the complex systems, methods, and standards LBSs rely on. For professionals in the areas of network operators, providers, and vendors, the book provides a compact and comprehensive overview of positioning methods, location protocols, and service platforms, which otherwise so far have only been covered in a jungle of many interrelated standardization documents and recommendations. Nevertheless, it has not failed to name these documents and reference them at the appropriate text passages for further study. The book also focuses on lecturers and students of computer science, business and media information systems, and electrical engineering. For this target group, it covers the fundamentals of spatial reference systems, spatial databases and GISs, wireless communications and cellular networks, as well as positioning, and shows how these topics are utilized for and contribute to the operation of LBSs. Furthermore, application developers may profit from the overview of location protocols, interfaces, languages, and service platforms as well as from related examples demonstrating their usage.

The motivation for the "LBS book project" originated from the lectures on "Mobile Communications" and "Location-based Services" I gave at the Institute for Informatics of the Ludwig Maximilian University Munich (Elev 515 m N 48° 08' 58" E 11° 35' 40") and from research in this area. I received massive support from superiors and colleagues, without which it would not have been possible to write this book. I especially thank Prof. Dr. Claudia Linnhoff-Popien and her Mobile and Distributed Systems Group. Claudia always inspired and encouraged me in writing this book, gave me helpful hints for organizing the preparation and compilation of the manuscript, and provided valuable comments on many parts of it. I also wish to acknowledge my colleagues for the many rich discussions in the areas of location-based and context-aware services, which significantly contributed to the topics selected for this book, and, in the last phase of writing the manuscript, for taking up tasks that I did not manage in due time. I am especially very grateful to Georg Treu. Georg not only managed the proofreading of the manuscript and gave me valuable input for improvement, but also inspired many fruitful discussions for LBS research, which we often had after a long working day in the bar "Leib und Seele" (Elev 533 m N 48° 08' 43" E 11° 35' 36"). Furthermore, I wish to thank the Munich Network Management (MNM) team guided by Prof. Dr. Heinz-Gerd Hegering for its support and cooperation and for providing an inspiring working environment. I am also grateful for the comments on the manuscript I received from our students, who are dealing with LBSs in practical courses and master theses. Last but not the least, I would like to stress that writing a book is a long and cumbersome process that needs some guidance. I am very grateful to the publishing team of John Wiley & Sons, Ltd for providing me this guidance, their support during the entire writing process, and for their patience.

The book is accompanied by a dedicated web page that provides the reader with links, comments, and latest news in the area of LBSs as well as with an errata. It can be accessed at the following URLs:

> http://www.location-based-services.org/
> http://www.mobile.ifi.lmu.de/lbs/

Finally, I appreciate any comments, suggestions, and feedback about the book. Please use the email address given below to contact me.

Axel Küpper
Munich
mail@axel-kuepper.de

List of Abbreviations

2G, 3G, 4G	Second, third, and fourth generation of mobile networks
3GPP	3rd Generation Partnership Project
3GPP2	3rd Generation Partnership Project 2
A-GPS	Assisted GPS
ADT	Abstract Data Type
A-FLT	Advanced Forward Link Trilateration
ALI	Automatic Location Identification
AMPS	Advanced Mobile Phone System
ANI	Automatic Number Identification
AoA	Angle of Arrival
API	Application Program Interface
ASK	Amplitude-shift Keying
AT	Absolute Time
ATD	Absolute Time Difference
AuC	Authentication Center
AWT	Abstract Window Toolkit
BCCH	Broadcast Control Channel
BCH	Broadcast Channel
BOC	Binary Offset Carrier
BSA	Basic Service Area
BSC	Base Station Controller
BSS	Basic Service Set
BSS	Base Station Subsystem
BSSAP	Base Station Subsystem Application Part
BSSAP-LE	Base Station Subsystem Application Part LCS Extension
BSSGP	Base Station System GPRS Protocol
BSSI	Basic Service Set Identifier
BTS	Base Transceiver Station
CA	cell allocation
CAD/CAM	Computer Aided Design and Manufacturing
CAS	call-associated signaling
CBC	Cell Broadcast Center
CC	Country Code
CDC	Connected Device Configuration
CDM	Code Division Multiplexing
CDMA	Code Division Multiple Access
CGALIES	Coordination Group on Access to Location Information for Emergency Services
CGI	Cell Global Identity

CGI	Common Gateway Interface
CI	Cell Identifier
CIA	Confidentiality – Integrity – Authentication
CLDC	Connected Limited Device Configuration
CORBA	Common Object Request Broker Architecture
CoO	Cell of Origin
CPP	Common Profile for Presence
CPICH	Common Pilot Channel
CS	Commercial Service
DBMS	Database Management System
D-GPS	Differential GPS
DoA	Direction of Arrival
DoD	Department of Defense
DOP	Dilution of Precision
DRNC	Drifting Radio Network Controller
DTD	Document Type Definition
E-112	Enhanced 112
E-911	Enhanced 911
ECEF	Earth Centered Earth Fixed
EDGE	Enhanced Data Rates for GSM Evolution
E-FLT	Enhanced Forward Link Trilateration
E-OTD	Enhanced Observed Time Difference
EHF	Extremely High Frequency
EIR	Equipment Identity Register
ELF	Extremely Low Frequency
ELIS	Emergency Location Immediate Request
ELRS	Emergency Location Reporting Service
EPSG	European Petroleum Survey Group
ESME	Emergency Services Message Entity
ESNE	Emergency Services Network Entity
ESRD	Emergency Service Routing Digits
ESRK	Emergency Services Routing Key
ESS	Extended Service Set
ETSI	European Telecommunication Standards Institute
FCC	Federal Communications Commission
FCD	Floating Car Data
FDD	Frequency Division Duplexing
FDM	Frequency Division Multiplexing
FDMA	Frequency Division Multiple Access
FGDC	Federal Geographic Data Committee
FLT	Forward Link Trilateration
FSK	Frequency-shift Keying
GCS	Ground Control Segment
GDOP	Geometric Dilution of Precision
GEO	Geostationary Earth Orbit
GERAN	GSM/EDGE Radio Access Network
GGSN	Gateway GPRS Support Node
GIS	Geographic Information System
GML	Geography Markup Language
GMLC	Gateway Mobile Location Center

LIST OF ABBREVIATIONS

GMSC	Gateway Mobile Switching Center
GMSK	Gaussian Minimum Shift Keying
GMT	Greenwich Mean Time
GPS	Global Positioning System
GPRS	General Packet Radio Service
GSM	Global System for Mobile Communications
GSN	GPRS Support Node
GTD	Geometric Time Difference
GTP	GPRS Tunneling Protocol
HDOP	Horizontal Dilution of Precision
HEO	Highly Elliptical Orbit
HLR	Home Location Register
HOW	handover word
Hz	Hertz
IDL	Interface Definition Language
IDS	Integrity Determination System
IETF	Internet Engineering Task Force
IEEE	Institute of Electrical and Electronics Engineers
IMS	IP Multimedia Subsystem
IMSI	International Mobile Subscriber Identity
IMT-2000	International Mobile Telephony System 2000
ISDN	Integrated Services Digital Network
ISO	International Organization for Standardization
ITU	International Telecommunications Union
J2EE	Java 2 Enterprise Edition
J2ME	Java 2 Micro Edition
J2SE	Java 2 Standard Edition
JVM	Java Virtual Machine
KVM	Kilobyte Virtual Machine
LA	Location Area
LAC	Location Area Code
LAI	Location Area Identifier
LAN	Local Area Network
LBS	Location-based Services
LCAF	Location Client Authorization Function
LCCF	Location Client Control Function
LCCTF	Location Client Coordinate Transformation Function
LCF	Location Client Function
LCS	Location Service
LCZTF	Location Client Zone Transformation Function
LEO	Low Earth Orbit
LER	Location Enabler Release
LIF	Location Interoperability Forum
LLP	LMU LCS Protocol
LMU	Location Measurement Unit
LoS	Line of Sight
LSAF	Location Subscriber Authorization Function
LSBcF	Location System Broadcast Function
LSBF	Location System Billing Function
LSCF	Location System Control Function

LSOF	Location Systems Operation Function
LSPF	Location Subscriber Privacy Function
LSTF	Location Subscriber Translation Function
MAC	Medium Access Layer
MAP	Mobile Application Part
MCC	Mobile Country Code
MEO	Medium Earth Orbit
MIDP	Mobile Information Device Profile
MLP	Mobile Location Protocol
MMS	Multimedia Messaging Service
MNC	Mobile Network Code
MS	Mobile Station
MSC	Mobile Switching Center
MSISDN	Mobile Subscriber ISDN Number
MSN	Mobile Subscriber Number
MSRN	Mobile Station Roaming Number
NAD-27	North American Datum of 1927
NAICS	National American Industry Classification Scheme
NAVSTAR	Navigation Satellite Timing and Ranging System
NCAS	non-call-associated signaling
NDC	National Destination Code
NENA	National Emergency Number Association
NGS	National Geodetic Survey
NIMA	National Imagery and Mapping Agency
NMT	Nordic Mobile Telephone system
NNSS	Nearest Neighbor in Signal Space
OBU	On-Board Unit
OCXO	Oven-Controlled Crystal Oscillator
OGC	Open Geospatial Consortium
OMA	Open Mobile Alliance
OMC	Operation and Maintenance Center
OMS	Operation and Management Subsystem
OpenLS	Open GIS Location Services
OS	Open Service
OSA	Open Service Access
OTA	Over the Air
OTD	Observed Time Difference
OVSF	Orthogonal Variable spreading Factor
P3P	Platform for Privacy Preferences
PAN	Personal Area Network
PAP	Push Access Protocol
PC	Personal Computer
P-CCPC	Primary Common Control Physical Channel
PCE	Privacy Checking Entity
PCF	Positioning Calculation Function
PCH	Paging Channel
PCMCIA	Personal Computer Memory Card International Association
PCP	Privacy Checking Protocol
PDC	Personal Digital Cellular system
PDE	Positioning Determining Entity

LIST OF ABBREVIATIONS

PDN	Packet Data Network
PDOP	Position Dilution of Precision
PDP	Packet Data Protocol
PDTCH	Packet Data Traffic Channel
PDU	Packet Data Unit
PIDF	Presence Information Data Format
PIDF-LO	PIDF-Location Object
PLMN	Public Land Mobile Network
PMD	Pseudonym Mediation Device
PMM	Packet Mobility Management
POI	Point of Interest
POI	Privacy Override Indicator
POTS	Plain Old Telephone System
PPR	Privacy Profile Register
PPS	Precise Positioning Service
PRCD	Positioning Radio Coordination Function
PRN	pseudorandom noise
PRRM	Positioning Radio Resource Management
PRS	Public Regulated Service
PSAP	Public Safety Answering Point
PS	packet-switched
PSK	Phase-shift Keying
PSMF	Positioning Signal Measurement Function
P-TMSI	Packet TMSI
QAM	Quadrature Amplitude Modulation
QoS	Quality of Service
QPSK	Quadrature Phase-shift Keying
OTDoA-IPDL	Observed Time Difference of Arrival with Idle Period Downlink
RA	Routing Area
RAI	Routing Area Identifier
RANAP	Radio Access Network Application Part
RFID	Radio Frequency Identification
RIT	Radio Interface Timing
RLP	Roaming Location Protocol
RMI	Remote Method Invocation
RM-ODP	Reference Model for Open Distributed Processing
RMS	root mean square
RNC	Radio Network Controller
RNS	Radio Network Subsystem
RRC	Radio Resource Control
RRLP	Radio Resource Link Protocol
RSS	received signal strength
RTD	Real Time Difference
RTT	round trip time
SA	Selective Availability
SAR	Search and Rescue
SAS	Standalone SMLC
SCF	Service Capability Feature
SDH	Synchronous Digital Hierarchy
SDM	Space Division Multiplexing

SDMA	Space Division Multiple Access
SFN	System Frame Number
SGSN	Serving GPRS Support Node
SIC	Standard Industrial Classification
SIM	Subscriber Identity Module
SIP	Session Initiation Protocol
SLIR	Standard Location Immediate Request
SLPP	Subscriber LCS Privacy Profile
SLRS	Standard Location Reporting Service
SMLC	Serving Mobile Location Center
SMS	Short Message Service
SMS-C	SMS Center
SMSS	Mobile Switching and Management Subsystem
SN	Subscriber Number
SNDCP	Subnetwork Dependent Convergence Protocol
SNR	signal-to-noise ratio
SOAP	Simple Object Access Protocol
SONET	Synchronous Optical Network
SoL	Safety-of-Life Service
SPS	Standard Positioning Service
SQL	Structured Query Language
SRNC	Serving Radio Network Controller
SS7	Signaling System No. 7
SSL	Secure Socket Layer
SUPL	Secure User Plane
SV	Space Vehicle
SVN	Space Vehicle Number
TCH	Traffic Channel
TCXO	Temperature-Compensated Crystal Oscillator
TDM	Time Division Duplex
TDMA	Time Division Multiple Access
TDOP	Time Dilution of Precision
TLM	telemetry
TLRS	Triggered Location Reporting Service
TLS	Transport Layer Security
TTFF	Time To First Fix
TMSI	Temporary Mobile Subscriber Identity
TOW	time-of-week
UE	User Equipment
UERE	User Equivalent Range Error
UML	Unified Modeling Language
UMTS	Universal Mobile Telecommunications System
URA	UTRAN Registration Area
UTC	Coordinates Universal Time
UTM	Universal Transverse Mercator
UTRAN	UMTS Terrestrial Radio Access Network
VDOP	Vertical Dilution of Precision
VLR	Visitor Location Register
VoIP	Voice over IP
VPN	Virtual Private Network

LIST OF ABBREVIATIONS

W3C	World Wide Web Consortium
WAP	Wireless Application Protocol
WBMXL	Wireless Binary XML
WCDMA	Wideband CDMA
WGS-84	World Geodetic System of 1984
WIPS	Wireless Indoor Positioning System
WLAN	Wireless Local Area Network
WSDL	Web Services Definition Language
WSP	Wireless Session Protocol
WTP	Wireless Transaction Protocol
WWW	World Wide Web
XML	eXtended Markup Language

1

Introduction

1.1 What are Location-based Services?

Although *Location-based Services* (LBSs) have been an issue in the field of mobile communications for many years, there exists neither a common definition nor a common terminology for them. For example, the terms *location-based service*, *location-aware service*, *location-related service*, and *location service* are often interchangeably used. One reason for this dilemma might lie in the fact that the character and appearance of such services have been determined by different communities, especially the telecommunications sector and the ubiquitous computing area.

The *GSM Association*, which is a consortium of 600 GSM network operators, simply defines LBSs as services that use the location of the *target* for adding value to the service, where the target is the "entity" to be located (and this entity is not necessarily also the user of the service). This abstract definition raises of course the question of what a concrete added value is. The GSM Association (2003) presents three examples where the added value is given by the filtering of information (for example, selecting nearby points of interest), showing the location of a target on a map, or automatically activating the service when a target enters or leaves a predefined location. Another similarly abstract definition of LBSs is given by the *3rd Generation Partnership Project* (3GPP), which is an international federation of many national standardization authorities aiming at providing the specification for GSM and UMTS: an LBS is a service provided by a service provider that utilizes the available location information of the terminal (3GPP TS 23.271). Following these definitions, most of today's LBSs are realized as data or messaging services, for example, based on the *Wireless Application Protocol* (WAP), the *General Packet Radio Service* (GPRS), or the *Short Message Service* (SMS). However, they can also appear in conjunction with traditional telephony services or future interactive multimedia services, as well as with *supplementary services* like call forwarding, freephone, charging, and televoting. Another application area is *location based* or *selective routing*, where telephone calls or data are routed depending on the subscriber's current location. This is used, for example, to route emergency calls to an emergency response agency that is close to the emergency caller's current location.

Location-based Services: Fundamentals and Operation Axel Küpper
© 2005 John Wiley & Sons, Ltd

3GPP strictly distinguishes between LBSs and *location services*. The latter exclusively deals with the localization of target persons and objects and with making the resulting location data available to external actors. A location service does not imply the processing of location data in the sense of filtering or selecting location-dependent information or performing other high-level actions (as an LBS does); it is only responsible for the generation and delivery of location data. However, with this function, location services essentially contribute to the operation of LBSs and can be regarded as an important subservice of them. Without a location service, an LBS user would have to enter location data manually, which would be a cumbersome procedure, especially when doing it by using a mobile device with limited user interface while on the move. Thus, LBSs and location services mostly appear in conjunction.

In research, LBSs are often considered to be a special subset of the so-called *context-aware services* (from where the term *location-aware service* has its origin). Generally, context-aware services are defined to be services that automatically adapt their behavior, for example, filtering or presenting information, to one or several parameters reflecting the context of a target. These parameters are termed *context information*. The set of potential context information is broadly categorized and, as depicted in Figure 1.1, may be subdivided into personal, technical, spatial, social, and physical contexts. It can be further classified as *primary* and *secondary contexts*. Primary context comprises any kind of raw data that can be selected from sensors, like light sensors, biosensors, microphones, accelerometers, location sensors, and so on (Schmidt and van Laerhoven 2001). This raw data may be refined by combination, deduction, or filtering in order to derive high-level context information, which is termed *secondary context* and which is more appropriate for processing by a given context-aware service.

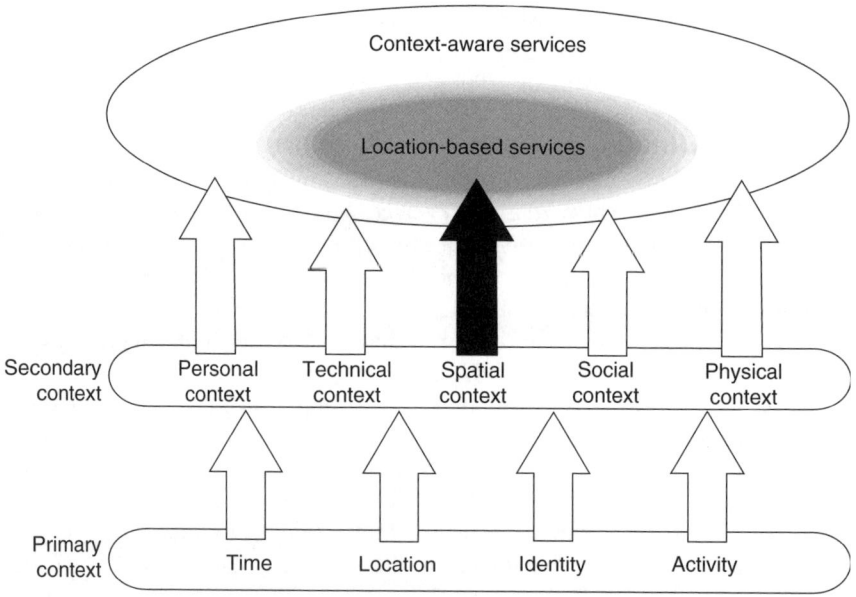

Figure 1.1 Context-aware and location-based services.

INTRODUCTION 3

As can be derived from Figure 1.1, LBSs are always context-aware services, because location is one special case of context information. In many cases, the concept of primary and secondary contexts can also be applied to LBSs, for example, when location data from different targets are related or the history of location data is analyzed to obtain high-level information such as the distance between targets or their velocity and direction of motion. Therefore, there is no sharp distinction between LBSs and context-aware services. And, in many cases context information that is relevant to a service, for example, information such as temperature, pollution, or audibility are closely related to the location of the target to be considered. Hence, its location must be obtained first before gathering other context information. A detailed introduction to the ideas and fundamentals of context-awareness is given in Dey and Abowd (1999) and Schmidt et al. (1999).

LBSs can be classified into *reactive* and *proactive LBSs*. A reactive LBS is always explicitly activated by the user. The interaction between LBS and user is roughly as follows: the user first invokes the service and establishes a service session, either via a mobile device or a desktop PC. He then requests for certain functions or information, whereupon the service gathers location data (either of himself or of another target person), processes it, and returns the location-dependent result to the user, for example, a list of nearby restaurants. This request/response cycle may be repeated several times before the session is finally terminated. Thus, a reactive LBS is characterized by a synchronous interaction pattern between user and service. Proactive LBSs, on the other hand, are automatically initialized as soon as a predefined location event occurs, for example, if the user enters, approaches, or leaves a certain point of interest or if he approaches, meets, or leaves another target. As an example, consider an electronic tourist guide that notifies tourists via SMS as soon as they approach a landmark. Thus, proactive services are not explicitly requested by the user, but the interaction between them happens asynchronously. In contrast to proactive LBSs, where the user is only located once, proactive LBSs require to permanently track him in order to detect location events.

In order to make the idea behind LBSs more clear, the following section presents a broad range of application scenarios.

1.2 Application Scenarios

The scenarios presented here are subdivided into economical initiatives, which are carried out by operators and providers to raise the attractiveness of their networks and data services and thus to increase the average revenue per user, and public initiatives, which are introduced by governments for supporting or fulfilling sovereign or administrative tasks.

1.2.1 Business Initiatives

The main motivation for offering LBSs is to gain revenue by increasing the average airtime per user, selling location information to third parties, and offering services tailored to the special needs of mobile users. A provider may either realize and offer LBSs on its own initiative or it may enter into business relationships with other actors, for example, from trade and commerce or the automobile industry, and realize and offer services on behalf of them. These relationships are defined by more or less complex business models, which are the subject of intensive research in the areas of business sciences and consulting (see, for example, (Agrawal and Agrawal 2003)). This section presents some classical examples of LBSs, which result from such business initiatives. However, actually there is a much broader range of LBSs, which cannot be completely covered here.

1.2.1.1 Enquiry and Information Services

The simplest and so far the most widespread type of LBSs are *enquiry* and *information services*, which provide the mobile user with nearby *points of interest* such as restaurants, automated teller machines, or filling stations. Upon request, the user is either automatically located by the mobile network or, if appropriate positioning technology is missing, he must explicitly enter his current location. Furthermore, he must specify the points of interest, for example, whether he would like to receive a list of all nearby restaurants or filling stations, and the desired maximum distance between his current position and the points of interest. The request is then passed to a service provider, which assembles a list of appropriate points of interest and returns it to the user. Thus, this type of service is basically an extension of the Yellow Pages for showing only entries of local relevance. In today's networks, these services are usually accessed over SMS, WAP, or I-mode. In some cases, they are combined with navigation facilities for guiding the user to the points of interest of his choice along the shortest route.

1.2.1.2 Community Services

Community services enable users that share common interests to join together in a closed user group (*community*) and to interact among each other via chat, whiteboards, or messaging services. In the recent years, the WWW has created various occurrences of these services supporting a broad and heterogeneous range of communities in such areas as cooking, traveling, family, computer, and eroticism. What is common to most of them is that users have to fix their nicknames, age, gender, domiciles, and other personal data in profiles that are matched against each other in order to support the mutual detection of users with similar interests. A very popular category is the so-called *Instant Messaging*, where users can assemble a buddy list of their favorite acquaintances. If a user is registered with the service, he can observe which of his buddies is also on line and can immediately enter into contact with him. This feature is commonly referred to as the *presence* feature.

Like for many mobile services, a breakthrough in mobile community services has not yet taken place, the reason for which is certainly the lack of convenient user interfaces for mobile devices. However, their extension with location-based features represents an obvious way to make them popular in mobile networks too. Typical functions are to show a user the current location of his buddies or to alert him if one of his buddies stays close by. Location-based community services are much more sophisticated and more difficult to realize than, for example, the enquiry and information services presented earlier. They require a permanent tracking of their members and sophisticated mechanisms for saving their privacy. The different characteristics of mobile community services and the chances they provide for users as well as for mobile operators are discussed in Fremuth et al. (2003).

INTRODUCTION

1.2.1.3 Traffic Telematics

The area of traffic telematics aims to support car drivers with a set of manifold services relating to their vehicles. It includes but is not limited to navigation, the automatic configuration of appliances and added features within the vehicle, diagnostics of malfunctions, or the dissemination of warning messages. The most widespread application so far has been navigation, which is enabled by *On-Board Units* (OBU) installed in the cars. On the basis of the current location, which is derived via GPS (Global Positioning System), the OBU guides the driver to the desired target by giving either vocal instructions or displaying the route graphically. The guidance is based on map material that is loaded from a local CD/DVD-ROM inside the OBU. More sophisticated versions of these systems are equipped with GSM/GPRS units and can thus keep the driver up-to-date with information from a remote server, including information on, for example, the latest traffic jams, weather conditions, and road works. On the basis of this information, it becomes possible to recommend alternative routes. The navigation services can be combined with several useful features. For example, Scharf and Bayer (2002) present a system that includes a number of services around parking lots, ranging from registering and tariffing parking lots, guiding the driver to the reserved lot, and the exchange of parking lots among drivers.

A hot topic in research is the wireless intervehicle communication, which relies on short-range communication technologies like WLAN or Bluetooth and which enables the exchange of warning messages, local traffic situations, or the position of filling stations in an ad hoc manner. The content of the messages originates from different sources, above all from sensor technology inside the vehicles. Data delivered by these sensors is subsumed under the term *floating car data* and comprises such parameters as the vehicle's speed, direction, and position. To derive high-level information, for example, like the aforementioned traffic situation, the floating car data must be refined in several steps and maybe even combined with the data received from other vehicles, before disseminating it to nearby vehicles. Intervehicle communication is a complex matter, which poses a number of strong requirements on the systems' reliability, security mechanisms, routing protocols, and positioning technologies. It is usually not considered to be a classical LBS, but adopts a number of similar technologies and mechanisms. The interested reader can get additional information from the works of Enkelmann (2003), Lochert et al. (2003), and Kosch et al. (2002).

1.2.1.4 Fleet Management and Logistics

While traffic telematics is concerned with supporting single, autonomous vehicles, fleet management deals with the control and coordination of entire fleets of vehicles by a central office. Typical target groups are freight services, public transportation, and emergency services. Location-based systems for fleet management are able to request the position of vehicles, display it on a map, determine the distance between different vehicles of a fleet as well as between a vehicle and its destination, and so on. On the basis of this information, the central office can dynamically delegate new orders and predict the arrival time of deliveries at the destination.

LBSs can also serve to support each form of logistics. As stated in Jakobs et al. (2001), the distribution of goods is no longer about moving cargo from A to B, but a complex process including sorting, planning, and consolidation of goods along a supply chain, which is usually composed of a sequence of different means of transportation. With the technologies of LBSs, it becomes possible to support faster transportation, different transportation modes, and the development of fallback scenarios in case of failures.

1.2.1.5 Mobile Marketing

Mobile marketing is a new kind of sales approach that helps manufacturers and service agencies to promote their products and services by interacting with consumers through their mobile devices. The contact with a consumer is usually established by using technologies such as SMS, *Multimedia Messaging Service* (MMS), or WAP, where the first one is the most popular "media channel" till date. Unlike conventional campaigns in television, newspapers, and journals, mobile marketing enables to select the target group of a certain product or service very accurately by evaluating the user profiles that reflect a customer's interests in products and services and possibly even his buying patterns in the past. In addition, it enables a high degree of interactivity between consumers and the agencies carrying out a campaign.

The consequent step forward is to make mobile marketing location-based in that the consumer is provided with information about products and services of local relevance. For example, a consumer might be informed about the special offers of a shop by sending a message to his mobile device just at the moment he is passing the shop. This message might contain information about products or services available on the spot as well as additional benefits like coupons or allowances. However, it must be stressed that mobile marketing in general and the location-based version in particular will only gain acceptance if consumers do not feel harassed by incoming advertisement messages, which turns out to be a very serious problem with the E-mail service in the Internet. Advertisement messages should be delivered to a consumer only if they are in accordance with his interest profile, and it must be possible to conveniently cancel a subscription either permanently or temporarily. For advertisement messages that are delivered depending on the user's location, it must also be guaranteed that they do not distract the user while carrying out activities that need concentration such as car driving, for which appropriate mechanisms, either in the network or in the mobile devices, are yet to be developed. Further information about location-based marketing can be found in Ververidis and Polyzos (2002).

1.2.1.6 Mobile Gaming

In recent years, mobile devices have developed from rudimentary mobile phones to sophisticated mobile computers with high-resolution multicolor displays, high-speed processors, and several megabytes of storage. Hence, devices with these capabilities can not only be used for making phone calls, but they are also very attractive to be used as mobile consoles for playing games, which are either preinstalled or which can be dynamically loaded over the air from a service provider against a fee. A very popular application is the interactive games that allow remote users to share the same session and to enter into a real-time competition, for example, in a football game or a race. The games are accessed via the mobile device and the data needed to organize and maintain a distributed game session are communicated over a cellular network.

Another occurrence is the location-based mobile games, where the virtual and real worlds merge and the current locations of users become an essential aspect of the play.

INTRODUCTION

An example is *Can you see me now?*, where on-line players have to catch professional players who run through real city streets, and the on-line players are equipped with a mobile device for tracking the runner and communicating with the game server (Benford et al. 2003). In Japan, another popular game is *Mogi*, where players have to cruise the streets of a city to collect virtually hidden treasures. The mobile device indicates the hiding places of treasures on a map, and the players have to move to this place in the real world as fast as possible before the treasure is collected by another player. For a comprehensive and thorough introduction to the business and technical aspects of the mobile entertainment industry, see the publications of the EU project mGain (2003).

1.2.1.7 Value-added Services

Value-added or *supplementary services* are terms originating from the traditional telecommunications domain and refer to enhancements of basic services, especially speech telephony (see (Magedanz and Popescu–Zeletin 1996)). More prominent examples are *call forwarding*, *freephone*, *split charging*, and *televoting*. Actually, positioning capabilities of a network can also be seen as a value-added service as they are, in many cases, offered as enhancements to other services. However, they can also serve to enable a more intelligent and flexible use of conventional supplementary services. For example, location-based call forwarding (or selective routing) means that incoming calls addressed to a user's mobile device are automatically rerouted to a nearby fixed terminal. Some operators have also implemented location-dependent charging and allow their customers to determine a so-called *homezone*, that is, a certain geographic area of some size from where they can make calls at special tariffs or even free of charge. Location-based supplementary services like these are predominantly based on proprietary solutions developed by the operators for marketing purposes. Unlike any other (supplementary) services in the telecommunications sector, they are not subject to standardization currently.

1.2.2 Public Initiatives

In many countries of the world, governments and authorities have recognized the potentials of new communication systems such as the Internet and use them for supporting and fulfilling sovereign and administrative tasks. Obviously, the new technical possibilities for tracking and locating people by mobile communication systems have inspired many governments to think of new services for various national purposes, ranging from fighting against crime and emergency services to collecting tolls. While some of these initiatives go along with legal mandates that require network operators to implement the required functions, others may be realized through the so-called *public-private partnerships*, that is, contracts between the government and operators, which have been negotiated according to the rules of free market economy. Although these initiatives do not fall into the category of conventional LBSs (and, in most cases, are hardly experienced by the citizens as such), the underlying mechanisms are nevertheless the same as those used for LBSs. Therefore, public initiatives in the aforementioned areas turned out to be very important driving forces for a broad commercial introduction of LBSs. In the following section, the most important examples of national activities in this field are reflected.

1.2.2.1 Enhanced Emergency Services

Emergency services represent a very obvious and reasonable application area where the deployment of location technology makes sense. In many cases, persons calling a so-called *emergency response agency* (e.g., police, fire) are unable to communicate their current location or they simply do not know it. While in many cases the address of a caller can be easily determined when the emergency call is made over the fixed telephone network, rescue workers are faced with serious problems when locating callers from mobile networks. This is worse, for example, in the United States, where about 50% of all emergency calls are increasingly made from mobile phones.

To cope with this problem, the *Federal Communications Commission* (FCC) in the United States passed a mandate in 1996 that obligated mobile operators to locate the callers of emergency services and to deliver their geographic position to the so-called *Public Safety Answering Point* (PSAP), the office where emergency calls arrive. According to the emergency number 911 in the United States, this mandate is known as *Enhanced 911* (E-911). The mandate also defines an accuracy standard that goes far beyond what is possible with the standard mechanisms of location management in cellular networks and therefore requires enhancement of existing network infrastructures. To give operators enough time for these enhancements, the FCC ruled that the introduction of E-911 be carried out in two phases:

- **Phase I.** In the first phase of E-911, it was required to derive a caller's location from the coordinates of the serving cell site from where the emergency call has been made. Typical 2G networks like GSM were designed for cell radii of several tens of kilometers (although in urban areas typical cell radii were in the range hundreds of meters), and hence the accuracy of this location technology was rather poor. In addition, the operators were obligated to forward the caller's telephone number to the PSAP, a feature called *Automatic Number Identification* (ANI), thereby allowing the PSAP to call back if the call is unintentionally interrupted. Phase I was scheduled to be completed by April 1998.

- **Phase II.** The second phase ruled that the operators be able to locate a caller accurately within 50 to 100 m in 67% and 150 to 300 m in 95% of all emergency calls, depending on the location technology used. As this accuracy standard was hard to meet by the cell-based approach mentioned earlier, complex enhancements on the network infrastructure became necessary. The E-911 mandate required the operators to begin the network enhancements not later than October 2001 and to finish them by December 2005. By this point in time, network operators must meet the aforementioned accuracy standard in their whole coverage area.

While all operators in the United States had no problem in meeting the deadline of Phase I, they were and still are faced with serious problems with the realization of Phase II, and that is why the FCC changed the conditions of Phase II several times and postponed the deadline. One reason for the problem was that many operators decided at a very early stage to implement premature positioning technology. When making the first field trials, it soon became apparent that especially in rural areas the prescribed accuracy standard was missing. As a result, one operator after the other decided to switch to alternative positioning technologies, thereby missing the initial deadline of Phase II, which ruled the

INTRODUCTION

starting of network enhancements by October 2001. However, it was common consensus in the meantime that all participants, not only the operators but also the FCC, had dramatically underestimated the complexity of positioning technology, which was very unproven when the FCC issued the E-911 mandate in 1996.

The introduction of enhanced emergency services is also an issue in many other countries. The EU launched activities for *Enhanced 112* (E-112) in 2000 and founded the *Coordination Group on Access to Location Information for Emergency Services* (CGALIES) (see (CGALIES 2002)). The intention of this group is to investigate and prepare for the introduction of location-based emergency services in all countries of the EU and to coordinate investments and implementation details among all actors involved in this initiative. However, the commitments for operators are less restrictive than in the United States. For the time being, the EU has merely issued a recommendation defining several features of E-112, but has neither fixed a schedule for an introduction nor defined any accuracy standards so far. Rather, the operators are required to merely locate emergency callers as accurately as possible. In Asia, the situation is very different. While many countries have not passed any regulatory arrangements so far, in Japan and Korea the counterpart of the American E-911 is expected to go into operation around 2005 or is even already available in some regions.

1.2.2.2 Toll Systems

In many countries, drivers have to pay tolls for using roads, for example, highways and streets in cities as well as tunnels and bridges. It is common practice, for many hundreds or even thousands of years, as long as people have to pay tolls anyhow, that the access to chargeable roads is controlled by the local staff on the spot and that the toll has to be paid to this staff either directly before or after passing the road. Many countries are still practicing this approach. Alternatively, in some countries the drivers have to buy vignettes that are valid for a certain duration of time, usually a month or a year. While the former approach causes horrible congestions on the roads, which is bad as the traffic volume is permanently increasing, the vignette suffers from the fact that a driver cannot be charged depending on the covered distance.

To cope with these problems, many countries have launched activities that aim at recording the usage of roads and collecting tolls electronically. Some systems developed so far require that each vehicle is equipped with an OBU, which exchanges data over the air with fixed control stations located along the roads. The data transfer is usually based on infrared or microwave and happens exactly at the moment a vehicle passes a control station. In other systems, control stations take a picture of passing vehicles and electronically analyze the license plate number by image recognition.

Unfortunately, each country having installed such a system or planning to do so has followed its own design decisions, and that is why often the systems of different countries are incompatible with each other, that is, OBUs cannot be deployed across national boundaries. Even the EU, which usually leaves out no efforts to harmonize everything (ranging from the size of eggs and tomatoes to more fundamental aspects such as a common currency) has not managed to coordinate the toll activities in its member states. As a result, Italy, Austria, Germany, and the United Kingdom, to name only a few, are building up or just operating proprietary systems. Nevertheless, the EU has recognized this necessity and founded a coordination group that prepares a stepwise migration path from existing systems toward a common, Europe-wide toll system, which, however, is expected to go into operation not before 2007.

While Austria has successfully launched its toll system in 2004, Germany experienced massive problems a year prior to this while trying to establish such a system for charging trucks on highways. In contrast to Austria, where the system works on the basis of microwaves as described earlier, the German government favored a satellite-based system: the OBU in a truck detects the usage of highways via GPS and determines the covered distance. The resulting data is then passed for billing purposes to a toll center via a public GSM network. In addition, hundreds of control bridges were installed on German highways in order to take pictures of trucks without OBUs for analyzing their license plate numbers. When the system was to be launched in August 2003, it turned out that the 210,000 OBUs that had just been installed in trucks suffered from profound defects. Because this was only one of many flaws, the system could not go into operation before January 2005, but even when it became operational, it had only a reduced functional range.

1.3 LBS Actors

LBSs are an interorganizational matter. Many actors are participating in the operational and nonoperational realization of LBSs. Generally, an *actor* is here defined as an autonomous entity like a person, a company, or an organization. An actor adopts one or several roles, which characterize either the functions it fulfills from a technical point of view or the impacts it exerts on LBSs from an economical or regulatory point of view. Accordingly, the roles listed in Figure 1.2 are classified into operational and nonoperational roles.

The operational actors are represented by the roles of the LBS provider, the user, the target, the network operator, the position originator, the location provider, and the content provider. Actors operating in these roles are cooperating during the execution of an LBS and request or provide subservices of the LBS. Each of them maintains a technical infrastructure ranging from single mobile devices (users and targets) to server farms (LBS, location, and content providers) and large-scale, complex cellular networks (network operators). The interaction between these roles during service operation happens over reference points consisting of protocols and connectivity services offered by various networks. Often, the technical realization of reference points is determined by *Service Level Agreements* (SLAs), which are adopted between the participating actors for fixing quality of service and accounting conditions.

Apart from a few exceptions, nonoperational actors are not engaged in the technical operation of LBSs. Rather, they have an indirect impact in that they dictate the economical or regulatory circumstances of LBS operation or influence it by the definition and adoption of technical standards. For example, trade and commerce may be interested in utilizing LBS technologies for mobile marketing and may authorize developers or service providers for creating adequate applications, and the automobile industry is specifically interested in furnishing their vehicles with appliances for navigation and fleet management. On the other hand, the government may have a direct influence in that it strictly regulates the utilization of location data by law, either for saving the privacy of individuals or vice versa, for purposes of lawful interception. Furthermore, the appearance of LBSs also strongly depends on the availability of standards from the responsible standardization committees, and on the acceptance and adoption of these standards by the vendors of network infrastructure.

INTRODUCTION

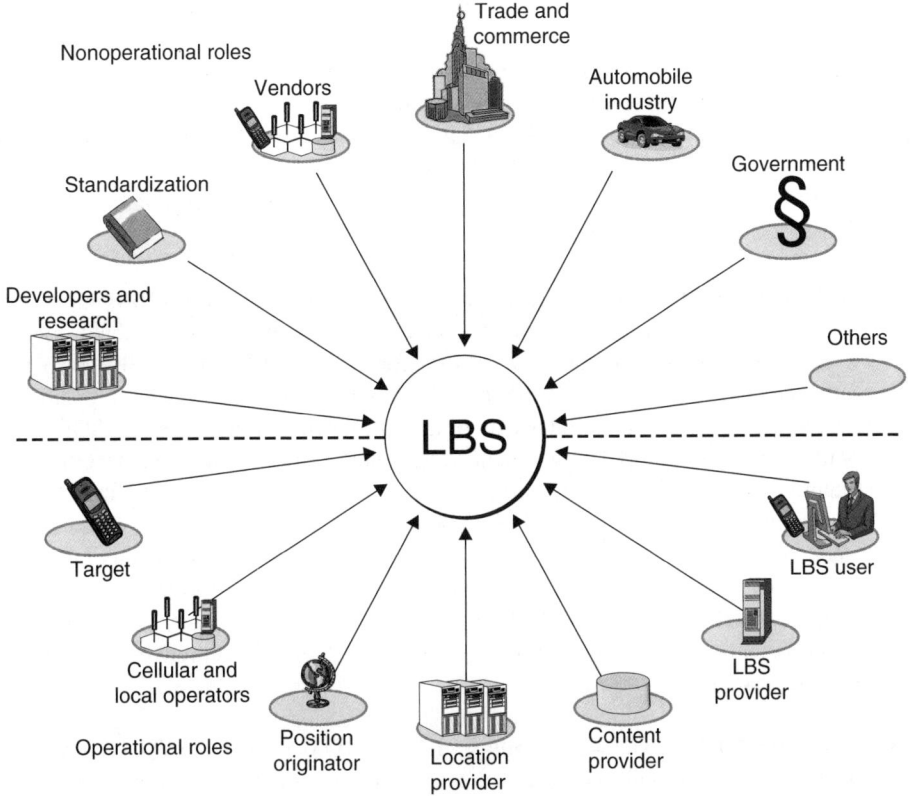

Figure 1.2 Roles for LBSs.

Finally, it should be mentioned that operational actors also affect nonoperational aspects, which especially holds for network operators, and significantly decide about success or failure of a certain technology.

Apart from standardization committees (which will be presented in the next section), the focus of this book is on the operational actors. In later chapters, their interaction will be explained by means of a supply chain, and the corresponding subservices, APIs, and protocols will be introduced.

1.4 Standardization

The acceptance and success of LBSs is essentially based on the availability of appropriate standards, which fix the interfaces, protocols, and APIs for supporting the cooperation between the actors introduced before and during LBS operation. Without the existence of standards, these actors would have to communicate over proprietary protocols and technologies, which would prevent competition, open service markets, and success of LBSs in general. Standards are considered and adopted by vendors and manufacturers of mobile network equipment and mobile devices as well as by the developers and programmers of applications and services. They guarantee a seamless interworking between equipment and software originating from different sources and in this way enable the use of services that are technically independent of a certain operator or provider. For example, the success of GSM was significantly determined by the possibilities of global roaming between different networks, which would not have been possible without the existence of international standards and their adoption by hundreds of mobile operators worldwide.

The technical realization and appearance of LBSs is mainly determined by the standardization works of the following consortia and committees:

- **3GPP and 3GPP2.** The *3rd Generation Partnership Project* (3GPP) is an international collaboration of several national standards bodies and focuses on the production of technical recommendations and reports for GSM and UMTS (3GPP Web site). In the context of LBSs, the work of 3GPP is of major relevance for positioning technologies in and with cellular networks and related location services. 3GPP specifications are continually being enhanced with new features. These enhancements are structured and coordinated according to *releases*. Today's GSM/UMTS networks are mainly operated on the basis of Releases 98 and 99, while work on Releases 4, 5, and 6 is still in progress. If not otherwise stated, descriptions in this book refer to Releases 99, 4, and 5. The *3rd Generation Partnership Project 2* (3GPP2) is a similar collaboration (3GPP2 Web site). It consists of North American and Asian standards bodies and deals with the specification of cdma2000 networks and related technologies.

- **OMA and Parlay.** The *Open Mobile Alliance* (OMA) and the *Parlay group* are collaborations of mobile operators, device and network manufacturers, information technology companies, and content providers (see (OMA Web site) and (Parlay Group Web site)). Their aim is less on the development of network solutions, but rather on the delivery of technical specifications for application-level and service frameworks. OMA is primarily concerned with the creation of end-consumer platforms like WAP or the emerging push-to-talk services, while Parlay develops open APIs for enabling third party service providers to access network functions of operators. Both groups support the development of LBS applications by providing protocols and APIs for the exchange of location data between different actors.

- **IETF and W3C.** In the past, the development of LBSs was primarily motivated by the telecommunications sector. However, with increasing mergers of classical telecommunications and Internet-based services, the necessity for supporting location-related technologies has also been recognized by consortia like the *Internet Engineering Task Force* (IETF) and the *World Wide Web Consortium* (W3C) (see (IETF Web site) and (W3C Web site)). These groups develop solutions that form the basis for specifications of the consortia mentioned earlier and specify special protocols for integration of location data into Internet or Web-based applications.

- **OGC.** The *Open Geospatial Consortium* (OGC) focuses on the specification of open standards for geospatial and location-based applications (OGC Web site). It is developing a broad range of standards for location languages, the transformation between different formats of location representation, the support of spatial databases and Geographic Information Systems (GIS), and middleware solutions for supporting the rapid creation and deployment of LBSs.

The specifications and standards produced by these consortia will be extensively covered and explained throughout this book.

1.5 Structure of this Book

This book is organized into three parts, thereby covering the diverse demands and interests of different target groups of readers like students and lecturers, professionals as well as developers.

The first part deals with the basic fundamentals the LBS operation relies on. It starts with an overview of different categories of location information in Chapter 2. A special focus of this chapter is on spatial reference systems, including coordinate systems, datums, and map projections, which are of importance for interpreting the results of positioning methods and for the processing of location information. The latter aspect is then covered in Chapter 3, which gives an insight into the principles and working of spatial databases and GISs. They are used for relating the positions and locations of target persons with geographic content, for example, in order to compose routing information or to find nearby points of interests. The chapter explains the underlying data models and shows how to represent spatial objects and topological relationships and how to query them. Chapter 4 explains the fundamentals of wireless communications, which are required for understanding the principles behind positioning in later chapters. It provides an overview of the basics of radio signals, introduces their representation in the frequency and time domain, and shows how to transfer data over a wireless link by the use of modulation and demodulation. Furthermore, the chapter describes the different phenomena radio signals are exposed to during their propagation, and finally gives an overview of multiplexing and multiple access schemes. The last chapter of the first part gives a general introduction into the working of cellular networks and location management.

The second part of the book is dedicated to positioning. Because it is a very complex matter and can be realized in several ways and by numerous, very different methods and systems, this part starts with an introduction to the fundamental aspects of positioning in Chapter 6. This chapter at first identifies the parameters measured during positioning, the general procedures, as well as the required infrastructures, and provides a scheme for classifying positioning methods according to different criteria. Subsequently, it introduces the basic positioning methods, which are proximity sensing, lateration, angulation, dead reckoning, and pattern matching. Finally, the chapter concludes with an overview of range measurements as needed for lateration and of error sources that may affect the accuracy of position fixes. The following chapters then deal with the implementation of positioning in the different systems. Chapter 7 introduces positioning by satellites, including the Global Positioning System (GPS), Differential GPS, and the future Galileo system. Chapter 8 explains positioning in and with cellular networks and covers the methods standardized by 3GPP for GSM and UMTS networks, among them Enhanced Observed Time Difference (E-OTD), Uplink Time Difference of Arrival (U-TDoA), and Observed Time Difference of Arrival (OTDoA). Furthermore, it illustrates how to increase accuracy and reduce latency of GPS by support of GSM/UMTS signaling, which is referred to as Assisted GPS, and provides an overview of positioning methods in cdmaOne and cdma2000 networks. The chapter concludes with a comparison of the cellular methods presented. Finally, Chapter 9 concludes the second part with an overview of positioning in indoor environments, including WLAN fingerprinting, RFID technology, and GPS indoor receiver technology.

The third part covers the operation of LBSs. Chapter 10 highlights interorganizational matters in that it identifies the different roles of actors participating in LBS operation and presents different scenarios of their interaction along a supply chain. After that, the chapter gives an overview of different querying and reporting strategies for exchanging location information between these actors. Of important concern when realizing interorganizational LBSs is the privacy protection of individuals, for which the chapter introduces the concepts of privacy policies and anonymization. Chapter 11 then deals with the concrete realization of the general concepts presented earlier. It highlights the working of location services in GSM and UMTS networks and explains different approaches and protocols for gathering location data and making it accessible for an LBS. A special focus is on the realization of emergency services, which is demonstrated by means of E-911 from the United States. Chapter 12 deals with middleware concepts for supporting the rapid creation and deployment of LBSs. As there is a lack of corresponding implementations and products in this field, it identifies at first the basic requirements and demands of an LBS middleware, and then presents available specifications and discusses their pros and cons. Finally, the third part and the book conclude with an outlook for the next generation of LBSs in Chapter 13.

Part I

Fundamentals

2

What is Location?

When dealing with LBSs it is first important to be clear about the meaning of the term "location". Although most people would claim that they are very familiar with the concept of location, as it, besides time, is one of the major quantities determining our everyday life, it is useful to have a closer look at it and to distinguish between different categories of location information. This chapter provides an overview of these categories. The main focus is on the concept of spatial locations that appear to the developers and users of LBSs in the form of coordinates. This chapter introduces the different ways to express a spatial location in a coordinate system and explains the idea behind datums, which are used for modeling the surface of the Earth. Furthermore, an introduction to the fundamentals of map projections is given, which are needed for projecting the bended surface of the Earth on a planar map.

2.1 Location Categories

Basically, the term "location" is associated with a certain place in the real world. When people make appointments, they usually agree to meet at a certain place, such as the airport, a bar, or the office. In other cases, they must inform the place of their residence, for example, in order to receive written correspondence or official directives. What these examples have in common is that location denotes a place of an object of the real world, and hence these kinds of locations belong to the class of *physical locations*.

Most recently, with the propagation of what people sometimes call the *cyberspace* or the *global village*, that is, the Internet, there is reason to introduce another concept of location. The Internet has created a broad range of new applications that have dramatically changed the way people receive information or interact among each other. In this area, the term "location" has sometimes another meaning and refers to a virtual meeting place, for example, a web site, a chat room, or a field shared by several players of a distributed computer game. These locations are typically denoted as *virtual locations*.

LBSs predominantly refer to physical locations, and do not know the concept of virtual locations, maybe apart from a few exceptions, for example, in the areas of mobile

Location-based Services: Fundamentals and Operation   Axel Küpper
© 2005 John Wiley & Sons, Ltd

gaming and augmented reality where physical locations are mapped on or mixed with virtual locations.

The category of physical locations can be further broken down to the following three subcategories that are relevant for creating and using LBSs:

Descriptive locations. A descriptive location is always related to natural geographic objects like territories, mountains, and lakes, or to man-made geographical objects like borders, cities, countries, roads, buildings, and rooms within a building. These structures are referenced by descriptions, that is, names, identifiers, or numbers, from where this category of location has derived its name. Thus, descriptive location is a fundamental concept of our everyday life, which is used by people for arranging appointments, navigation, or delivering goods and written correspondence to well-defined places. Without having organized our real-world environment and infrastructure according to well-defined descriptions of geographical objects, people would meander without orientation.

Spatial locations. Strictly speaking, a spatial location represents a single point in the Euclidean space. Another, more intuitive term for spatial location is therefore *position*. It is usually expressed by means of two- or three-dimensional coordinates, which are given as a vector of numbers, each of it fixing the position in one dimension. In contrast to descriptive locations, positions are not used in our everyday life, because people prefer to orientate in terms of geographical objects instead of using coordinates. However, spatial location is indispensable for professional applications like aviation or shipping, which depend on the availability of highly precise and accurate location information. The concept of spatial locations also provides the basis for surveying and mapping of descriptive locations.

Network locations. Network locations refer to the topology of a communications network, for example, the Internet or cellular systems like GSM or UMTS. These networks are composed of many local networks, sometimes also referred to as *subnetworks*, connected among each other by a hierarchical topology of trunk circuits and backbones. Service provisioning in these networks assumes that the location of the user's device with respect to the network topology is known. This is achieved by network addresses that contain routing information, in combination with directory services, for mapping numbers, identifiers, or names of another scheme onto the network address. For example, in the Internet a network location refers to a local network which is identified by means of its IP address. In mobile networks, on the other hand, a network location is related to a base station a mobile terminal is currently attached to.

LBSs may be based on all three categories of locations (see Figure 2.1). The target persons of an LBS are pinpointed by positioning, for which many different methods exist. Some of them deliver a spatial location, for example, GPS, while others provide a network location, for example, Cell-Id, or a combination of both. Once the location of a target has been derived, it needs to be further refined for being processable by the LBS. This especially concerns the deriving of an appropriate descriptive location that is meaningful to the LBS user.

WHAT IS LOCATION?

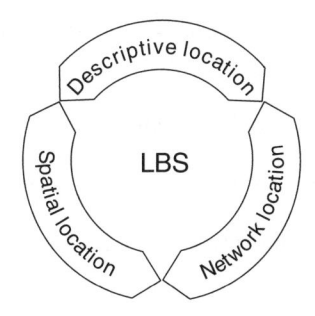

Figure 2.1 Location categories.

Hence, an important function of LBSs is the mapping between the different categories of locations. If positioning delivers a spatial or network location, it must often be mapped onto a descriptive location in order to be interpretable by the respective LBS user. On the other hand, a descriptive location might be transferred into a spatial location in order to relate it with other locations, for example, as it is required for distance calculations. In another example, it might be necessary to translate a spatial or descriptive location into a network location to support location-based routing, for example, in the context of enhanced emergency services like E-911.

The mapping is usually accomplished by spatial databases and Geographic Information Systems (GIS), which will be introduced in the next chapter, while the concept of network locations will be explained in the context of cellular networks and location management in Chapter 5. The remainder of this chapter deals with all aspects of spatial location.

2.2 Spatial Location

Spatial locations or positions are based on well-defined *reference systems* that subdivide a geographic area, in most cases the entire Earth, into units of a common shape and size. A spatial reference system consists of the following elements:

- a coordinate system,
- a datum, and
- a projection (if location is represented on a map).

The following sections provide an introduction to these elements and present an overview of the most common reference systems used for LBSs.

2.2.1 Coordinate Systems

A coordinate system is used to reference a certain spatial location by a vector of numbers that are called *coordinates*. The coordinates of an object refer to its position either measured by distance or by angle with regard to two or three axes of the coordinate system, depending on whether the position is to be fixed on a plane or in space. In geometry, a coordinate system belongs either to the class of *Cartesian* or *Ellipsoidal coordinate systems* and can be uniquely defined by the following elements:

1. its *origin*, which represents the intersection of two or three axes,

2. its *scale*, which represents the subdivision of the axes in common units, and

3. its *orientation*, which fixes the directions of the axes on a plane and in space, respectively.

In order to interpret location information in a consistent way, one has to agree on a well-defined coordinate system that fixes these components.

2.2.1.1 Cartesian Coordinate Systems

Cartesian coordinate systems are certainly known to most readers from mathematics. They are used to express the spatial location by specifying its distances to the axes. The axes are labeled X and Y in a two-dimensional and X, Y, and Z in a three-dimensional coordinate system. They are arranged at right angles to each other, that is, they are said to be mutually orthogonal. Each pair of axes defines a plane, which is labeled according to the axes spanning this plane, for example, XY-plane. The axes intersect at the origin, and the origin coordinates are usually defined to be $(0, 0)$ and $(0, 0, 0)$ respectively.

An example of such a Cartesian coordinate system has the cumbersome name *Earth Centered, Earth Fixed X, Y,* and *Z* (ECEF). It is used by many positioning technologies such as GPS to express a certain position in three dimensions on a global, worldwide basis. The term "Earth Centered" results from the fact that the origin is defined to be at the center of the mass of the Earth, which is also called the *geocenter* of the Earth. The term "Earth Fixed" means that the axes are fixed with respect to the Earth and hence rotate with the Earth. Figure 2.2 illustrates the orientation of the ECEF coordinate system.

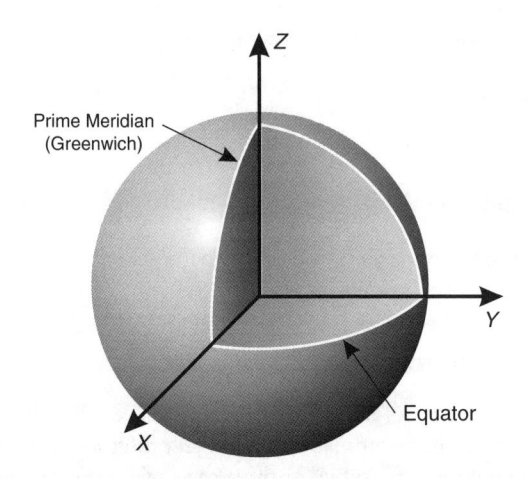

Figure 2.2 Earth Centered, Earth Fixed X, Y, Z.

WHAT IS LOCATION?

The Z-axis corresponds to the rotation axis of the Earth and is hence defined by the line that is given by the geocenter and the North Pole. The X-axis intersects the equator at a point that is referred to as the *Prime Meridian* or the *Greenwich meridian* (see explanations below). The Y-axis is then defined relative to the X-axis in that it completes a right-handed orthogonal system. The XY-plane is also called the *equatorial plane*. The scale of ECEF is usually given in meters, which is, by the way, a linear measure that was defined in the eighteenth century and was then intended to be the ten millionth part of the distance from the equator to the North Pole. The coordinates of a position in an ECEF system then look like this:

$$(X, Y, Z) = 4{,}176{,}671.786 \text{ m}, \ 857{,}107.334 \text{ m}, \ 4{,}728{,}338.964 \text{ m})$$

ECEF is a very convenient system for calculating *line-of-sight* distances between two positions, for example, between a satellite and an object located at the surface of the Earth, and hence it has been adopted by GPS for estimating positions. However, its major drawback is that it does not reflect the bended surface of the Earth and hence is not a very meaningful way to represent positions at the Earth's surface or to determine distances along the bended Earth's surface. For instance, the coordinates of the previous example do not indicate whether the position is at the surface of the Earth, at some height above the Earth, or within the Earth. Also, ECEF cannot be applied for map projections.

2.2.1.2 Ellipsoidal Coordinate Systems

A more convenient way to express spatial location is to use an ellipsoidal coordinate system that models the Earth's surface as an ellipsoid (in contrast to the predominant opinion, the shape of the Earth does indeed not correspond to a sphere). A *reference ellipsoid* describes the shape of the Earth by fixing an equatorial and a polar radius. The origin of an ellipsoidal coordinate system is given by two reference planes, both arranged orthogonal to each other and crossing the geocenter. The horizontal plane corresponds to the equatorial plane and is unambiguously fixed in this way. The vertical reference plane includes the Earth's rotation axis and hence intersects North and South Poles. By this definition, its orientation in East–West direction is not unambiguously fixed, and, as will be described later, needs to be further restricted.

A position at the surface of the Earth is then represented by the angles between the reference planes and the line passing from the geocenter to the position (see Figure 2.3). The *latitude* ϕ is defined as the angle between the equatorial plane and the line, that is, it represents the North–South direction measured from the Equator, while the *longitude* λ describes the angle between the vertical plane and the line, that is, it reflects the East–West direction of a position with regard to the vertical reference plane. Another term for latitude is *parallel* (as lines of latitude run parallel to the Equator), while lines of longitude, which run perpendicular to the Equator, are often denoted as *meridians*. The concept of latitude and longitude merely allows the expression of a position located at the Earth's surface. However, for many applications, such as aviation, it is also important to know the height or *altitude* z of a position from the surface of the Earth. The *geodetic height* is used to express the altitude, and it is defined as the distance between a position and the reference ellipsoid along the perpendicular to the ellipsoid.

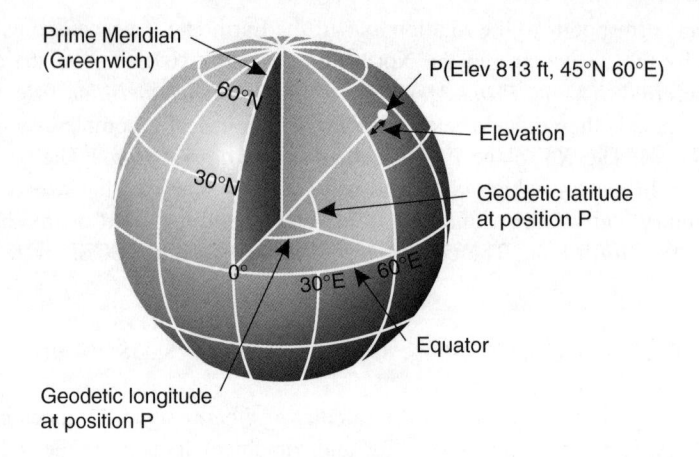

Figure 2.3 Latitude, longitude, and height.

While the altitude can be expressed by any length unit such as meters or feet, the unit of latitude and longitude is given in degrees (°). The latitude has a value range of −90° to +90° (or N90° to S90°), where −90° and +90° correspond to the South and North Poles of the Earth. The longitude ranges from −180° to +180° (or W180° to E180°). The next smaller unit of degree is the minute ('). Each degree comprises 60 min, and a minute is then further subdivided into 60 s ("). Because lines of latitude are arranged as parallels, they have always a constant surface distance on Earth. For example, one second (1") of latitude approximately corresponds to 30 m everywhere on Earth. In contrast, lines of longitude converge at the poles, and hence their surface distance depends on the latitude where this distance is measured. One second of longitude at the equatorial plane is also approximately 30 m, at 45° of latitude it is approximately 22 m, and at the poles the distance is exactly 0 m. Note that these surface distances refer to an altitude of 0 m, that is, they are measured at the surface of the reference ellipsoid.

Table 2.1 Notations for latitude, longitude, and height

Notation	Example
Degrees, minutes, seconds of lat. and long.	Elev 505 m N 48° 08' 57" E 11° 35' 47"
Tenth of seconds of latitude and longitude	Elev 505 m N 48° 08' 57.2" E 11° 35' 47.6"
Hundreds of thousands of degrees	Elev 505 m N 48. 14,922° E 11.596,83°

For many applications, the precision of seconds is too coarse grained, and hence it is common practice to render position data more precisely by additional decimal places of a second, for example, to tenths or hundredths of a second. It is also common practice to express position data without units of minutes and seconds and to use a number of decimal places of the degree unit instead. Table 2.1 lists the three most common notations for latitude, longitude, and height. The positions given in the table refer to the same position as the ECEF coordinates presented in the earlier example.

So far, there remains one open question: the origin of latitudes are unambiguously fixed by the equatorial plane, but where exactly is zero longitude, that is, the so-called *Prime Meridian*? The answer to this question goes back to the late nineteenth century and was primarily motivated by the efforts to set an international time standard, which acts as a reference on a worldwide basis for measuring time. Before this, almost every town in the world used its own reference time, and there were no conventions that fixed the beginning and end of a day or how long an hour should exactly last. In 1884, the *International Meridian Conference* in Washington DC decided on a Prime Meridian that passes through the Royal Greenwich Observatory in Greenwich, England. This decision was the result of a voting among 41 delegates from 25 nations. It was mainly influenced by the fact that at this time the United States had already chosen Greenwich to be the reference for their national time zone system and that the bigger part of the world's commerce still depended on sea charts that used Greenwich as Prime Meridian. Since then, the time zones of the world have been arranged on the basis of *Greenwich Mean Time* (GMT) (which has been renamed as *Coordinated Universal Time* (UTC) later), and the meridian that passes through Greenwich is used as the Prime Meridian, which is hence also called *Greenwich Meridian* (see Figure 2.3).

2.2.2 Datums

For a long time, the human race believed that the Earth was a perfect, regular sphere, an assumption that was first postulated by the Greek mathematician and philosopher Pythagoras around 500 B.C. It was not before 1687 when the English scientist Sir Isaac Newton concluded from extensive investigations that planets must have the shape of an oblate ellipsoid of revolution. Newton argued that planets would be perfect spheres only if they did not rotate. The ellipsoidal shape was the result of a slight "flattening out" due to rotation and gravitation. Since then, studies of *geodesy* created many models approximating the shape of the Earth, which were commonly referred to as *geodetic datums*. A datum defines the size and shape of the Earth as well as the origin and orientation of the coordinate system that is used to reference a certain position. It is classified as follows:

- **Horizontal datums.** These datums approximate the Earth's shape as a *reference ellipsoid* (or, at earlier times, as a sphere). The ellipsoid defines the origin and orientation of the coordinate system. Horizontal datums are the essential prerequisite for referring a position with regard to the surface of the Earth and are hence also the basis for projecting the Earth's surface on a planar map.

- **Vertical datums.** They are based on the determination of a *geoid*, which always reproduces the Earth's mean sea level more or less accurately (but in either case much more accurately than a reference ellipsoid). Vertical datums primarily serve to determine the height of a certain position with regard to mean sea level, for which the geoid acts as a benchmark for origin and orientation.

The following sections provide an introduction to the fundamentals of both types of datums.

2.2.2.1 Horizontal Datums

The reference ellipsoid of a horizontal datum is defined by its equatorial radius a and its polar radius b (see Figure 2.4). The former is also denoted as *semimajor axis*, the latter as *semiminor axis*. When fixing a reference ellipsoid, it is common practice to use the parameters a and the flattening f, where f is defined as follows:

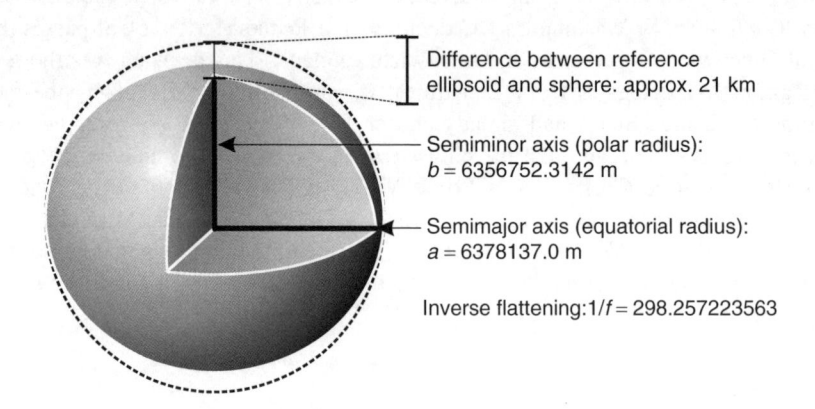

$$f = \frac{a-b}{a} \qquad (2.1)$$

Figure 2.4 Parameters of a reference ellipsoid.

The difference between the semi axes a and b of a typical Earth ellipsoid differ by approximately 21 km, and hence the magnitude of f is very small. Therefore, the inverse flattening $1/f$ is commonly used to specify an Earth ellipsoid. Table 2.2 lists a number of popular reference ellipsoids used in the last two centuries. Most of them have been named after their generators.

Table 2.2 Examples of reference ellipsoids

Ellipse	Year	Semimajor axis [m]	1/Flattening
Airy	1830	6,377,563.396	299.3249646
Everest	1830	6,377,276.345	300.8017
Bessel	1841	6,377,392.155	299.1528128
Clarke	1866	6,378,204.400	294.9786982
Clarke	1880	6,378,249,136	293,465
International (Hayford)	1924	6,378,388.0	297,0
WGS-72	1972	6,378,135.0	298.26
WGS-84	1984	6,378,137.0	298.257223563

Unfortunately, the actual shape of the Earth does not correspond to an ellipsoid, and hence horizontal datums can be merely seen as a rough approximation. The Earth's surface shows a number of irregularities like hills and mountains as well as rifts and dales. Even if the entire Earth is entirely covered with water, it would hardly match the shape of an exact ellipsoid, which is due to gravitational anomalies and external gravitational forces being responsible for tides. For a long time, datums were therefore tailored to a special local region only, for example, a country or a continent. These *local datums* model the shape of the respective regions very accurately, but would show significant deflections if using them for other regions. With the emergence of satellite surveying in recent years, it is possible to fix *global datums* that provide a reasonable approximation for the entire Earth.

Global and local datums primarily differ in their relationship with the geocenter and the center of the reference ellipsoid. For global datums, the ellipsoid center corresponds to the geocenter, while for local datums it is shifted by a certain amount that is required to align the ellipsoid with the surface of the respective local region as accurate as possible. Table 2.3 lists the reference ellipsoids used for different local datums as well as their shift with regard to the geocenter.

Table 2.3 Selected local datums and their reference ellipsoids

Datum	Ellipsoid	Datum shift (D_x, D_y, D_z)
NAD-27 (Central America)	Clarke 1866	(0 m, 125 m, 194 m)
NAD-27 (Alaska)	Clarke 1866	(−5 m, 135 m, 172 m)
European 1950	Hayford	(−87 m, −98 m, −121 m)
Tokyo	Bessel 1841	(−148 m, 507 m, 685 m)

The difference between global and local datums is also illustrated in Figure 2.5. A local datum is always fixed by a *datum origin*, which normalizes the reference ellipsoid with regard to a certain position on Earth. All subsequent positions are then computed from this origin. For example, the origin of the *North American Datum of 1927* (NAD-27) is located at Meades Ranch in Kansas (which is also the center of the United States) and the origin of the *European Datum* is located at Potsdam in Germany.

Although local datums are still in use today, there is an increasing tendency to derive position data by the use of global datums, which is due to satellite-based positioning technologies like GPS that provide worldwide coverage. The most common global horizontal datum used today is that of the *World Geodetic System of 1984* (WGS-84), a system that also defines a vertical datum and is used for GPS (see Table 2.2 for its parameters).

2.2.2.2 Vertical Datums

An important magnitude when fixing a datum is the mean sea level, which is defined as zero altitude and which is averaged over temporally varying sea levels in a local area. These variances are results of wave actions and external gravitational forces being responsible for tides. Unfortunately, approximating the Earth's shape by using the mean sea level as a reference does not lead to a uniform, ellipsoidal figure. This is due to the fact that the sea level exactly conforms to the gravitational field of the Earth, which is all but constant everywhere. Rather, this field shows local gravitational anomalies, and hence the sea level, as seen from a global point of view, has local dales and knolls and may deviate from a fairly well-approximated reference ellipsoid by up to 100 m. This gives rise to another,

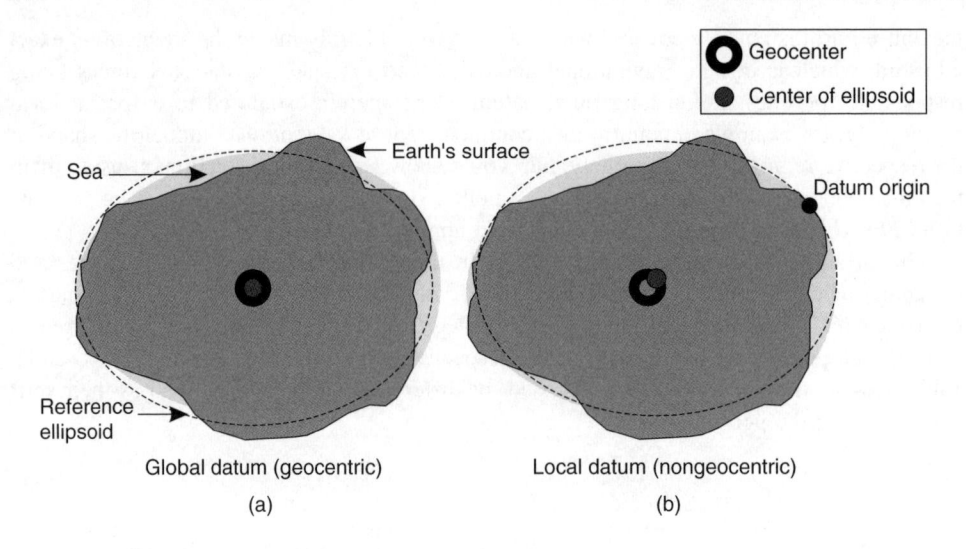

Figure 2.5 Global and local datums.

more precise approximation of the Earth's shape by using a *geoid*. A geoid is defined to be the surface of fixed equigravitational potential value, which coincides on average with mean sea level. Given this definition, the gravitational potential at each point of the geoid surface has the same amount, and it is normalized in a way that it approximately matches with the mean sea level. Figure 2.6 illustrates the differences between the surface of Earth, a reference ellipsoid, and a geoid.

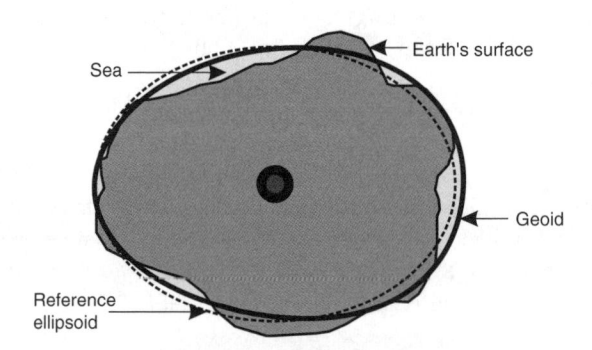

Figure 2.6 Difference between geoid and reference ellipsoid.

Having fixed a reference ellipsoid and a geoid, three different classes of heights can be distinguished (see also Figure 2.7):

- **Orthometric height.** This is the height of a position normal to the geoid. It is often also considered to be the height above mean sea level.
- **Ellipsoid or geodetic height.** This height refers to the normal distance between a position above the Earth's surface and the reference ellipsoid.

WHAT IS LOCATION?

- **Geoid height.** This height expresses the normal distance between the reference ellipsoid and the geoid.

Generally, the geoid height is relatively small compared to the dimension of Earth. In WGS-84, which defines both a reference ellipsoid and a geoid, it is in the range of tens of meters, and the maximal difference between them is approximately 100 m.

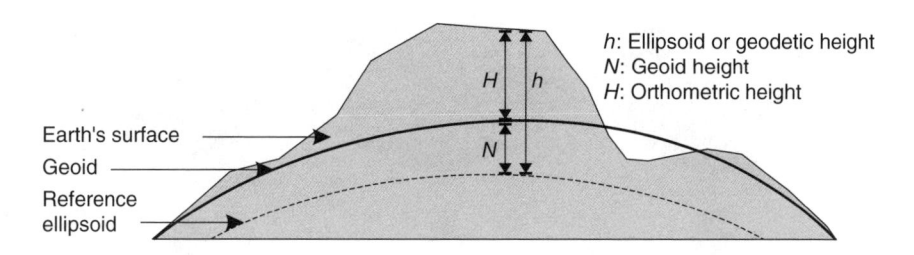

Figure 2.7 Ellipsoid, orthometric, and geoid heights.

Usually, a GPS receiver in the first step determines its position using the ECEF Cartesian coordinate system, and subsequently transforms the derived coordinates into latitude, longitude, and ellipsoid height under consideration of the WGS-84 ellipsoid. In the last step, the orthometric height can be derived from that if the geoid height is known for that particular position. For that purpose, the WGS-84 geoid is usually given by a 10 by 10 degree grid, which specifies the geoid height for all coordinates that meet at multiples of 10 degrees of latitude and longitude. The geoid height at a special position can then be determined by performing a linear interpolation on the closest grid points and its geodetic heights. The error caused by this interpolation can be neglected, as it is comparatively small with regard to other positioning errors imposed by GPS (see Chapter 7). To improve the results of interpolation, it is also possible to use a more fine-grained grid, for example, of 15 by 15 min. Various grids of geodetic heights are published (and permanently updated) by the *National Imagery and Mapping Agency* (NIMA) (see (NIMA)).

2.2.3 Map Projections

Displaying the Earth's surface on an ellipsoid or geoid is usually not very useful for many applications like navigation. In many cases, it is more convenient to depict the bended surface of the Earth on a flat surface like a piece of paper (i.e., a map), a computer monitor, or the display of a mobile device. Map projections are used to represent the bended, three-dimensional surface of the Earth on a two-dimensional surface, the former one being the *projected* and the latter being the *projection surface*.

Although projections are based on complex mathematical models, it is common practice to imagine them as outlines of items like continents and oceans that are projected from the inner surface of the Earth onto the projection surface. The complexity of this process results from the fact that it is not possible to flatten a surface that is curved in all directions like an ellipsoid or a sphere without distortion. The projection surface, in contrast to this, is either flat or is curved in only one direction in a way that it can be transformed back into a flat surface without any distortion after the projection. According to the shape of the projection surface it is hence classified as *planar*, *conical*, and *cylindrical projections* (see Figure 2.8).

Planar projection   Conical projection   Cylindrical projection

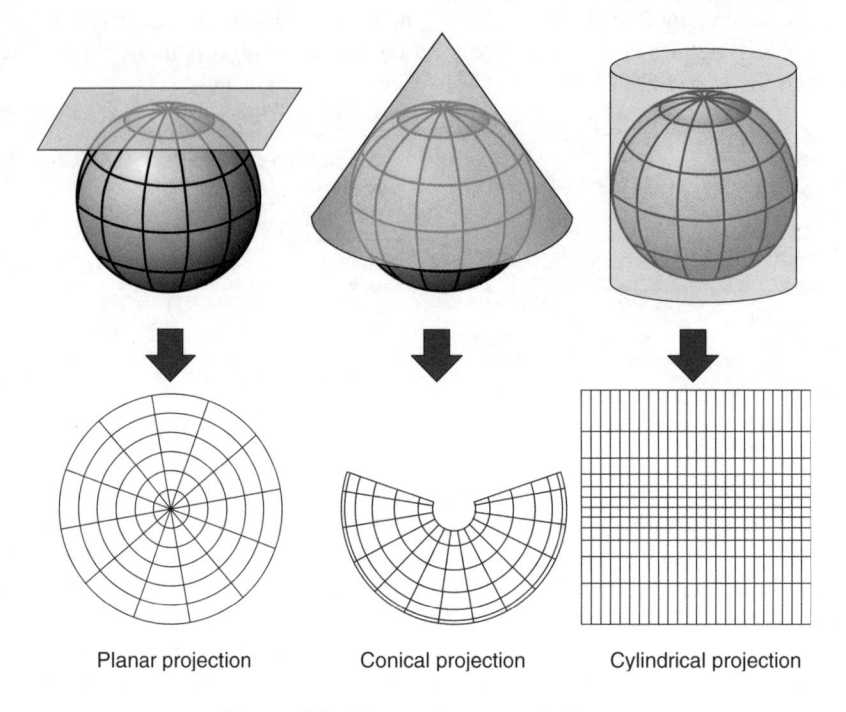

Figure 2.8 Types of map projections.

In a planar projection, the ellipsoidal surface is simply projected onto a plane, which is tangent to one point of the Earth. Figure 2.8 shows a planar projection where the tangent point coincides with the North Pole. However, the tangent point can be located at any point on Earth, and the resulting map represents a view of the Earth as it would appear from space directly above this point. Using a conic projection, the flat projection surface is transformed into a cone that is put over the ellipsoid at any point. After the projection, the cone can be unrolled into a flat sheet again with no deformations of the depicted items. The cylindrical projection works in a similar way, but uses the shape of a cylinder as projection surface.

Each of these projections can be further subdivided into several subclasses depending on the orientation of the projection surface with regard to the rotation axis of the Earth, which is commonly denoted as *aspect*. Figure 2.8 depicts only projections with a *normal aspect*, which are characterized by the fact that the axis of the projection surface is in conjunction with the Earth's rotation axis. Another classification is projections with a *transversal aspect*, where projection surface and rotation axis are aligned by an angle of 90°. Projections with any orientation other than 0° or 90° are subsumed under the term *oblique aspect*. Besides the different aspects, projections can also be classified according to the distance between the projection surface and the Earth. The projection surface may be arranged at some height above the Earth, or it may be tangent to the Earth at some point (in case of a planar projection) or at a circle along its surface (in case of conic and cylindrical projections). Some projections also intersect the Earth, resulting in one or two secant lines.

The variety of projection types, which might be somewhat confusing to the reader, goes back to the fact that each of them reflects the Earth's surface with a certain degree and combination of different distortions. Distortion is a result of flattening the curved surface of the Earth and occurs in one or the other way in each projection. The following list gives an overview of different kinds of distortions and explains their impact on the appearance of maps:

- **Areal distortion.** This type of distortion is given if different regions on a map show disproportional sizes with respect to each other. If a map preserves the areal relationships between structures, it relies on an *equal-area* projection. Equal-area projections are useful for thematic maps, that is, for comparing dimensions of different countries or continents.

- **Angular distortion.** Some projections reflect the shape of items corrupted, that is, twisted or bent compared to their real shape, and falls into the category of angular distortions. A map that does not suffer from angular distortion and preserves local shapes is said to be *conformal*. Conformal projections are very useful for navigation.

- **Scale distortion.** Scale distortions are given if proportions of distances with respect to each other are not shown correctly. All projection techniques show scale distortions of different amounts.

- **Distance distortion.** A projection causes distance distortions if the distance between any point on a map and a certain reference point is not preserved. A map is said to be *equidistant* if it does not show this kind of distortion. Equidistant projections are used to measure distances from a fixed position.

- **Direction distortion.** This kind of distortion refers to the *azimuth*, which is the angle between a point on a line and another point. Direction distortion is given if the azimuth between any point and a fixed reference point is not displayed correctly. Otherwise the projection is said to be *azimuthal*, which is especially useful for displaying polar regions.

Thus, different types of projections differ in the degree and combination of distortions they cause. A certain projection is usually developed for a particular application, for example, navigation or thematic maps, for which certain kinds of distortions should be avoided while others are less bothering. The most important distortions are areal and angular distortions, which can occur independent of each other, that is, there exist projections that are either conformal or equal-area maps. Scale distortions, on the other hand, cannot be avoided at all and appear with any kind of projection, but with different amounts.

Map projections are a complex matter that can hardly be covered in this book in detail. The following sections therefore deal with those kinds of projections that are of major relevance in the context of LBSs.

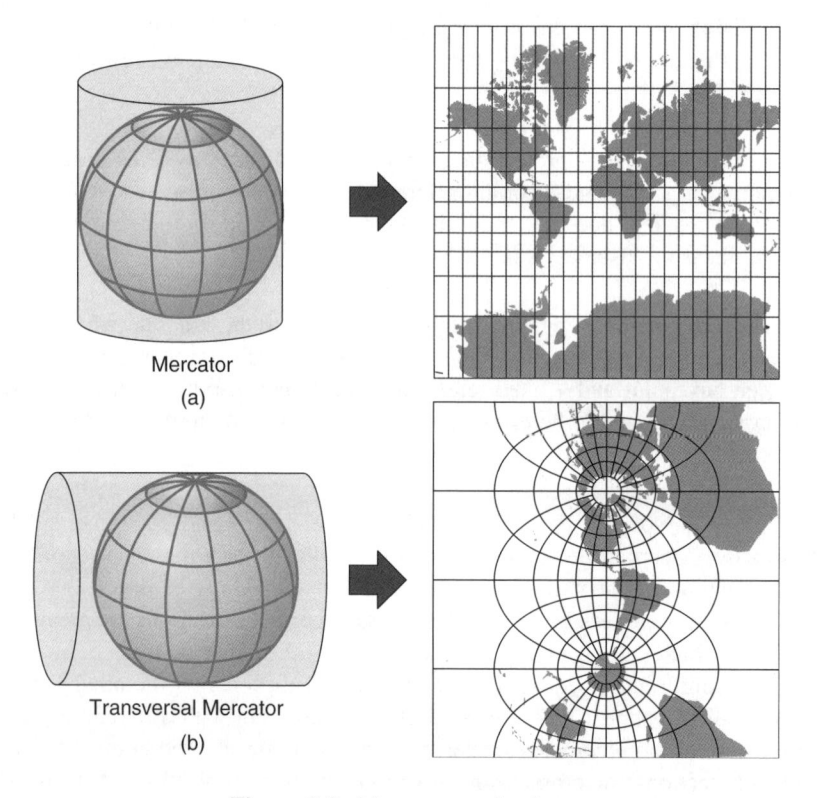

2.2.3.1 Mercator Projections

Most sea charts in the sixteenth century did not account for the then recently proven fact that the Earth was round and hence reflected its surface all but accurately. Gerhardus Mercator, a Flemish mathematician and cartographer, therefore developed a cylindrical projection in 1568 which provided a more appropriate mean for navigation. Since then, the *Mercator projection* is the basis for many other projection techniques, and still appears in many variants today.

The original Mercator projection is depicted in Figure 2.9(a). It is a normal cylindrical projection, where the projection surface is wrapped around the Earth in that it builds a tangent line with the Equator. Meridians and parallels are straight lines being arranged orthographic to each other. Mercator is a conformal projection that preserves local shapes and directions, that is, the angle between two lines on a map accurately reflects this angle in the real world. It is thus a meaningful instrument for navigation. However, conformity is paid by tremendous areal and scale distortions. As an example consider Greenland, which on a Mercator map adopts an area greater than South America, although its actual size is only 1/8 of that of South America. Commonly speaking, Mercator displays only items near the Equator without distortions, and increasingly distorts them the farther away they are located from the Equator. As a result, the pure Mercator projection

Figure 2.9 Mercator projections.

is not suited for mirroring countries with large North–South extent, such as Argentina or Brazil.

A prominent variant of Mercator is the *Transverse Mercator projection*, which was developed by Johann Heinrich Lambert in 1772. In contrast to the original Mercator, the cylinder is turned by 90° with respect to the Earth's rotation axis and hence has a common tangent line with a selected meridian, which is referred to as the *central meridian*, see Figure 2.9(b). This projection results in a map where meridians and parallels are depicted as curved lines (apart from the central meridian and the two meridians intersecting with the central meridian at 90°). The projection is also conformal and preserves shapes along the central meridian very accurately. However, while Mercator distorts structures in North–South direction, the Transversal Mercator suffers from the same problem in East–West direction.

2.2.3.2 Universal Transverse Mercator

Today, cartographers still fall back on the Transverse Mercator projection for developing maps. In many countries, it is applied with special parameters that have been fixed either arbitrarily or with the aim to reproduce the geographic regions of the respective countries in an appropriate manner. As a result, there is a coexistence of hundreds of national maps and a lack of international standardization making position data from different maps hard to compare with one another. However, an international system of increasing popularity is the *Universal Transverse Mercator* (UTM), which has its origin in the military area and serves as a common, nation-independent system with a common worldwide coverage for NATO members. At the same time, like GPS, it is also of major importance for civil applications.

UTM is a two-dimensional Cartesian coordinate system that comprises a series of Transversal Mercator projections, each of them being limited to a certain size in East–West direction, thereby avoiding large East–West distortions. Figure 2.10 illustrates the composition of a UTM map. The cylinder is rotated in steps of 6° of longitude around the Earth, thereby acquiring 60 zones of 6° longitude. In order to avoid large distortions near the poles, the zones are artificially restricted to a range of S80° to N84°, thereby guaranteeing that South America, Africa, and Australia on the one hand and parts of Greenland on the other are still covered.

Expressing coordinates in UTM takes some getting used to. For each zone two coordinate systems with different origins are defined, one for the Northern and another for the Southern Hemisphere, thereby avoiding coordinates of negative numbers. The scale is usually given in meters, and a zone covers approximately 900,000 m in East–West and 20,000,000 m in North–South direction. For the Northern Hemisphere, the origin of the coordinate system is aligned at the Equator and 500,000 m westward of the zone's central meridian. For referencing positions on the Southern Hemisphere this coordinate system is shifted by 10,000,000 m southward (see also Figure 2.10).

For unambiguously referencing a position on Earth, it is also required to specify the zone this position belongs to. For this purpose, the zones are serially numbered, starting with zone 1, which ranges from W180° to W172°, and ending with zone 60, which ranges from E172° to E180°. Figure 2.11 shows the resulting UTM world map.

Unfortunately, two different notations have been established for referencing positions in UTM, one being preferred in the civil world and the other being predominantly in use

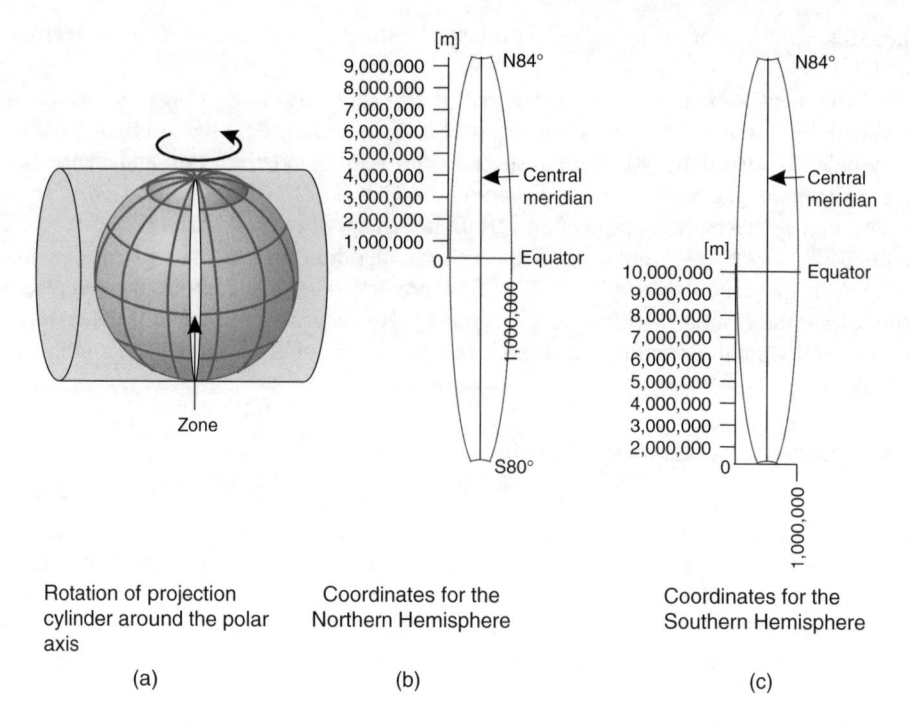

Figure 2.10 Composition of UTM coordinate system.

for military applications. A position expressed in terms of the civil notation looks like this:

32N – 0693,145 – 5336,147

The first number specifies the UTM zone and the hemisphere the position belongs to. The character "N" refers to a position on the Northern Hemisphere, whereas "S" denotes a position on the Southern Hemisphere. The indication of the hemisphere is important to interpret the following numbers with regard to the correct coordinate system. The first number specifies the position's distance in meters eastward from the ordinate of the coordinate system, which is called *easting*. The second number indicates the distance northward from the abscissa, which is denoted as *northing*.

For the military notation, each zone is subdivided into 20 horizontal bands of 8° latitude, starting with band C, which goes from S80° to S72° and ending with band X, which ranges from N72° to N80° (see also Figure 2.11). Each zone is then further broken down into a matrix of squares, each of them covering an area of 100 sq km and being represented by a combination of two letters. For example, the position given earlier appears in the military notation as follows:

32U – PU – 93,145 – 36,147

where the first term denotes the zone and band of the position. The second term specifies the square. The first letter encodes multiples of 100 km of the easting and the second one of the northing. Thus, these letters replace the first two places of easting and northing compared to the civil notation. The remainder is then given by the following two numbers.

WHAT IS LOCATION?

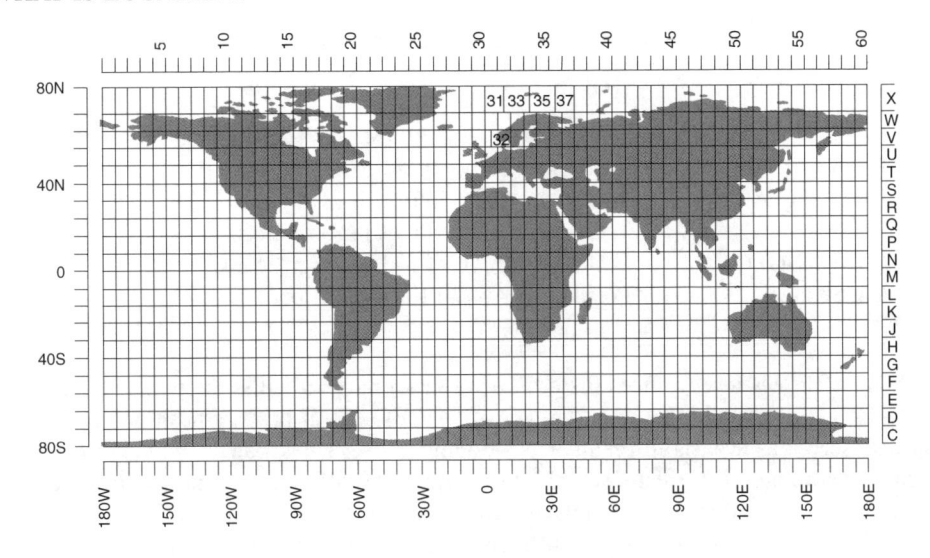

Figure 2.11 UTM world map.

The assignment of letters to squares happens according to an algorithm, which will not be covered here. The interested reader can get additional information on this from the web site of the NIMA (NIMA Web site). Also note that sometimes civil and military notations are mixed. In this case, the first term corresponds to the military notation, specifying zone and band, directly followed by easting and northing according to the civil notation. This scheme, for example, is often used by GPS receivers.

The major advantage of UTM is that it is a rectangular coordinate system, which allows to derive very accurately the distance between two positions from a map (assuming that the positions fall into the same zone). These distance measurements are less affected by scale distortions than those achieved with other map projections, which is due to the comparatively small East–West dimension of zones. Also, UTM has been designed to further reduce the remaining small distortions for larger regions of a zone: the scale along central meridians is uniformly reduced to 0.9996, which produces two lines of zero distortion 180 km eastward and westward from the central meridian (and a scale of approximately 1.003 at the zone boundaries).

As indicated earlier, UTM is an international scheme for maps. Several countries use very similar systems, for example, the British National Grid or the German Gauß-Krüger coordinate system. These systems slightly differ from UTM in the dimensions of zones and the way zones and subunits of zones are referenced. A detailed introduction to the topic of map projections is given by Snyder (1987) and Robinson et al. (1995).

Table 2.4 gives an overview of the different coordinate notations (and the associated datums) one is primarily faced with when building LBSs.

2.3 Conclusion

To draw a conclusion, it is of major importance to consider the underlying datums of position data and be aware of the fact that geographic content derived from maps may

Table 2.4 Overview of notations for coordinate systems

Coordinate system	Datum	Example
ECEF X, Y, Z	–	4,176,671.786, 857,107.334, 4,728,338.964
Lat. Long. Alt.	WGS-84	Elev 505 m N 48° 08' 57" E 11° 35' 47"
Lat. Long. Alt. (dec.)	WGS-84	Elev 505 m N 48.149,22° E 11.59683°
UTM	WGS-84	32N – 0693,145 – 5336,147
UTM (NATO)	WGS-84	32U – 0693,145 – 5336,147
GaußKrüger	Potsdam	4–470,107 – 5334,656

suffer more or less heavily from distortions. In the context of LBSs the preferred datum is WGS-84, which is due to the fact that GPS is one of several favorite positioning methods and also serves as a basis for cellular positioning. Nevertheless, developers of LBSs may be faced with many other datums and a broad range of different map projections, because geographic content that is to be related with position data derived from positioning has often been recorded on the basis of different datums and map projections.

In general, datum transformations are a complex matter for which the datum parameters must be known. In addition, they impose inaccuracies on position data, depending on the precision of source and target datums. A complete list of local datums used in the different regions of the world, the parameters of their ellipsoids, and their shifts with regard to WGS-84 can be obtained from the web site of the NIMA (NIMA Web site). In the context of LBSs, it has become a practice in recent years to tag each kind of position data and geographic content by means of a scheme maintained by the *European Petroleum Survey Group* (EPSG). This group has created a database of all local and vertical datums, map projections, and coordinate systems together with their parameters and has tagged each of them according to a numbering scheme. For example EPSG:4326 represents the WGS-84 reference system, while EPSG:4267 represents NAD-27. These numbers are used by many protocols and interfaces that the LBSs rely on. The complete database can be downloaded from the web site of EPSG, see (EPSG).

Often, it is required to transform position data from one datum to another, for example, when displaying coordinates derived by GPS on a map that has been surveyed with NAD-27. A web site that provides software for datum transformations and a number of useful links on this topic is hosted by the National Geodetic Survey (NGS) (see (NGS Web site)). Furthermore, the OGC has released a specification for *Coordinate Transformation Services*, which provide open interfaces for transforming between different coordinate systems and datums, and which is a valuable source for LBS developers (see (OGC 2001)).

3

Spatial Databases and GIS

Spatial location or position information represents an appropriate means for exactly pinpointing an object on Earth. Most of the positioning methods used in the area of LBSs, for example, GPS, deliver spatial location information as a direct result from the measurements of one or several observables and subsequent calculations. However, spatial location is not an intuitive approach that is clearly understood in all situations. For example, delivering a target's position to the LBS user as *N 48° 21' 17" E 11° 47' 15"* is less meaningful than the simple statement that the target person currently resides at the airport in Munich, Germany. Similar problems appear when setting two positions in relation, for example, for determining the distance between the LBS user and a selected point of interest and deriving the expected traveling time from that. A simple approach would deliver only the *line-of-sight distance* between both positions, but it would be more convenient for the user to get the *shortest route distance* in a road or public transportation network, preferably in combination with the shortest route displayed on a map and additional navigation assistance. As these examples demonstrate, it is inevitable to care for the mapping between spatial and descriptive location information as well as for maintaining and deriving relationships between locations in general.

Spatial databases and *Geographic Information Systems* (GISs) are the essential key technologies for fulfilling these tasks. While they cover a broad range of applications, for example, in the areas of surveying, mapping, and transportation, in the context of LBSs they are important for indicating the positions of one or several targets with respect to geographical content like borders of cities and countries, road networks, or buildings. They are used for mapping spatial location onto meaningful descriptive location information and vice versa, which is referred to as *geocoding* and *reverse geocoding* respectively, as well as for creating digital maps and routing information, or for finding nearby points of interest.

This chapter gives a general overview of spatial databases and GISs as it is necessary for the understanding of LBSs, rather than providing an in-depth knowledge of all related background technologies, research and related challenges, or even concrete products. Starting with a definition of spatial databases and GIS, the chapter explains the underlying data models and shows how to represent spatial objects and topological relationships. It introduces the three common database approaches GISs rely on and the concept of features and

---

Location-based Services: Fundamentals and Operation Axel Küpper
© 2005 John Wiley & Sons, Ltd

themes. Furthermore, the method for querying geographic content and relating it with the position data of LBS targets is demonstrated by an example. Finally, the chapter concludes with an overview of the algorithms of computational geometry and an introduction to the Geography Markup Language (GML).

3.1 What are Spatial Databases and GIS?

The question in this heading is much easier to answer for spatial databases than for GIS. In general, a database is created, organized, and maintained by a *Database Management System* (DBMS), which is a software executed on a server for fulfilling these tasks. In this sense, a spatial DBMS focuses on the efficient storage, querying, and optimization of spatial data. The range of existing GIS definitions, on the other hand, is very broad and partly controversial. Often, GISs are misconstrued as a mere geographic tool for producing maps. A more complete and comprehensive definition stems from (Rigaux et al. 2002), which characterizes the functions of a GIS as follows: "It stores geographic data, retrieves and combines this data to create new representations of geographic space, provides tools for spatial analysis, and performs simulations to help expert users organize their work in many areas, including public administration, transportation networks, military applications, and environmental information systems." Another definition is given by (McDonnell and Kemp 1996), which describes a GIS as "a computer system for capturing, managing, integrating, manipulating, analyzing, and displaying data which is spatially referenced to Earth."

In this context, it is important to note that a spatial DBMS forms an integral part of a GIS that is used for maintaining spatial data, and a GIS offers functions that go far beyond that of a spatial DBMS. Spatial DBMSs are also required for many other applications that process spatial data, such as *Computer Aided Design and Computer Aided Manufacturing* (CAD/CAM), graphical user interfaces, and augmented reality. The main differences between GISs and those applications are that GISs have to cope with larger volumes of data, model geography, and produce new information by combining the existing ones (Thurston et al. 2003).

On the market, several GIS products are available, such as ArcInfo and ArcView (both from ESRI), Tigris from Intergraph, or Smallworld. Some of them are bundled with a spatial DBMS, which can often be optionally replaced by other systems, while others are available as extensions to relational DBMSs. And, there emerge various systems that have been specifically tailored to the special needs of LBSs, such as IntelliWhere from Intergraph, miAware from MapInfo, and ArcIMS from ESRI. However, the following sections do not cover the particularities of these products, but provide an overview of the common mechanisms and structures of GISs.

3.2 Geographic versus Spatial Data Models

For understanding the idea behind GISs, it is necessary to distinguish between two levels of abstraction, which are depicted in Figure 3.1. The upper layer in a GIS, the so-called *geographic data model*, provides a conceptual view of geographic content in terms of units called *features*. A feature represents a real-world entity, for example, a building, road, river, country, or city. It consists of a *spatial component*, which fixes its location, shape,

SPATIAL DATABASES AND GIS

and topological relationship with other entities, and a *description*, which provides nonspatial information about the entity, for example, the name of a city or road, or the population of a country. Each feature has a well-defined set of operations, which is tailored to the type of real-world entity it represents. For example, a feature representing a road may provide an operation to request its length, while that of a city may offer an operation for obtaining its area size. The operations are used by applications for manipulating or requesting spatial or descriptive information of a feature and relating it to those of others. Thus, the geographic data model represents the interface between the GIS and the application.

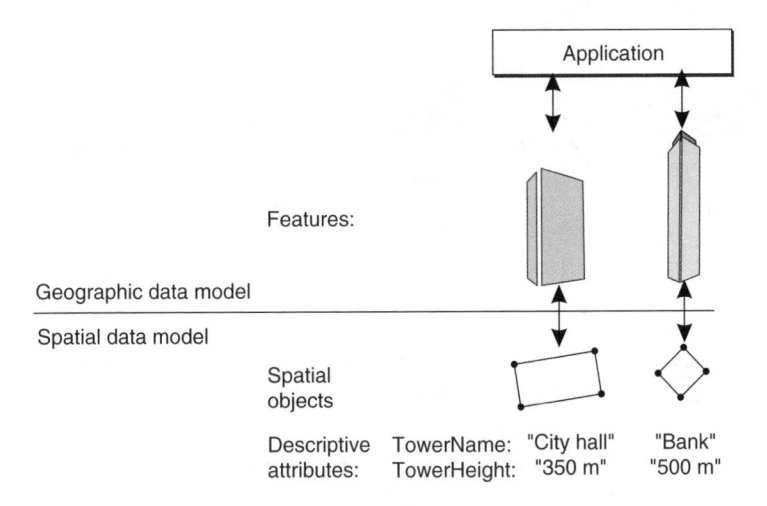

Figure 3.1 Layered GIS approach.

The lower layer deals with all aspects of the physical data management, including storage, query processing, and optimization, as well as concurrency and recovery. A particular concern in GISs is the representation of the spatial components of features, which are called *spatial objects*, as well as their combination with *descriptive attributes* for maintaining their descriptions. Spatial objects can be represented in numerous ways. Generally, it is classified into two representation categories *raster mode* and *vector mode* (which will be described in the following text), but for each mode various further modeling approaches exist. Likewise, the combination of spatial objects and descriptive attributes can be done in several ways. In the context of GISs, the lower layer coping with these issues is referred to as *spatial data model*.

The motivation behind the distinction between geographic and spatial data models is to hide the complexity of the internal data representation from the application. As mentioned earlier, features provide a well-defined set of operations to the application, but the way the features implement these operations is hidden from the application. Thus, it becomes possible to access geographic content in a unified way, for example, by using standardized query languages, without coping with the particularities of the respective implementation.

The following section gives an overview of representations of spatial objects, followed by another section introducing how to operate on features.

3.3 Representing Spatial Objects

Two main representation modes for spatial objects exist, which are referred to as *raster* and *vector modes*. Figure 3.2 shows features describing buildings, roads, and rivers (a), and the representation of their spatial objects in raster (b) and vector modes (c). Note that in most implementations spatial objects are preferably modeled in the 2-D or 2.5-D Euclidean space, because a representation of complex 3-D objects requires much more storage and computing resources and is not needed by most applications anyway, at least not in the area of LBSs. In 2.5-D, the height of a spatial object is modeled as a function of its two-dimensional coordinates.

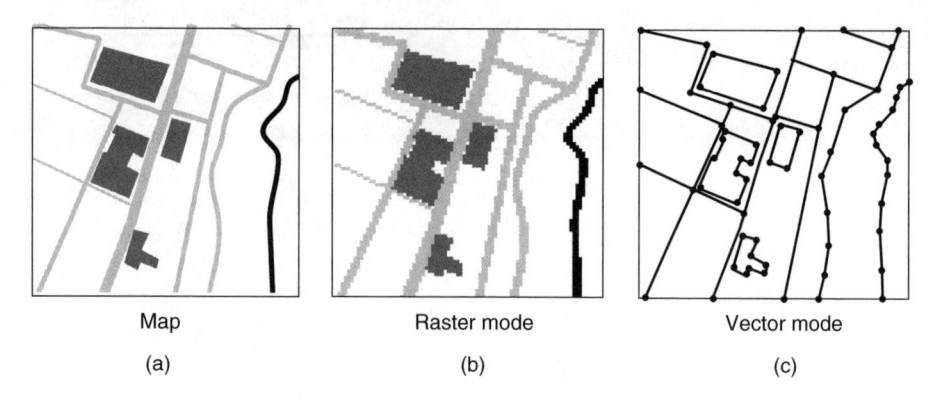

Map   Raster mode   Vector mode
(a)   (b)   (c)

Figure 3.2 Raster and vector mode.

The following section provides a brief introduction to the raster mode, followed by a more detailed explanation of the vector mode, which is much more relevant for supporting LBSs.

3.3.1 Raster Mode

A representation of spatial objects in raster mode can be best compared with a bitmap image, which consists of a number of *pixels* organized in a grid of rows and columns. Basically, in most cases, raster data is indeed derived from images, for example, satellite images, which serve as a basis for observing weather phenomena, vegetation in a certain geographic region, polar winds, or electromagnetic radiation. The analysis of such images and its conversion into raster data is also referred to as *tessellation*.

Following the principle of bitmap images, a spatial object is represented by a collection of pixels. The shape of a spatial object is reflected by the special arrangement of adjacent pixels, and its position is implicitly given by the integer coordinates of these pixels within the grid. Thus, an important measure for the quality of raster data is the resolution of the grid. A geographic area of fixed size is subdivided into a number of rows and columns, and this number defines the area in the real world covered by each pixel. The higher the number of rows and columns, the more fine grained is the resolution, but more memory is needed for storing the raster data in the database and more time is required for processing it. Another drawback is that the transmission of high-resolution

SPATIAL DATABASES AND GIS

raster data may be very bandwidth intensive, which is especially a concern in mobile communications.

Similar to bitmap images, a pixel can be assigned a value. As depicted in Figure 3.2(b), pixels describing road features have been assigned a value that corresponds to a light gray, building features are represented by dark gray, and river features by black. This shows us how descriptions can be attached to spatial objects in raster mode. Instead of explicitly linking spatial objects to attributes stored separately, the descriptions are implicitly given by the values of pixels. The possible value range of a pixel together with its semantics are referred to as *raster attribute*. The number of bits needed for representing a raster attribute is referred to as depth, and, besides rows and columns, forms the third dimension of raster data and consequently has an impact on its storage needs.

The raster mode serves well for reflecting simple descriptions such as the temperature or rainfall at a certain point and also for highlighting the progression of raster attribute values between neighboring points. However, it is not suitable for maintaining more complex descriptions like road names or opening hours of points of interest like restaurants and shopping malls. Therefore, the raster mode is primarily deployed for supporting professional applications, for example, in the areas of meteorology or pollution control, rather than for consumer mass market applications like LBSs, for which the vector mode is almost always preferred.

3.3.2 Vector Mode

In vector mode, the spatial objects are specified by means of coordinates of a reference system, which, as explained in Chapter 2, consists of a coordinate system, a datum, and, if position is expressed in 2-D, a projection. Basically, it is of no relevance which reference system is used, assuming that it is clearly referenced, for example, by means of EPSG, and that different spatial objects are based on the same reference system if spatially interrelating them.

The simplest spatial object that can be built this way is a single *point*, which might represent a person, a vehicle, or a point of interest. The data representation a single point relies on is simply that of a vector that represents the position with regard to the axes and the origin of the coordinate system. As depicted in Figure 3.3, the point object is used as a primitive to build more complex objects. A straight *line* is modeled by the interconnection of two endpoints (a), and a *polyline* is represented by an ordered list of points (b). The points are called *vertices* and the connection of a pair of vertices is called *edge*. The first and last elements of the list mark the polyline's endpoints. Examples of real-world entities represented by polylines are typically roads, traffic lines, or rivers. If the entities are bent or winding, their shapes can be represented by a concise set of consecutively connected vertices, which are arranged close enough to each other so that the resulting approximation meets the demands of the respective application. A polyline is said to be *complex* if nonconsecutive edges of it intersect (c). It is also possible to model a linear object as a composition of several polylines (d).

Surfacic objects are modeled as polygons, as shown in Figure 3.4. They are classified into *simple* and *complex* polygons. A polygon is said to be simple if it consists of a single, nonintersecting boundary (a, b). Otherwise, that is, if it has a boundary intersecting with itself (c) or defined by an outer and one or several inner boundaries (d), it is called *complex*.

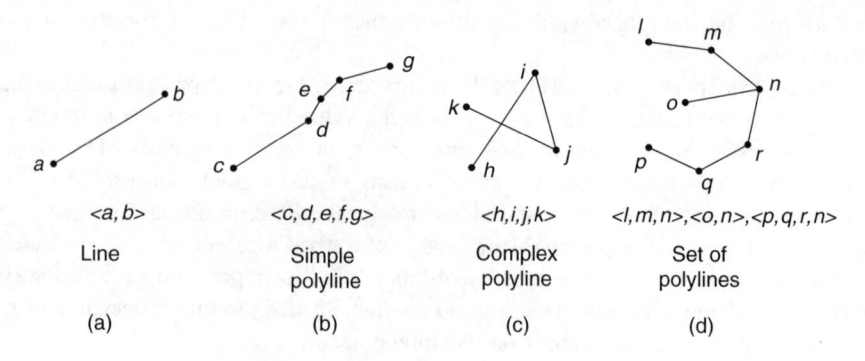

Figure 3.3 Linear spatial objects.

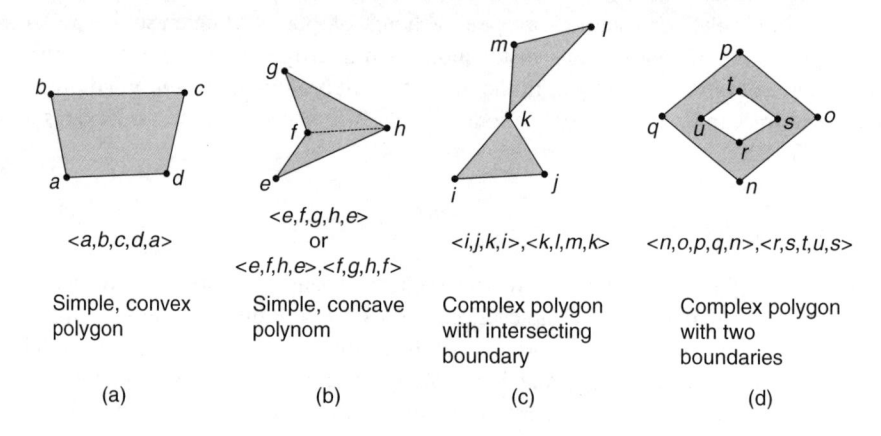

(a) <a,b> — Line
(b) <c,d,e,f,g> — Simple polyline
(c) <h,i,j,k> — Complex polyline
(d) <l,m,n>,<o,n>,<p,q,r,n> — Set of polylines

Figure 3.4 Surfacic spatial objects.

(a) <a,b,c,d,a> — Simple, convex polygon
(b) <e,f,g,h,e> or <e,f,h,e>,<f,g,h,f> — Simple, concave polynom
(c) <i,j,k,i>,<k,l,m,k> — Complex polygon with intersecting boundary
(d) <n,o,p,q,n>,<r,s,t,u,s> — Complex polygon with two boundaries

A simple polygon is basically a closed polyline and can thus be described by a list of vertices with its first and last elements being the same. For performing geometrical computations, it is important to distinguish between *convex* and *concave* polygons. A polygon is said to be convex if it has no internal angles greater than 180° (a); otherwise, it is called *concave* (b). Generally, algorithms of computational geometry, for example, the *point-in-polygon* or the *polygon-intersection* check, are much easier for convex than for concave polygons and can be performed faster and with less overhead. Therefore, it might be useful to partition a concave polygon into several convex ones (b), or, alternatively, into trapezoids or triangles. However, each kind of such partitioning also goes along with some computational overhead, which is not necessarily compensated by a reduction of complexity of subsequent geometrical computations. Similar to partitioning, complex polygons can be subdivided into simple ones and stored as sets of lists.

As can be derived from these explanations and from Figures 3.3 and 3.4, in a simple approach, spatial objects can be represented as lists of vectors. However, in this rudimentary approach it is not possible to distinguish between the different types of polylines and polygons. Therefore, a common approach is to use dedicated data representations that

SPATIAL DATABASES AND GIS 41

impose various constraints on how to specify a particular spatial object. For example, representations for simple polygons may rule that the first and last vector of a list refer to the same point and that edges must not intersect. Other constraints may prescribe at which vertex to start, with the description and the order in which the vertices have to appear in the list. In Figure 3.4, the vertices of polygons (a) and (b) appear in clockwise order, while those of (c) and (d) are listed counterclockwise. Another concern is the representation of topological relationships among several spatial objects, which is considered in the following section.

3.3.3 Representing Topological Relationships

Topological relationships, for example, adjacency or intersection, can be derived during the runtime of the respective application by performing operations on the spatial objects stored in the database, or they can be explicitly modeled when setting up the database. The former approach is obviously a more dynamic one and provides more flexibility. However, it may cause delays in the execution of the application because it often requires complex calculations on spatial objects and thus binds considerable computational resources. This is of particular concern for time-critical applications as given by LBSs, where users only accept delays in the range of some seconds. The explicit modeling, on the other hand, is a static approach that provides less flexibility but causes only little computational overhead. As a consequence, queries to the spatial database can be processed very efficiently and fast.

This section introduces three models for the explicit modeling of topological relationships, which are called *spaghetti*, *network*, and *topological models*, and which are compared in Figure 3.5. In the figure, as well as in Table 3.1, the pointed brackets <> represent a list of elements, while square brackets [ ] denote a tuple of elements.

In the spaghetti model, each spatial object is stored entirely independent of any other one, which basically results in the representation of lines, polylines, and polygons as ordered lists of vectors as depicted in Figure 3.5(a). The advantage of this model is that it is very simple and the GIS user can easily add and remove spatial objects to and from the system without worrying about topological relationships between them. However, the latter aspect is simultaneously a drawback, because relationships are only implicitly available in this way and have to be detected during runtime. For example, searching polygons that are adjacent to another given polygon makes it necessary to perform a check on common boundaries for each polygon in the database, which might be a time-consuming process. Likewise, it is cumbersome to check which polygons or polylines share a common point. In addition, the spaghetti model is very sensitive with regard to inconsistencies between spatial objects, which especially appear if point coordinates originate from different sources or need to be updated. And, the update of a point has to be explicitly done in all spatial objects in which it is a part.

In order to cope with such problems, it is customary to identify components that are shared between several spatial objects and to model them explicitly. This is done in both the network and the topological model. As suggested by its name, the network model specifically fosters applications for establishing, managing, and analyzing any kind of network, for example, road, public transportation, or computer networks. Its main motivation is therefore to cover the relationships between points and polylines. The concept of polylines as a list of homogeneous point objects is replaced by another component referred to as *arc*. An arc has well-defined starting and ending points, which are called *nodes*, as well as a number

42   SPATIAL DATABASES AND GIS

Spaghetti model
(a)

Lines:
l_1:<[0,6],[3,9],[8,8]>
l_2:<[6,0],[8,8]>
l_3:<[0,0],[8,8]>

Polygons:
p_1:<[0,6],[3,9],[8,8],[6,0],[0,6]>
p_2:<[0,0],[0,6],[6,0],[0,0]>

Network model
(b)

Nodes:
n_1:[[0,0],<a_5>]
n_2:[[0,6],<a_2>]
n_3:[[8,8],<a_2,a_3,a_5>]
n_4:[[6,0],<a_3>]
n_5:[[3,3],<a_5,a_6,a_7,a_8>]

Arcs:
a_2:[n_2,n_3,<3,9>]
a_3:[n_3,n_4,_]
a_5:[n_1,n_3,_]

Polygons:
p_1:<[0,6],[3,9],[8,8],[6,0],[0,6]>
p_2:<[0,0],[0,6],[6,0],[0,0]>

Topological model
(c)

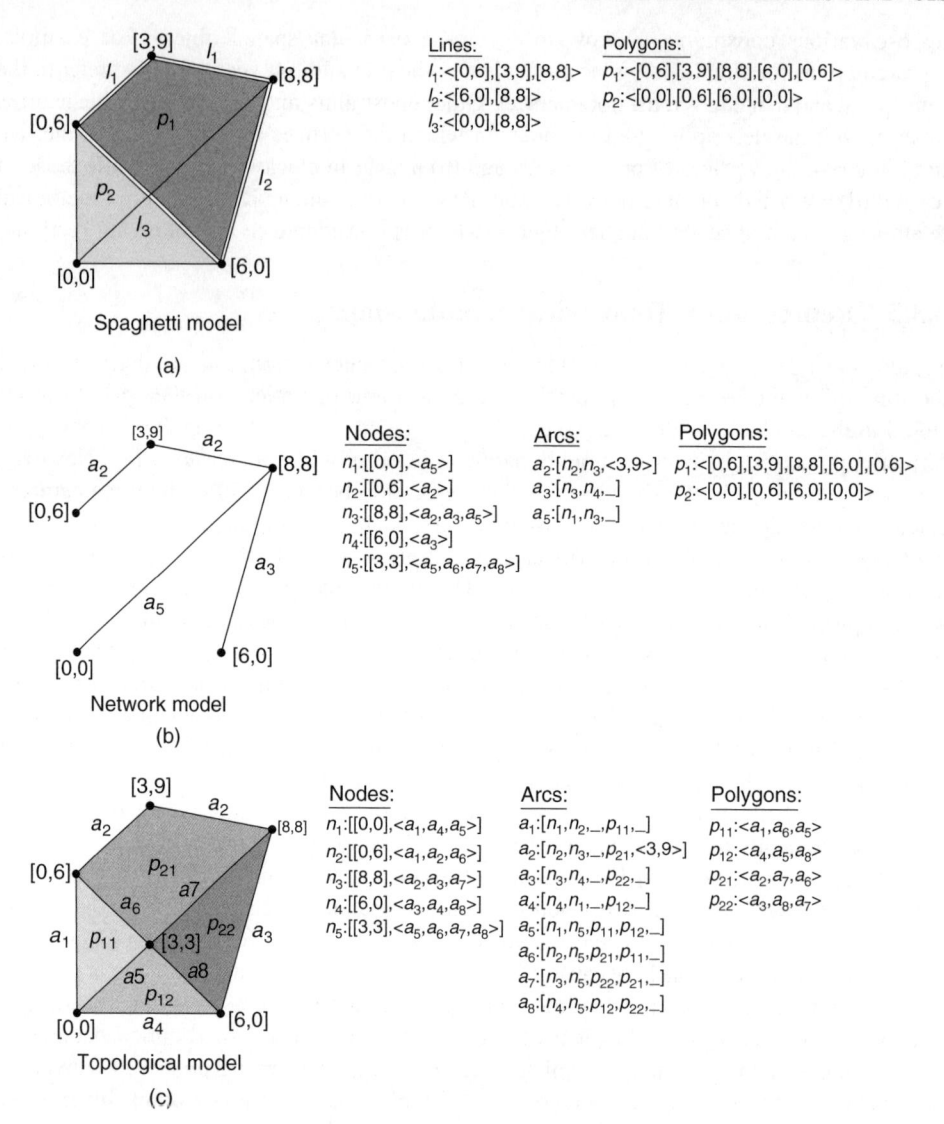

Nodes:
n_1:[[0,0],<a_1,a_4,a_5>]
n_2:[[0,6],<a_1,a_2,a_6>]
n_3:[[8,8],<a_2,a_3,a_7>]
n_4:[[6,0],<a_3,a_4,a_8>]
n_5:[[3,3],<a_5,a_6,a_7,a_8>]

Arcs:
a_1:[$n_1,n_2,$_,p_{11},_]
a_2:[$n_2,n_3,$_,p_{21},<3,9>]
a_3:[$n_3,n_4,$_,p_{22},_]
a_4:[$n_4,n_1,$_,p_{12},_]
a_5:[n_1,n_5,p_{11},p_{12},_]
a_6:[n_2,n_5,p_{21},p_{11},_]
a_7:[n_3,n_5,p_{22},p_{21},_]
a_8:[n_4,n_5,p_{12},p_{22},_]

Polygons:
p_{11}:<a_1,a_6,a_5>
p_{12}:<a_4,a_5,a_8>
p_{21}:<a_2,a_7,a_6>
p_{22}:<a_3,a_8,a_7>

Figure 3.5 Explicit Modeling of Topological Relationships.

of points in between. Starting and ending nodes may be isolated points or they may reflect intersections with other arcs. As can be derived from Table 3.1, a node contains a list of all arcs meeting at it. Polygons are still modeled as lists of points. This model allows to rapidly navigate through a network: starting at a given node, it is easily possible to derive the arcs interconnected by it and, in turn, to find out at which nodes these arcs terminate, and so on.

The topological model extends this approach in that it also reflects adjacent relationships between polygons. If an area is entirely covered by nonoverlapping polygons without any gaps, each edge is the common boundary of two polygons. The basic idea is therefore to

Table 3.1 Data representations of spatial objects (Rigaux et al. 2002)

	Spaghetti model	**Network model**	**Topological model**
point	[x:real,y:real]	[x:real,y:real]	[x:real,y:real]
node	–	[point,<arc>]	[point,<arc>]
polyline	<point>	–	–
arc	–	[start:node, end:node, <point>]	[start:node, end:node, polygon, polygon, <point>]
polygon	<point>	<point>	<arc>

extend the arc specification in that, besides starting and ending nodes and a list of points, it also contains links to both these polygons. A polygon is then no longer specified as a list of points, but as a list of arcs as demonstrated by means of the notation listed in Table 3.1. In this way, it is very easy to perform checks on adjacency. Furthermore, relationships do not suffer from inconsistencies (as in the spaghetti model) because a polygon can only be composed by a number of arcs shared with other polygons. Updates are generally performed on points or nodes and are automatically adopted by all associated spatial objects; it is not required to update them separately.

However, the topological model also imposes some constraints in that intersections between edges are always represented as nodes, which results in some drawbacks. For example, an edge crossing a polygon always subdivides it into two polygons, and the partial overlapping of two polygons results in three disjoint polygons. In other words, the topological model always produces a *planar* graph, which is in contrast to representations in the spaghetti or network model, which may be *nonplanar*. While for some applications this feature is desired, it is disadvantageous for others. For instance, spatial objects existing in the real world, such as the district of a city, may be decomposed into several objects without any semantics, for example, if the district is subdivided into two spatial objects by the crossing of a road.

3.3.4 Database Approaches for Spatial Objects

Generally, there is a distinction between relational and object-oriented DBMSs for storing conventional, nonspatial data. In a relational database, data is organized in tables where each row of a table represents a record or entry and each column represents an attribute. A record has to be well defined in that it should contain a fixed number of attributes, and each attribute must be of a certain, simple data type such as *integer*, *float*, *character*, or *string*. It is possible to relate records of different tables such that each record of a table is identified by a unique key that forms one attribute of the record, and to use this key as a reference in another table. Relational databases are very popular today, one reason being certainly the availability of standardized query languages like the *Structured Query Language* (SQL). However, they are also known to suffer from bad performance characteristics, especially when managing very large amounts of data and relations between numerous tables.

Object-oriented DBMSs, on the other hand, adopt the basic mechanisms of object-oriented modeling such as classes, methods, encapsulation, and inheritance for organizing data. The counterpart of a record is an object that encapsulates data of simple or complex types and provides a number of methods for accessing and manipulating them. Object-oriented DBMSs provide much better performance than relational ones, but it also takes more time and skills to design them. Another drawback is that no standardized query language exists. Some DBMS products are therefore based on a hybrid approach that adopts the benefits of both relational and object-oriented DBMSs while avoiding their drawbacks. These systems are known under the term *object-oriented relational DBMSs*. Most GISs operate on basic or extended relational DBMSs for representing spatial objects and descriptive attributes, and hence, in the subsequent explanations, object-oriented DBMSs are not further considered.

Rigaux et al. (2002) identify the three following approaches to represent spatial objects by relational DBMS (see also Figure 3.6):

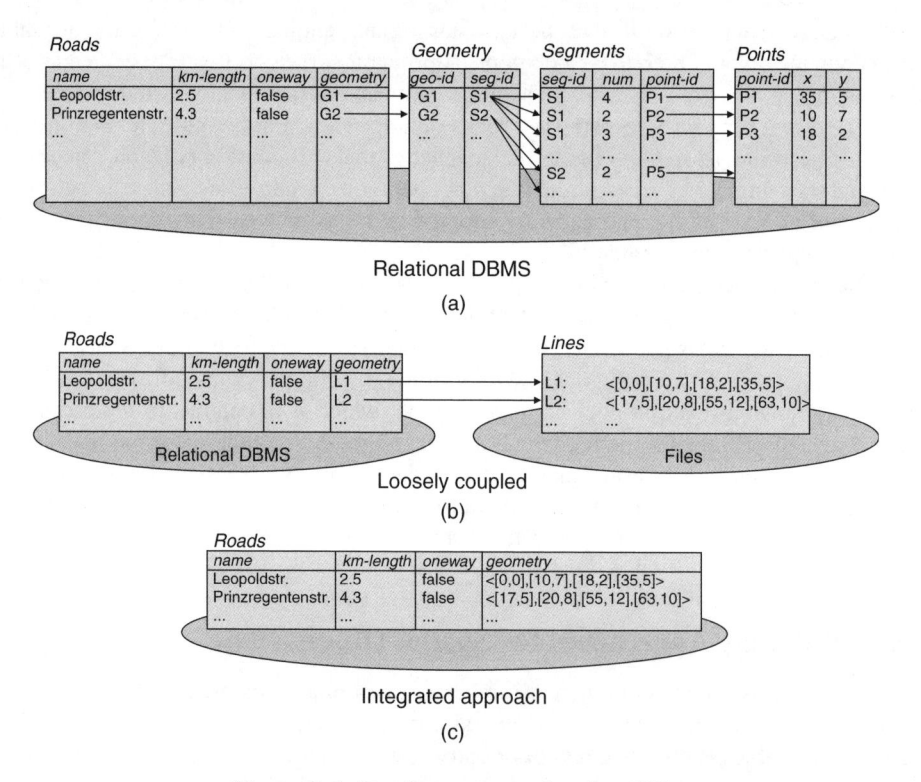

Figure 3.6 Database approaches for GIS.

Representation by relational DBMSs. The most rudimentary approach is to model spatial objects by a conventional relational DBMS as shown in Figure 3.6a. The drawback of this approach is that no dedicated data types for representing spatial objects are available, but only simple types like *integer* and *string*. Thus, spatial objects must be composed by relations of records between numerous tables. Another problem arises from the fact that the number of points in spatial objects is usually not fixed, and hence it might be necessary to represent each of them by several records at once. The benefit of this approach is that conventional DBMSs and query languages like SQL can be used, but it suffers from difficulties in managing and processing data as well as from bad performance due to the linking of many records.

Loosely coupled approach. The loosely coupled approach consists of two subsystems, one for storing descriptive attributes and another for maintaining spatial objects (see Figure 3.6). The first one is given by a conventional relational DBMS, while for the latter a file system is used. Spatial objects are stored in a certain format in files, and they are linked to the respective descriptive attributes in the relational DBMS via internal identifiers. Although this approach is more suitable for storing and processing spatial objects than the sole use of relational DBMSs, it suffers from difficulties concerning recovery techniques, querying, and optimization.

Integrated approach. Extended relational DBMSs provide an integrated approach by introducing new spatial data types like points, polylines, and polygons, as well as modified querying languages, which foster and ease the storage and query of spatial objects and their combination with descriptive attributes. As a result, it is possible to handle spatial data very efficiently and optimize query processing. This integrated approach is illustrated in Figure 3.6(c).

Most of today's commercial GIS products adopt the loosely coupled approach, for example, ArcInfo from ESRI or Tigris from Intergraph. For other relational DBMSs, for example, Oracle8i and Postgres, extensions are available for following the integrated approach.

3.4 Features and Themes

As mentioned earlier, the geographic data model provides another level of abstraction on which geographic content is organized by features, each of it representing a real-world entity by a spatial component and a description. The data representation of spatial objects and descriptive attributes, which forms the spatial data model, is hidden from the respective application. Instead, the application interacts with geographic content over well-defined operations of features that are tailored to the special type of real-world entity that the feature represents.

The appearance and structure of features as well as the topological relationships among them are fixed by *themes*. Homogeneous features describing the same type of real-world entity always belong to the same theme and must therefore be compliant with the structure, the operations, and the associated semantics prescribed by the theme. The concept of themes enables to compose spatial content by a kind of overlay concept, where the features of each theme are arranged on a dedicated layer. Depending on the demands of the respective application, it is then simply possible to overlay layers of different themes in order to detect topological relationships between their features. This is illustrated in Figure 3.7.

The figure shows three layers of different themes that might be used for supporting a location-based sightseeing application. All layers cover the same geographic region with the same benchmark. Layer (1) reflects the road network of a city, Layer (2) contains public

Layer 1: road network
(a)

Layer 2: public park areas
(b)

Layer 3: points of interest
(c)

Combination of layers
(d)

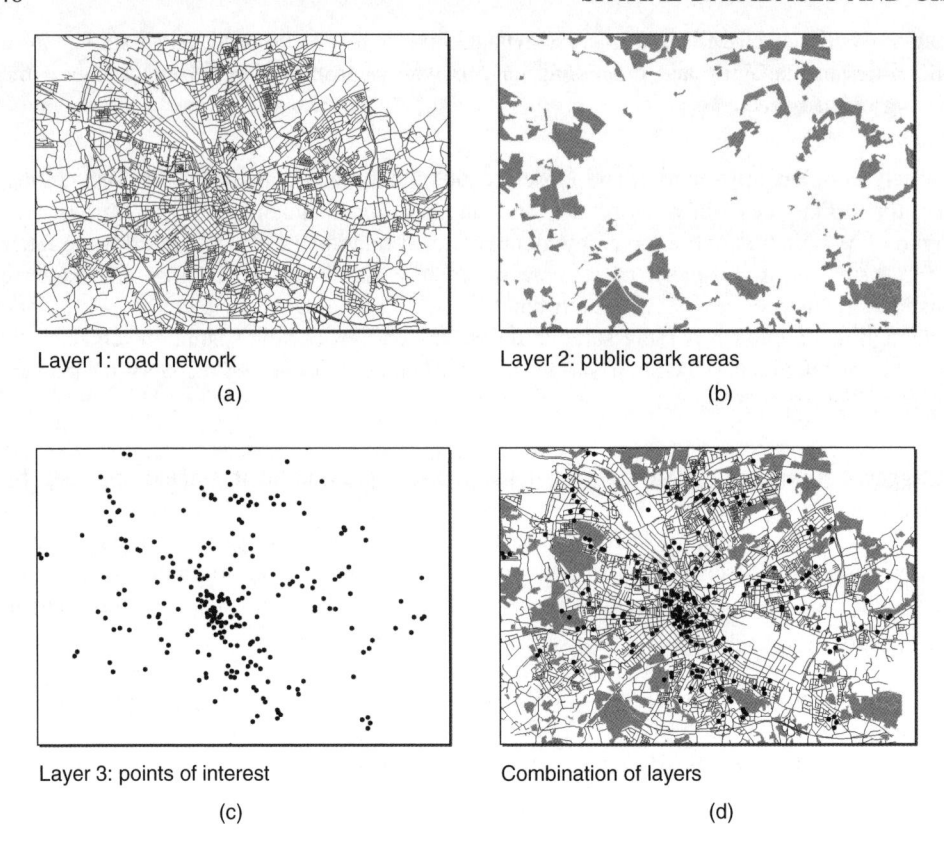

Figure 3.7 Layers and combination of layers.

park areas of that city, and Layer (3) contains points of interest such as churches, memorials, and other historic sites. Figure 3.7(d) shows the overlay of all three layers. Depending on the functionalities of the sightseeing service, different layers or different combinations of layers can be used for relating the LBS user's position with geographic content. In a simple approach, the user could be provided with a list of nearby points of interest, for example, being in a range not larger than 500 m to his own position. Optionally, the line-of-sight distance to each point of interest could be displayed. If desired, the user can also receive a list of nearby public park areas. For a more sophisticated sightseeing service, it is possible to derive nearby points of interest in a first step, and, in a second one, to determine the shortest path together with its distance between the user's position and the selected points of interest. This latter function requires to overlap Layers (1) and (3) and to apply some algorithms for detecting the shortest path. The path determined this way together with the path distance is then returned to the user. This principle can be simply extended in order to realize a number of service options. For example, an additional layer covering public transportation lines could be used to offer navigation with regard to a public transportation network and its stations instead of displaying the footway through the road network.

3.4.1 Conceptual Schemes

By the use of themes, features can be typified and must therefore follow the structure prescribed by the respective theme they belong to. This structure is usually called a *schema*, which is a term that stems from relational DBMSs. Here, a schema defines the tables, the fields in each table, and the relationships between fields and tables. In analogy to this, a schema in GIS defines a feature's description and spatial component. In addition, it also fixes topological relationships to other themes, which, for example, allows for checking whether a feature is spatially contained in the feature of another theme. The resulting model of themes and their relationships is called a *conceptual schema*.

As an example, consider the conceptual schema that is defined as an UML class diagram in Figure 3.8. It contains themes and relationships needed for realizing an LBS that delivers nearby points of interest to the user. Rather than matching the user's coordinates with the positions of all points of interest contained in the GIS, the schema allows to first check in which city and district the user currently resides. For this purpose, the themes City and District contain a spatial component Area, which describes the borders of a city and district respectively as polygons. Both themes are interconnected by an aggregation relationship. In a first step, the user's coordinates are checked for containment against the geographical areas of all cities in the GIS. If the city is found, this process is repeated for all districts of that city.

City	District	PointOfInterest
CityName *string* Area *polygon*	DistrictName *string* Area *polygon*	PoIType *string* PoIName *string* Address *string* Position *point*

City ◇—— District ▶ contains ——● PointOfInterest

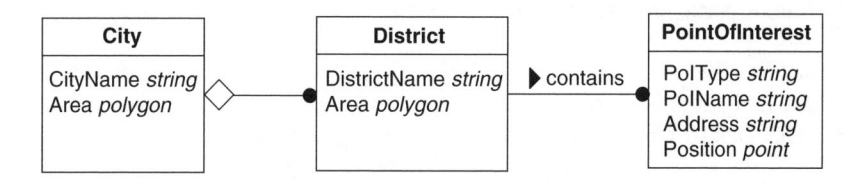

Figure 3.8 Conceptual schema example.

District, in turn, is connected to the PointOfInterest theme by the contain relationship. This theme fixes a number of descriptions for each point of interest according to the attributes PoIType (e.g., restaurant, church, memorial, etc.), PoIName, and Address. These attributes can be used to select points of interest according to the user's preferences. In addition, as the user may currently reside at the borders of a district, it is also possible to return the points of interest of neighboring districts. This is managed by an operation for checking pairs of polygons on adjacency as will be explained in the next section. Optionally, the user could also be supplied with the line-of-sight distances between his own location as derived from positioning and the positions of the points of interest as derived from the Position attribute.

Alternatively, the GIS may maintain a more sophisticated and complex conceptual schema describing the geometry and topology of roads and buildings, which would then replace the Address attribute. In this way, it would also be possible to determine the route distance between the user and the restaurants, instead of the line-of-sight distance.

3.4.2 Operations

The idea behind features is to model them as so-called *Abstract Data Types* (ADTs) that provide a well-defined set of operations that are used to obtain spatial or descriptive information about a feature or to spatially relate it with others. Applications do not directly access a feature's internal structure, but only by using the operations that it offers. The special focus in GIS is the definition of spatial ADTs. Examples of basic types are the *region ADT* for representing a geographic area, the *line ADT* for linear real-world entities like roads, and the *point ADT* for small-sized entities like persons or vehicles. From an internal point of view, they are represented by spatial objects (points, lines, polylines, or polygons) of the chosen spatial data model, but from the point of view of the application, they are only visible as ADTs. Point, line, and region ADT may serve as a basis for building more sophisticated ADTs. Examples may be a *road ADT* based on the line ADT, or a *city* or *building ADT* based on the region ADT. These derived ADTs adopt the operations and the attributes of their basic ADTs and may be extended with additional operations dedicated to the special characteristics of the corresponding real-world entity, for example, asking for the population of a city.

In general, an operation is defined by its name and its signature, where the signature determines the types of its arguments and results. GISs predominantly offer operations with spatial ADTs as arguments and Boolean, scalar, or spatial results. The following list gives an extract of many possible operations, which has been adapted from (Rigaux et al. 2002) and which is of particular importance to LBSs:

- *PointInRegion*: $point \times region \rightarrow bool$. Tests whether a given point entity is contained in a geographic region. The point ADT may represent an LBS user for which the GIS checks whether it currently stays in a given city.

- *Overlaps*: $region \times region \rightarrow bool$. Tests whether two geographical areas overlap.

- *Intersection*: $region \times region \rightarrow region$. Computes and returns the intersection of two geographical areas.

- *Meets*: $region \times region \rightarrow bool$. Tests two given geographical areas on adjacency. Can be used for selecting points of interest in the user's current and all neighboring districts of a city.

- *AreaSize*: $region \rightarrow real$. Returns the area size of a geographical area, for example, the size of the country the user stays in.

- *PointInLine*: $line \times point \rightarrow bool$. Tests the intersection between a point and a line. May be used to derive the user's location in terms of a station or line of a public transportation network.

- *ShortestRoute*: $point \times point \rightarrow line$. Returns the shortest route between two points, for example, as needed for navigation services.

- *Length*: $line \rightarrow real$. Calculates the length of a line, for example, a route in a navigation application.

SPATIAL DATABASES AND GIS

- *Distance*: *point* × *point* → *real*. Determines the line-of-sight distance between two points, for example, target and LBS user.

As can be derived from these examples, spatial operations are used to obtain some information about a feature (e.g., size or length), to generate new spatial components, or to test one or several features on the fulfillment of a certain property, which is also denoted as *predicate*. Generally, it is distinguished between *metric*, *direction*, and *topological predicates*. Metric predicates are those that test whether one or several spatial components fulfill a property with regard to a certain metric, for example, whether the size of a geographic area is less than a predefined value or whether the distance between two points exceeds a predefined value. Direction predicates check the orientation of an object with regard to another one, for example, in terms of "northwards" or "southwards". Finally, topological predicates investigate two objects with respect to topological relationships like containment or adjacency. The theory behind them is explained in the next section.

3.4.3 Topological Predicates

According to (Rigaux et al. 2002), topological relationships among spatial objects are characterized by the fact that they are invariant when these objects are translated, rotated, or scaled. They are used for testing, for example, whether a given object is contained in another one, or whether two objects overlap or meet. "Contain", "overlap", and "meet" are examples of topological predicates that describe the kind of topological relation between a pair of objects. Figure 3.9 gives some examples of topological predicates commonly used.

In order to avoid incorrectness or incompleteness in the semantics of such predicates, it is necessary to formally define them. For a formal definition, a number of parameters have to be taken into account, for example, the dimension of the involved spatial objects and their general structure (e.g., lines versus polylines, simple versus complex polygons, and so on). In order to cope with the resulting case differentiations, the formal definition is usually given by the *9-intersection model*. In this model, each spatial object is represented as a set of points in the Euclidean plane R^2. For a given point set A, it is distinguished between the interior A^0 of A, its boundary δA, and the exterior (or complement) A^{-1} of A. The point sets of two spatial objects A and B are then interrelated by building all nine intersections between these point sets in order to find out how they are topologically related. The model is usually represented by the 9-intersection matrix:

$$I_9(A, B) = \begin{bmatrix} A^0 \cap B^0 & A^0 \cap \delta B & A^0 \cap B^{-1} \\ \delta A \cap B^0 & \delta A \cap \delta B & \delta A \cap B^{-1} \\ A^{-1} \cap B^0 & A^{-1} \cap \delta B & A^{-1} \cap B^{-1} \end{bmatrix} \quad (3.1)$$

For each intersection described by these matrix elements it is distinguished between the empty set $\emptyset$ and the nonempty set $\neg\emptyset$, which leads to $2^9 = 512$ possible combinations. In order to reduce this number, it is possible to impose some constraints on the spatial objects to be considered. For example, assuming that A and B are two-dimensional, coherent, and simple polygons (i.e., without holes), the exterior does not need to be taken into account, which results in the following *4-intersection matrix*:

$$I_4(A, B) = \begin{bmatrix} A^0 \cap B^0 & A^0 \cap \delta B \\ \delta A \cap B^0 & \delta A \cap \delta B \end{bmatrix} \quad (3.2)$$

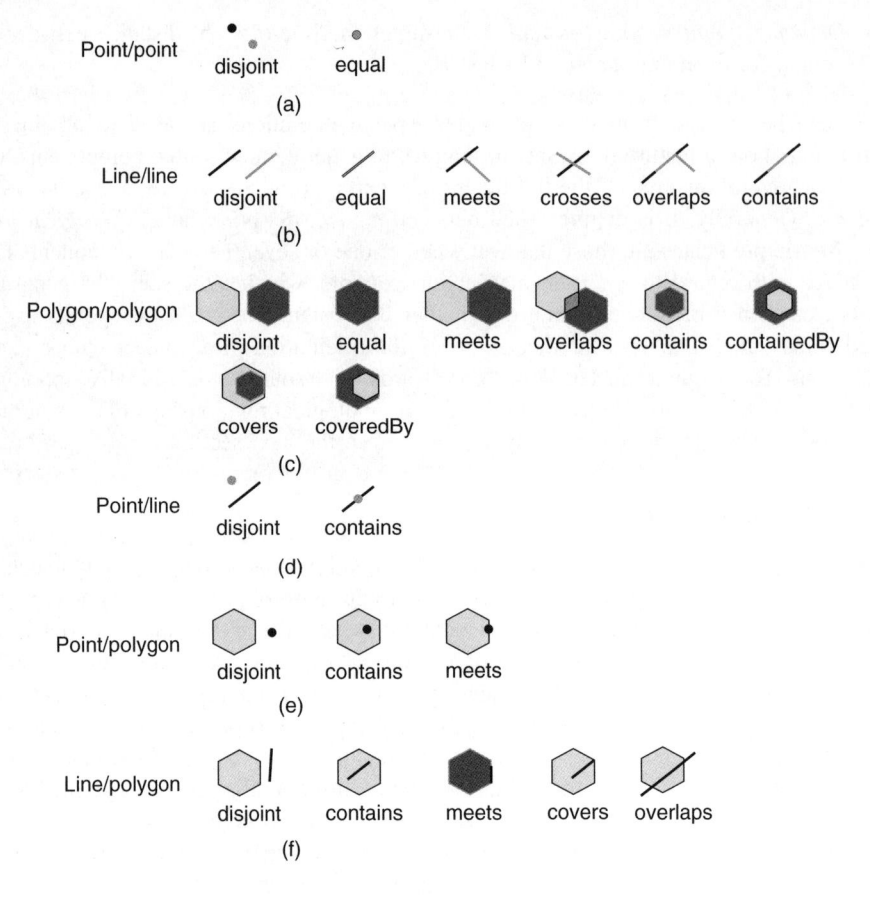

Figure 3.9 Topological predicates.

This matrix reduces the possible combinations to $2^4 = 16$, which are listed in Table 3.2. Some of them make no sense as they do not describe any possible topological relationship. For example, assuming that the boundaries of two polygons do not intersect and that their interiors do not intersect their boundaries, it is then not possible that their interiors intersect (see third row of the table). Removing such contradictory combinations leads to the following eight predicates between a pair of two-dimensional, coherent, and simple polygons: *disjoint*, *meet*, *equal*, *containedBy*, *cover*, *contain*, *coveredBy*, and *overlap*. The appliance of these predicates is illustrated in Figure 3.9(c).

The 4-intersection model is thus a very straightforward model, but is only valid under the constraints mentioned earlier. Relations between objects of different dimensions (≤ 2) as well as between complex polygons must be considered by means of the 9-intersection matrix. More details on that can be found in (Egenhofer 1989) and (Egenhofer and Franzosa 1991).

3.4.4 Queries

This section demonstrates the querying of a spatial, relational DBMS by means of an example that is based on the conceptual schema of cities, districts, and points of interest

SPATIAL DATABASES AND GIS

Table 3.2 Topological predicates (Egenhofer and Franzosa 1991)

	$\delta A \cap \delta B$	$A^0 \cap B^0$	$\delta A \cap B^0$	$A^0 \cap \delta B$
A disjoint B	$\emptyset$	$\emptyset$	$\emptyset$	$\emptyset$
A meets B	$\neg\emptyset$	$\emptyset$	$\emptyset$	$\emptyset$
–	$\emptyset$	$\neg\emptyset$	$\emptyset$	$\emptyset$
A equals B	$\neg\emptyset$	$\neg\emptyset$	$\emptyset$	$\emptyset$
–	$\emptyset$	$\emptyset$	$\neg\emptyset$	$\emptyset$
–	$\neg\emptyset$	$\emptyset$	$\neg\emptyset$	$\emptyset$
A containedBy B	$\emptyset$	$\neg\emptyset$	$\neg\emptyset$	$\emptyset$
A coveredBy B	$\neg\emptyset$	$\neg\emptyset$	$\neg\emptyset$	$\emptyset$
–	$\emptyset$	$\emptyset$	$\emptyset$	$\neg\emptyset$
–	$\neg\emptyset$	$\emptyset$	$\emptyset$	$\neg\emptyset$
A contains B	$\emptyset$	$\neg\emptyset$	$\emptyset$	$\neg\emptyset$
A covers B	$\neg\emptyset$	$\neg\emptyset$	$\emptyset$	$\neg\emptyset$
–	$\emptyset$	$\emptyset$	$\neg\emptyset$	$\neg\emptyset$
–	$\neg\emptyset$	$\emptyset$	$\neg\emptyset$	$\neg\emptyset$
–	$\emptyset$	$\neg\emptyset$	$\neg\emptyset$	$\neg\emptyset$
A overlaps B	$\neg\emptyset$	$\neg\emptyset$	$\neg\emptyset$	$\neg\emptyset$

as introduced in Figure 3.8. The notation used for creating tables and queries have been adopted from (Rigaux et al. 2002), which also presents an analogous approach for querying spatial object-oriented DBMSs. At first, it is necessary to determine the relational schema the subsequent queries are based on:

```
CreateTable City (
    CityName string,
    Region polygon,
    Key(CityName))

CreateTable District (
    DistrictName string,
    Region polygon,
    CityName string,
    Key(CityName))

CreateTable PointOfInterest (
    PoIType string,
    PoIName string,
    Address string,
    Position point,
    DistrictName string,
    Key(PoIName))
```

Thus, we operate on the three tables `City`, `District`, and `PointOfInterest`, which contain records representing the corresponding features. The aggregation and containment relationships shown in Figure 3.8 are modeled by attributes that link the different tables. For example, the attribute `DistrictName` in the `PointOfInterest` table refers to that record in the `District` table, which represents the district the respective point of interest lies in. The `District` and `City` tables are interrelated in a similar way. In addition, it is assumed that the user is represented by a data tuple `LBStarget` that, among others, contains an attribute `Position` which carries its coordinates as derived from positioning.

The following queries are based on SQL syntax (see (Wilton and Colby 2005) for an introduction). The first query is used to obtain the city in which the user currently resides:

```
select CityName
from City
where PointInRegion(LBStarget.Position,Region)
```

The query makes use of the *PointInRegion* operation that has been introduced in Section 3.4.2. For all records stored in the `City` table, the point-in-region check is performed. Assuming that the spatial components of cities do not overlap, the query returns at most one city. If no result is returned, the user may reside in a rural area that is not covered by any of the records stored in the table. The current district of the user can be obtained in a similar way.

The following query returns a list of points of interest, together with their types, names, and addresses that are located in the district where the LBS user currently resides and that match with the type the user has specified in `LBStarget.SelectedPoIType` (e.g., "restaurants"):

```
select PoIType,PoIName,Address
from PointOfInterest
where DistrictName=(
    select DistrictName
    from District
    where PointInRegion(LBStarget.Position,Region))
and LBStarget.SelectedPoIType=PoIType
```

The query contains an encapsulated `select` statement for deriving the current district of the user. Once this district is obtained, the containing points of interest are matched against the desired type of point of interest. If this query does not succeed, the LBS user may initiate an additional query to search for appropriate points of interest in neighboring districts. Such a query may then appear as follows:

```
select PoIType, PoIName, Address
from PointOfInterest
where DistrictName=(
    select d2.DistrictName
    from District d1, District d2
    where PointInRegion(LBStarget.Position,d1.Region)
    and Meets(d1.Region,d2.Region))
and LBStarget.SelectedPoIType=PoIType
```

In this query, the current district of the user is first obtained by the `PointInRegion` operation as before, and the resulting district is then checked on adjacency with other districts by using the `Meets` operation introduced in Section 3.4.2.

In a similar fashion, it is possible to search for other items, locate the target in a road or public transportation network, calculate shortest routes, or create maps, assuming that the respective GIS supports the corresponding operations. The following section demonstrates the realization of such operations by means of an example.

3.5 Algorithms of Computational Geometry

Some of the operations on ADTs mentioned before can be simply realized on the basis of the explicit representation of topological relationships as delivered by the network or topological model, for example, checks on adjacency or intersection. However, if information about one or several spatial objects is requested that have not been explicitly modeled before, it is necessary to derive it during runtime of the application. As an example, consider the current position of an LBS user that is to be interrelated with geographic content, for example, in the sense of reverse geocoding, maybe dependent on the user profile and other time-dependent circumstances. In this case, spatial operations have to fall back on algorithms of computational geometry that calculate the desired results during runtime.

In general, an algorithm can be described as a detailed, finite sequence of actions performed on some data in order to solve a problem. In the special case of computational geometry, algorithms are needed for operating on geometric or spatial objects such as points, polylines, and polygons introduced earlier. They are not only of relevance for spatial databases and GISs, but also for realizing geographic user interfaces or CAD/CAM applications. With regard to the spatial objects introduced in this chapter, examples of such algorithms are:

- the *point-in-polygon check* for testing whether a point lies within a polygon,

- the *polyline-intersection check* for detecting intersections between polylines,

- the *polygon-intersection check* for detecting intersections between polygons,

- the *polygon-intersection computation* for calculating a new spatial object from the intersection of two polygons,

- the *area-size computation* for calculating the area of a polygon,

- the *length computation* for calculating the length of a line or polyline,

- the *shortest path computation* for calculating the shortest path between two points in a network.

An important design goal when developing algorithms is to minimize their complexity. The complexity is commonly expressed as the upper bound of the number of actions performed by the algorithm depending on the input size. Algorithms of computational geometry are generally considered to be of comparatively high complexity, which is a significant problem for real-time applications like LBSs. These applications impose very high time constraints on algorithms, because their users typically expect to obtain results within a few seconds. This is worse when the system has to serve a high number of users simultaneously, which results in additional system requirements on scalability, load balancing, and data consistency. Therefore, research in the area of computational geometry is permanently concerned with the improvement and optimization of existing algorithms and the development of new ones in order to speed up calculations or even to find an algorithm that is optimal for a given problem.

Since there exist numerous algorithms in the area of computational geometry and the theory behind them is very extensive, it is not possible to exhaust this matter in this book. However, in order to give an understanding of the principal working, the point-in-polygon check is demonstrated here as an example. It is of particular importance for reverse geocoding and thus for LBSs, for example, for checking whether the derived position of an LBS target is located within a city.

The working of the point-in-polygon check for simple polygons is illustrated in Figure 3.10. It assumes that a ray r is drawn in one direction, for example, parallel to the x-axis, starting from the point to be tested. The algorithm then simply counts the number of crossings between the ray and all edges of the polygon. If this number is even, for example, as for point p_1, then the point lies outside the polygon. Otherwise it lies inside, for example, point p_2 in the figure.

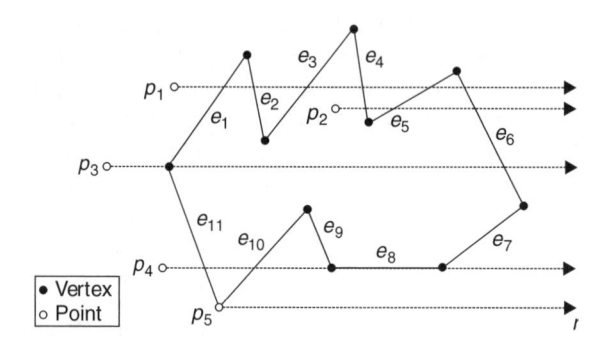

Figure 3.10 Point-in-polygon check.

However, the algorithm requires some exceptional cases to be taken into account. The first one is illustrated by point p_3, whose ray exactly intersects with one vertex of the polygon. As a result, the algorithm would deliver three intersections, that is, between the ray r and the edges e_1, e_6, and e_{11}, and come to the wrong conclusion that p_3 lies within the polygon. The next problem is demonstrated by means of point p_4, whose ray turns out to be collinear with the edge e_8. The algorithm would count intersections of r with the edges e_7, e_8, e_9, e_{10}, and e_{11}, and again derive the wrong conclusion that p_4 lies within the polygon. The last problem concerns point p_5, which coincides with the vertex interconnecting e_{10} and e_{11}. In this case, the algorithm would wrongly conclude that the point lies outside the polygon. These exceptional cases can be solved by applying the following rules:

- Count an edge e as an intersection only if there exists at least one endpoint of e strictly above r (this refers to p_3).

- Do not count edges being collinear with r (this refers to p_4).

- If the point is on one edge of the polygon, it lies within the polygon (this refers to p_5).

SPATIAL DATABASES AND GIS

An algorithm for checking whether a point p lies within a simple polygon P in pseudocode may then appear as follows:

```
begin
    count=0
    r=horizontal ray containing P
    for (i=1 to n)
        if (Meet(p,edge[i])) then return(true)
        if (Intersect(r,edge[i]) and not Collinear(r,edge[i])) then
            if (Above(edge[i],r)) then count:=count+1
    end for
    if (count is odd) then return(true)
    otherwise return(false)
end
```

The algorithm terminates with the Boolean result "true" if p lies within P, otherwise returns "false". It makes use of four other algorithms (Meet, Intersect, Collinear, Above) for covering the four exceptional cases mentioned earlier. The input size of the algorithm is given by the number n of edges to check, from which its complexity in terms of computation time can be derived. As each edge is subject to a constant number of tests, it is of linear complexity, that is, $\Theta(n)$.

3.6 Geography Markup Language

Spatial databases and GISs provide mechanisms for storing and processing location information. However, the operation of LBSs essentially relies on that the location information is exchanged between several actors of the LBS supply chain identified in Chapters 1 and 10. This includes the exchange of complex geographic content, for example, between content providers or between content and LBS providers.

The *Geography Markup Language* (GML) is a specification of the OGC for exchanging geographic content. It is based on the *OpenGIS Abstract Specification*, which is a common model of geography, also proposed by the OGC, and which provides directives, rules, and terminologies for modeling and expressing geographic content (OGC 1999). This specification is not only of relevance to GML, but in the meantime is being adopted by the majority of vendors in the area of spatial databases and GIS.

Similar to GIS, GML organizes geographic content as features. Each feature is made up of a number of so-called *properties*. Roughly speaking, it can be distinguished between *geometry properties*, which represent the location and shape of the feature, *topology properties*, which represent connectivity or adjacency with other features (similar to the network and topological models introduced earlier), *temporal properties*, which describe the feature's temporal behavior, and *descriptive properties*, which carry nonspatial information.

GML is a complex framework of XML schema definitions as illustrated by the UML class hierarchy in Figure 3.11. The root of the hierarchy is an abstract GML object, which establishes the general GML model and syntax and which serves as a basis for feature, geometry, and other schema definitions. Note that the figure shows a class hierarchy, that is, the relationships between the schemas are of type inheritance and do not cover containment interdependencies resulting from XML <include> or <import> statements. For example, a feature may contain one or several geometry objects for reflecting the location and shape of the feature. Developers can make use of the schemas in Figure 3.11 either by adopting and extending them by inheritance, for example, in order to define extended features meeting the demands of the respective applications, or by including them in their own schema definitions. For a detailed overview of the class hierarchy shown in Figure 3.11, see (OGC 2003).

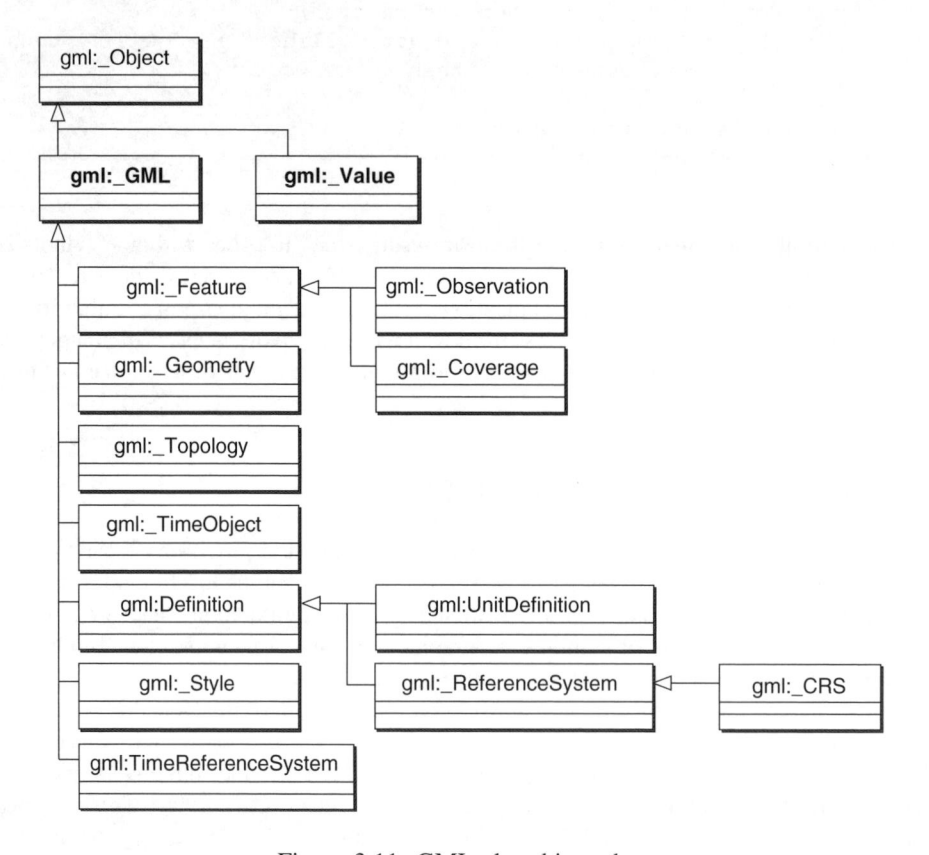

Figure 3.11 GML class hierarchy.

Spatial objects are modeled by the geometry schema in GML. Examples of supported geometry elements are:

- **Point.** Represents a 0-dimensional object.
- **Line String.** Describes a piecewise linear path by a list of coordinates connected by straight line segments.
- **Linear Ring.** Describes a closed, piecewise linear path as a list of coordinates interconnected by straight-line segments. The last coordinates must coincide with the first one and at least four coordinates are required.

SPATIAL DATABASES AND GIS

- **Polygon.** Represents a connected surface. May consist of one outer boundary (exterior) and one or several inner boundaries (interiors), each of it fixed by linear rings.

- **MultiPoints, MultiLineStrings, MultiPolygons.** Define collections of the respective elements.

- **MultiGeometry.** Provides a container for multiple, arbitrary geometry elements.

All of these geometries are based on lists of coordinates fixed by `<coordinates>`, `<coord>`, or `<pos>` elements, which slightly differ from each other in the syntax used to specify the coordinates. A point object is defined as follows:

```
<Point gid="p1" srsName="EPSG:4326">
    <coordinates>48:21:17N,11:47:15E</coordinates>
</Point>
```

The `gid` attribute attaches a tag to the object for referencing it in other objects, and the `srsName` attribute denotes the spatial reference system the coordinates refer to. In the example, this reference system is identified by EPSG, and the identifier "4326" refers to WGS-84. A polygon is composed by one or several linear rings as follows:

```
<Polygon gid="poly1" srsName="ownSRS">
    <outerBoundaryIs>
        <LinearRing>
            <coordinates>0,0 100,0 100,100 0,100 0,0</coordinates>
        </LinearRing>
    </outerBoundaryIs>
    <innerBoundaryIs>
        <LinearRing>
            <coordinates>10,10 10,40 40,40 40,10, 10,10</coordinates>
        </LinearRing>
    </innerBoundaryIs>
</Polygon>
```

Owing to simplicity, a proprietary reference system has been selected in this example. The other geometric objects identified in the previous list can be constructed in a similar fashion.

Geometric objects defined in this way can be embedded into feature objects. The corresponding feature schema provides a structure for combining geometric objects with descriptive and temporal attributes. The feature schema serves as a basis for defining tailored features meeting the demands of the respective applications, which are referred to as *application schemas*. The following example shows how the application schema for representing an LBS target may be constructed:

```
<element name="LBSTarget" type="ex:LBSTargetType"
    substitutionGroup="gml:_Feature"/>

<complexType name="LBSTargetType">
    <complexContent>
        <extension base="gml:AbstractFeatureType">
            <sequence>
```

```
            <element name="Name" type="string"/>
            <element name="Nickname" type="string"
                minOccurs="0" maxOccurs="unbounded"/>
            <element name="MSISDN" type="string"
                minOccurs="1" maxOccurs="unbounded"/>
            <element ref="gml:location">
        </sequence>
     </extension>
   </complexContent>
</complexType>
```

At first, it is declared that the application schema `LBSTarget` can be used as a substitute for the common `gml:_Feature` schema. The type definition of `LBSTarget` then inherits the structure of the common `gml:AbstractFeatureType` and attaches some elements to it needed for representing the LBS target person as a feature. For example, the LBS target must be denoted by an element `<familyName>` and optionally by one or several nicknames carried by a `<nickname>` element. To guarantee an unambiguous identification, it is necessary to provide at least one telephone number of the target in addition, which, in GSM terminology, is referred to as *Mobile Subscriber ISDN Number* (MSISDN). Finally, the target's position is contained within the `<gml:location>` element. An example of an LBS target feature may then appear as follows:

```
<LBSTarget>
    <Name>John Miller</Name>
    <Nickname>Netaholic</Nickname>
    <MSISDN>+491791234567</MSISDN>
    <MSISDN>+491739876543</MSISDN>
    <gml:location>
        <Point srsName="EPSG:4326">
            <coordinates>48:21:17N,11:47:15E</coordinates>
        </Point>
    </gml:location>
</LBSTarget>
```

Note that `<gml:location>` is an abstract element that allows embedding of any geometric object. If the exact coordinates of the target are known, for example, when derived by GPS positioning, the location may be contained in a point object. If, on the other hand, the location is only known in terms of the radio cell where it currently resides, then its location may be described by a linear ring that fixes the location and the shape of this cell.

As depicted in Figure 3.11, GML also provides schemas for observations and coverages. An observation feature describes the act of observing or measuring, for example, by a camera or sensor, a person, or other entities. The main difference of an ordinary feature is that it provides properties for fixing temporal aspects and results of the observation and measurement respectively. Coverage features, on the other hand, provide a means for specifying raster data, such as remotely sensed images, aerial photographs, soil distribution, or digital elevation models.

Additional mechanisms of GML include the grouping of several features in the so-called *feature collections*, for example, in order to model a city as a collection of road features, and the modeling of topological relationships between features by means of the topology schema shown in Figure 3.11. Other schemas enable the specification of one's own spatial reference systems (Coordinate Reference System (CRS) schema), the fixing of measurement units (unit definition schema), or the description of graphical styles for features, geometries, and topologies (style schema). The detailed specification of these schemas can be derived from (OGC 2003).

Finally, it should be noted that GML has experienced a significant restructuring in the last versions of its specification, which has resulted in a change of interdependencies between schema definitions and many new element definitions. The latter especially concerns the geometry schema, which has been extended with various new geometry objects that partially replace objects of earlier releases. However, the current release 3.0 also cares for backward compatibility with older version. The examples given in this section are based on GML 2.0.

3.7 Conclusion

The theory and technology behind spatial databases and GIS are very complex, and this chapter could only give a general understanding of this topic as it is required in the context of LBSs. Issues that could not be addressed are, among others, *constraint data models*, which represent an alternative to the classical relational and object-oriented models for managing spatial content, *indexing* for efficiently processing spatial queries as well as the particularities of commercial systems. A good overview of these topics as well as a more detailed coverage of the subjects addressed in this chapter can be found in (Rigaux et al. 2002).

In research, the so-called *spatio-temporal databases* are gaining more and more momentum in recent years. These databases allow the modeling and representation of mobile spatial objects based on the constraint data model, while the classical approaches introduced in this chapter rather focuses on the modeling of static objects. Although these technologies to a large extent have not yet found their way into concrete products, they might become important also in the area of LBSs within the next years. They provide facilities for efficiently storing, querying, and interrelating mobile spatial content, which would, for example, allow to observe LBS targets in a given geographical context over a longer period of time and consider the history of their movements during LBS operation (which is a sensitive matter with regard to privacy, as will be explained in later chapters). An overview of this emerging field can be found in (Erwig et al. 1999) and (De Caluwe et al. 2004).

The designers and developers of LBS applications can adopt the functions of spatial databases and GIS via *Application Program Interfaces* (APIs), perhaps of the commercial products mentioned at the beginning of this chapter. Alternatively, the OGC offers an open interface specification, which is termed *simple features* and is available for the *Interface Definition Language* (IDL) of CORBA, for SQL, and for Microsoft's OLE/COM technology (see (OGC 1998)). Some open source implementations based on these specifications are available from (Sourceforge.net). Another common approach is to integrate spatial databases and GIS in an LBS middleware, which provides access to them via subservices like directory, route, or map services. This will be explained in Chapter 12.

4

Basics of Wireless Communications

Today's wireless communications would not be possible without radio signals that are generated and emitted from a sender, propagate through the atmosphere, and are received and interpreted by a receiver. In the area of LBSs, there exist two applications for radio signals. First, as described earlier, they are needed for the wireless communication between a mobile terminal and a fixed network, which is accomplished by manipulating the signal's parameters depending on the data to be transferred through a process known as *modulation*. Second, they provide the basis for almost all positioning methods used for locating an LBS target. This is accomplished by utilizing some physical phenomena of radio signals, especially the velocity with which they travel or the attenuation or path loss they experience when traveling from the sender to the receiver.

For an understanding of the working of LBSs in general and positioning in particular, it is therefore useful to get an overview of the basics of wireless communications. This chapter provides an overview of the essentials of radio signals, introduces their representation in time and frequency domain, and shows how to modulate and demodulate data to and from them. Furthermore, it is explained how radio signals propagate and how they are influenced during propagation by certain circumstances. Finally, the chapter gives an introduction into the mechanisms of multiplexing and multiple access, which are needed for subdividing the air interface into a number of independent channels and for coordinating the access to these channels respectively.

4.1 Signals

In computing and communication technology, information is represented by binary *digital data*. This data appears as a sequence of bits, each of it taking on one of two discrete values, the binary 0 and the binary 1. In the earlier days of communication networks, information was encoded as *analog data*, which is characterized by a continuous value range at some interval. A typical example of analog data is human speech.

Location-based Services: Fundamentals and Operation Axel Küpper
© 2005 John Wiley & Sons, Ltd

Data defined in this way is hence an abstract entity that needs to be represented in the physical world in a certain manner. One way of representation is to use transistors, which are simple on/off switches controlled by electricity and where a single bit is represented by the state of the switch. Other ways are the usage of magnetic or optical storage media such as hard disks or compact disks. While these technologies have the aim to make data permanent and available for processing, the transfer of data between physically distributed entities is always based on *signals*, which are propagated over a *transmission medium*. In wired networks, transmission media are typically copper twisted pair, copper coaxial cable, and optical fiber. The transmission medium for wireless communications is always the atmosphere, the space, or water.

Like data, signals can be subdivided into *digital* and *analog signals*. Digital signals are composed by a sequence of well-defined voltage pulses, which change their values at discrete time intervals and which are passed along a wired medium in a certain pattern that reflects the bits to be transmitted. Some wired systems like Ethernet or ISDN make use of such voltage pulses to transmit data.

In contrast to this, signals in wireless communications are electromagnetic waves, which by their very nature are analog, that is, they have a continuously varying strength. As suggested by its name, an electromagnetic wave consists of an electric and a magnetic field as shown in Figure 4.1. The moving charge of a source generates an oscillating magnetic field, and this in turn generates another changing electric field. To put it simply, an electromagnetic wave is the product of the interaction between time-varying electric and magnetic fields, which propagate away from the emitting source.

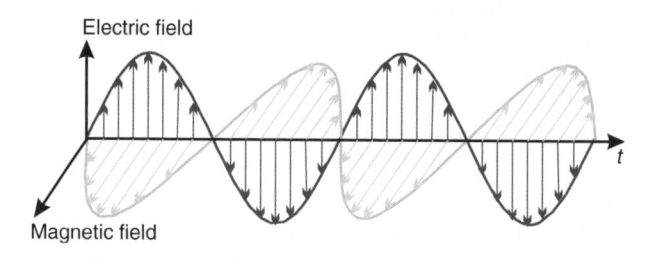

Figure 4.1 Composition of an electromagnetic wave.

Electromagnetic waves are produced and received by *antennas*. The receiving antenna converts radio signals from the surrounding environment into alternating current and delivers it to electronic equipment connected to the antenna, known as the *receiver*. Conversely, the transmitting antenna radiates alternating current delivered by a transmitter into the surrounding environment in the form of radio or microwave signals. For a bidirectional communication, as it occurs in most wireless communication systems considered in this book, the same antenna is usually used for both transmission and reception.

As can be derived from Figure 4.1, an electromagnetic wave in its simplest form is just a sine wave:

$$g(t) = A_t \sin(2\pi f_t t + \varphi_t) \tag{4.1}$$

Thus, a signal is defined by the following time-dependent parameters: the *amplitude* A, the *frequency* f, and the *phase* φ. The amplitude is the maximum strength of the signal,

BASICS OF WIRELESS COMMUNICATIONS

that is, its peak, and is typically measured in volts. The frequency is the rate at which the signal repeats, and it is expressed in *Hertz* (Hz), a unit which is defined as cycles per second. Finally, the phase denotes the relative position of a signal in time within a single period and is expressed in degrees.

If the velocity of a signal is known, which is usually close to the speed of light (see Section 4.2.3), it is possible to determine its *wavelength*, which is the distance in space between the corresponding phase of two consecutive cycles of the signal, for example, between the peak of one cycle and the peak of the next one. The wavelength is a function of a signal's frequency f and its velocity v:

$$\lambda = \frac{v}{f} \qquad (4.2)$$

Usually, frequency is an important measure when considering the amount of data that can be transferred by a signal in a given period of time. Although strongly correlated with frequency, the wavelength is the preferred unit when investigating propagation effects of signals, which are often a function of the relation between the wavelength and the dimension of obstacles. Also, it is an essential parameter in antenna design. Both will be described in subsequent sections.

4.1.1 Modulation

A pure sine wave does not carry any information in its pure form. In order to represent data, one or several of its three parameters A, f, and φ have to be varied over time at the transmitter, a process that is known as *modulation*. The receiver then scans the incoming signal and extracts the data from it, which is referred to as *demodulation*. As a sine wave manipulated in this way carries information, it is usually denoted as *carrier*, and the frequency of this wave is referred to as *carrier frequency*. Depending on whether the data that is to be modulated onto a signal is digital or analog, it is distinguished between *digital* and *analog modulations*, both of them being relevant to wireless communications.

Digital modulation, for which an alternative term is *shift keying*, is concerned with converting digital data into an analog signal by varying one or several parameters of a sine wave of a given carrier frequency. In general, it is distinguished between *amplitude*, *frequency*, and *phase-shift keying* (ASK, FSK, PSK). Figure 4.2 gives an overview of these modulation schemes. Using ASK, the amplitude of the carrier represents the two binary values 0 and 1. For example, an amplitude of zero, that is, the absence of the carrier, may encode the binary 0, while the presence of the amplitude may represent the binary 1, or vice versa. In FSK, the binary values are represented by two frequencies that are arranged directly above and below the carrier frequency. The lower frequency may represent the binary 0, and the upper frequency the binary 1, or vice versa. Finally, in PSK the phase of the carrier is shifted to represent data. A phase shift of 180° may encode the binary 1, and a phase shift of 0° the binary 0, or vice versa.

What is common to these shift-keying methods is that all of them define two signal states to represent data, and hence they are also referred to as *binary shift-keying methods*. In practice, more advanced modulation schemes are often used that define more than two signal states, in order to convert several bits at once. For example, in *Quadrature PSK* (QPSK), as used for most 3G networks, the phase of the signal is shifted by multiples

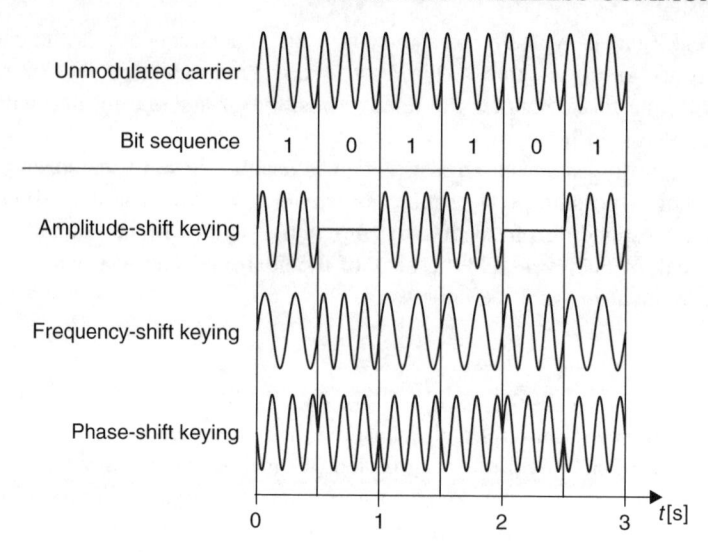

Figure 4.2 Digital modulation schemes.

of 90°, whereby it becomes possible to encode two bits by one signal change and which consequently results in an increased data rate. Another example is *Quadrature Amplitude Modulation* (QAM), which defines three different amplitudes and four phase shifts for each amplitude and hence allows the encoding of four bits at once.

Although it appears that in this way data rates can be arbitrarily increased by simply defining additional signal states, this is certainly not true. Such signals are very susceptible with regard to attenuation and distortion during transmission, and hence, with an increasing number of different signal states, it becomes very hard for the receiver to distinguish between them. The degree of attenuation and distortion is affected by many conditions. For wireless transmission, for example, these conditions are given by the distance between the transmitting and receiving antennas, or the arrangement of buildings between them. These effects will be discussed in the following text.

At this point, it is important to note that no transmission medium is capable of transmitting the exact shape of a modulated signal. During transmission, a signal is always attenuated and distorted, whereby the degree of attenuation and distortion always depends on the respective medium used and, especially in wireless communications, on circumstances like the arrangement of obstacles between transmitting and receiving antenna. Distortion has its origins in different kinds of *noise* and, in wireless transmission, also in the multipath propagation of signals (see Section 4.2). As a result of these effects, the receiver may experience massive problems in discriminating between the different signal states of a modulated carrier, which then results in errors of the interpreted data.

4.1.2 Representing Signals in the Frequency Domain

So far, we have considered signals as a varying amplitude depending on time, that is, we have observed the behavior of signals in the *time domain*. However, an essential parameter of a signal is the frequency or frequency range it adopts, and hence it is normal to analyze

BASICS OF WIRELESS COMMUNICATIONS

signals also depending on frequency or in the *frequency domain*. While the timely behavior of signals is clearly and intuitively understandable, the concept of the frequency domain is somewhat more complicated. As an example, consider again the time-domain representation of signals in Figure 4.2. The frequency of the unmodulated carrier can be simply derived by counting the number of peaks per second. However, a modulated carrier does not only consist of a single frequency but the signal is spread over a range of several frequencies by the process of modulation. This leads to the question of how many and which frequencies the amplitude-shifted signal in Figure 4.2, for example, comprises. In this case, the answer to this question cannot be given by simply counting peaks per time unit or by using any such parameter. Rather, the answer has to be found mathematically in that a signal is transformed from the time-domain to the frequency-domain representation.

The foundation for such a transformation goes back to the early nineteenth century and was constituted by the French mathematician Jean-Baptiste Fourier. He discovered that any periodic signal $g(t)$ of period T is composed of a number of sine and cosine waves and described this coherence with the following formula, which is also known as the *Fourier series*:

$$g(t) = \frac{A_0}{2} + \sum_{n=1}^{\infty}[A_n \cos(2\pi n f_0 t) + B_n \sin(2\pi n f_0 t)] \tag{4.3}$$

where f_0 is given by $f_0 = 1/T$ and denotes the frequency of a signal that repeats many times over. It is therefore also referred to as the *fundamental frequency* or *first harmonic* of the signal. As can be derived from that Equation, the fundamental signal is modified by summing up sine and cosine waves, whose frequencies are integer multiples of f_0, given by n, and which are therefore denoted as *harmonics* of the fundamental frequency. Important parameters of the harmonics are their amplitudes A_n and B_n. These coefficients can be determined by transforming Equation 4.3 as follows:

$$A_0 = \frac{2}{T}\int_0^T g(t)\,dt$$

$$A_n = \frac{2}{T}\int_0^T g(t)\cos(2\pi n f_0 t)\,dt \tag{4.4}$$

$$B_n = \frac{2}{T}\int_0^T g(t)\sin(2\pi n f_0 t)\,dt$$

Thus, the frequency components of a given periodic signal $g(t)$ can be determined by analyzing the voltage or current level of its harmonics. As an example, consider Figure 4.3, which shows the splitting of a periodic signal $g(t)$ into its harmonics and demonstrates this process in both the time and frequency domains. It is assumed that $g(t)$ is ASK-modulated as in Figure 4.2 and carries the bit sequence 10. The figure shows the signal's harmonics in the time and frequency domains (as derived from Equation 4.4) in the left and middle column, and the signal composed by the respective number of harmonics in the time domain (as derived from Equation 4.3) in the right column. The figure clearly demonstrates that by taking into account a sufficient number of harmonics the initial signal can be reconstructed more and more true to the original. Furthermore, it shows that some harmonics do not have any amplitude or voltage levels and therefore do not carry any data. The maximum voltage is carried by the sixth harmonic, which corresponds to the carrier signal that has been used for modulation.

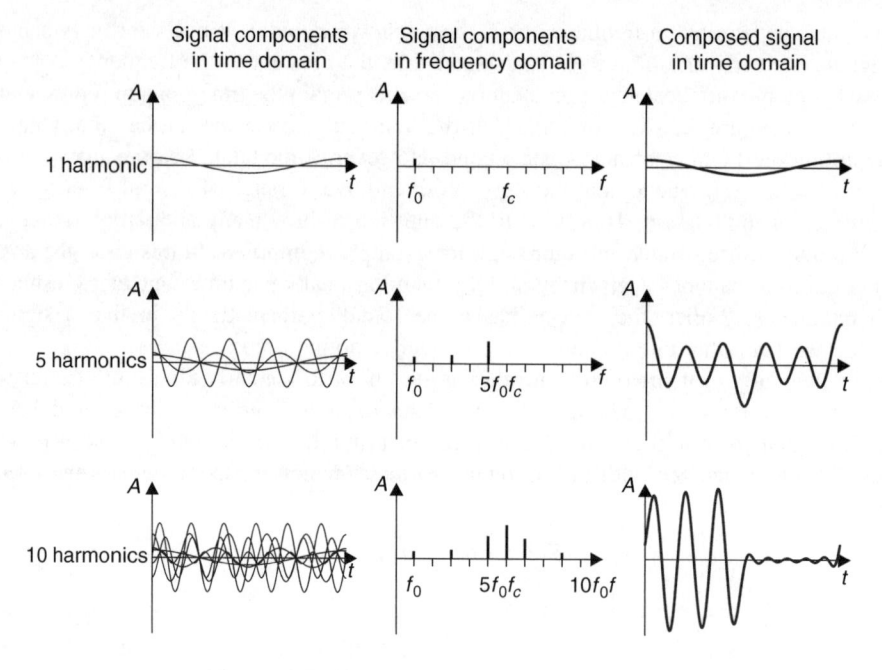

Figure 4.3 Frequency versus time domain.

As this example clearly demonstrates, a carrier signal of constant frequency is spread over many frequencies when modulating data onto it, and the Fourier series is a useful tool for determining the frequencies of such a signal. The range of frequencies adopted by a signal is called the *spectrum*, and the width of the spectrum is referred to as the *bandwidth* of the signal. However, so far, we have only considered periodic signals and hence implicitly assumed in our example that the ASK-modulated signal repeats its entire pattern over and over again. As a result, we have seen that such a periodic signal is made up of a discrete spectrum described by integer multiples of f_0. However, when considering aperiodic signals, the period T approaches infinity. As a result of this, the fundamental frequency f_0 goes to zero and hence the harmonics get closer and closer, thus merging into a continuous spectrum. This circumstance is the basis for the *Fourier transformation*, which is derived from the Fourier series and, in different variants, is a more appropriate method to analyze the spectrum of signals. The opposite direction, that is, the transformation from frequency to time domain, is referred to as *inverse Fourier transformation*. The theory behind these transformations is a complex matter and is beyond the scope of this book. The interested reader can find more information on this topic in (Brigham 1998).

4.1.3 Signal Spectrum and Bandwidth

Now that we have introduced the concepts of frequency domain, it is also necessary to deal with some aspects of the signal bandwidth. From a theoretic point of view, signals that result from digital modulation like the ASK-modulated carrier in our example adopt an infinite bandwidth, as they require an infinite number of harmonics to perfectly rebuild the shape

BASICS OF WIRELESS COMMUNICATIONS

of its original pattern (this is also reflected by the infinite number of sums in Equation 4.3). However, the bandwidth of signals is always limited by the physical properties of the underlying transmission medium, and because of these properties, the different frequency components of a signal are transmitted only up to a certain cutoff frequency. Above this cutoff frequency, signals become increasingly disrupted and are hard to interpret at the receiver or they are even not passed by the medium.

However, the bandwidth of signals is not only limited by the medium but is often also intentionally limited. The bulk of the energy is contained in a very narrow frequency band that concentrates around the carrier frequency. To reconstruct a sent signal at the receiver, it is sufficient to analyze only this limited frequency band. Frequency components above or below this band show only minor power levels and actually do not carry any data. These frequencies are also denoted as *side lobes*, and it is common practice to remove them before transmission with the use of *filters*. One reason for filtering offside lobes is that they might disrupt signals of neighboring frequency bands if the medium is shared among different users through frequency division multiplexing.

The term *bandwidth*, as used in most of the literature, is actually related to the *effective bandwidth* of a signal that comprises the data-carrying frequencies of a certain minimum energy level, and not to the *absolute*, often infinite bandwidth of a signal. This convention is also used for the remainder of this book, that is, the *bandwidth* of a signal means its *effective bandwidth*.

It is important to be clear about the fact that it requires a certain minimum bandwidth to transmit data at a certain rate. The relationship between bandwidth and data rate was first established by Harry Nyquist, who discovered in 1924 that a baseband signal of the limited bandwidth B can be fully reconstructed if it is sampled at a rate $v \geq 2B$ (Nyquist 1928). It is therefore sufficient to sample an incoming signal at the receiver with $v = 2B$. Sampling a signal faster than this rate would be useless as frequency components above B do not exist. On the basis of this relationship, the maximum data rate C that can be transmitted with a signal of given bandwidth B is determined by the following equation, which is also referred to as the *Nyquist theorem*:

$$C = 2B \log_2 M \; [bits/s] \quad (4.5)$$

where M denotes the number of signal states as defined by the modulation scheme. As an example, consider Figure 4.4, which shows the effective bandwidth of a signal that carries data with a given rate C and which concentrates around the carrier frequency f_c.

Unfortunately, the Nyquist theorem assumes a perfect transmission medium that does not suffer from attenuation and distortion, and, as already stated, does not exist. A strict application of this theorem means that it would be possible to increase the data rate C arbitrarily by just increasing the number of signal states M of the modulation scheme. In order to come to a more realistic model, Claude Elwood Shannon extended Nyquist's theorem in 1949 (Shannon 1949), and established the following relationship between bandwidth, data rate, and noise:

$$C = B \log_2(1 + SNR) \quad (4.6)$$

In general, *noise* is defined to be any unwanted signals that combine with and hence distort the signal intended for transmission. It is usually quantified as the *signal-to-noise ratio* (SNR), which states the relationship between signal and noise power. However, Shannon

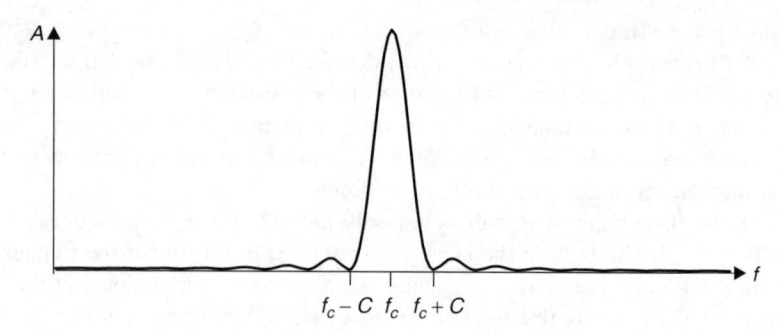

Figure 4.4 Bandwidth of a signal.

only considered the so-called *thermal noise* in his equation, which is of relatively continuous shape and results from the agitation of electrons in a conductor. He neglected other forms of distortion, for example, impulsive noise, which results from lightning or ignition, or intermodulation noise, which appears if the medium realizes different channels either on the same or on neighboring frequency bands. Therefore, Shannon's equation should rather be seen as an upper bound for the maximum data rate that can be realized on a channel of given bandwidth. In reality, this bound is hard to achieve, which especially holds for wireless transmission. The following section deals with the special characteristics of signal propagation in wireless networks.

4.2 Propagation of Radio Signals

Signals on different media behave in different ways, depending on the physical properties of the respective medium. As already mentioned, they are always subject to attenuation and different types of noise, which basically appear on all media but with different degrees. Nevertheless, there exist major differences between wired and wireless media. In wired communications, signals are always passed along a *solid* or *guided medium* such as coaxial wire or twisted pair, and therefore they are largely protected from negative impacts from outside the media. Wireless communications, on the other hand, is based on *unguided media* like the atmosphere, space, or water, and hence signals may be exposed to several sources of interference on their way from the transmitting to the receiving antenna. In addition, the way signals propagate also depends on their spectrum. This section deals with the different aspects of signal propagation in wireless communications and discusses their impacts on data transmission and positioning technology.

4.2.1 The Electromagnetic Spectrum

The propagation behavior of signals significantly depends on the range of frequencies they adopt in the electromagnetic spectrum. This spectrum, the notation used for different frequency bands, and the arrangement of different classes of signals are depicted in Figure 4.5.

The notation of frequency bands relevant to telecommunications was fixed by the *International Telecommunications Union* (ITU), which assigned names ranging from *Extremely*

BASICS OF WIRELESS COMMUNICATIONS

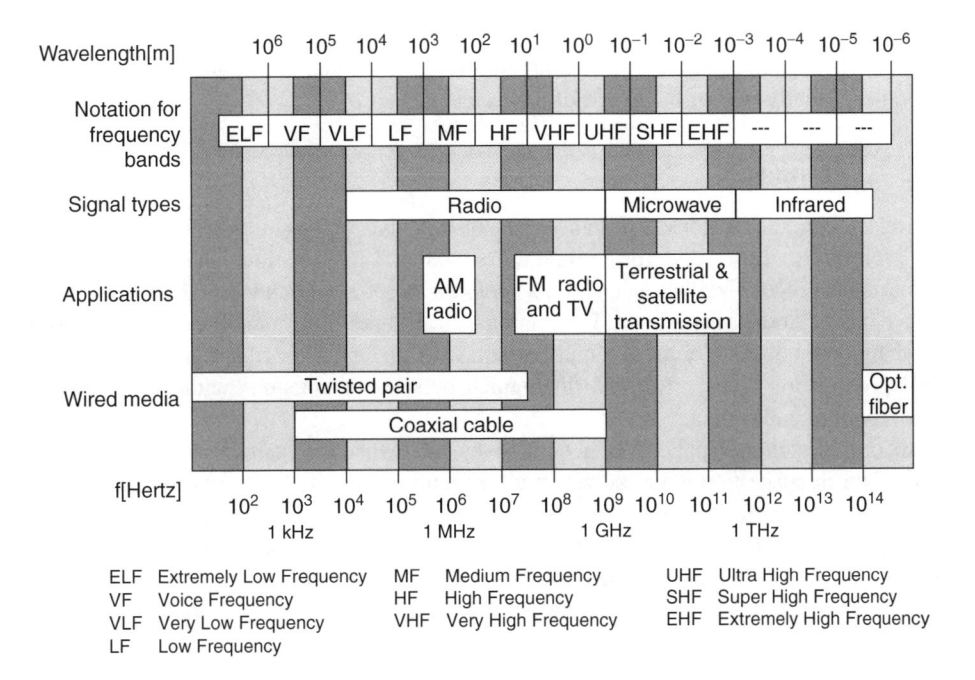

Figure 4.5 The electromagnetic spectrum for telecommunications (Stallings 2002a).

Low Frequency (ELF) up to *Extremely High Frequency* (EHF). With the allocation of additional bands for telecommunications in the upper area of this spectrum, this naming scheme was extended several times in the past, and certainly it will be further extended in future as this allocation goes on. The figure also clearly illustrates the typical frequency bands the wired and wireless media operate on and the applications they are used for. For wireless communications, signals are classified into *radio*, *microwave*, and *infrared signals*.

Radio signals are arranged in the frequency bands below 1 GHz and have wavelengths in the magnitude of meters and kilometers. Typical applications are radio and TV broadcast. Microwave signals adopt the frequency bands between 1 GHz and several 100 GHz and have wavelengths between a meter and a few millimeters. Nearly all communication systems, being terrestrial or satellite based, use microwave signals for exchanging data. Finally, infrared signals extend from several 100 GHz up to approximately 400 THz and hence have very short wavelengths about some millimeters or fractions of a millimeter. They are primarily used for short-range communications, such as television remote controls, or for the data exchange between handheld devices.

The classification into radio, microwave, and infrared signals are mainly constituted by the different propagation characteristics. Generally, radio signals are emitted omnidirectional from an antenna and can pass several hundreds of kilometers without being affected by obstacles, which makes them very attractive for radio and television broadcast. Their large coverage results also from the fact that they are reflected back and forth between the Earth's surface and the ionosphere. Microwave signals, on the other hand, can be transmitted as directional beams. However, their propagation may be increasingly obstructed by obstacles, but owing to the effects of multipath propagation, communication is possible even if no

LoS between sending and receiving antenna exists. Infrared signals are transmitted as highly directional beams and require that sending and receiving antennas are located very close to each other, usually within the range of some meters.

4.2.2 Antennas

In wireless communications, different types of antennas are used, which primarily differ from each other with respect to the directivity of signal propagation. The directivity of an antenna can be recognized by means of its *radiation pattern*, which is usually represented as an envelope of some shape around the antenna in a three-dimensional coordinate system. The shape symbolizes points of equal signal strength and is a function of the angle of signal propagation. Basically, it is distinguished between antennas with *omnidirectional* and *directional* radio patterns.

An omnidirectional pattern is generated by an *isotropic antenna*, which is simply a single point in space that emits power in all directions equally. The resulting radio pattern is hence a sphere with the antenna at its center and with equal signal strength at the surface of the sphere (see Figure 4.6(a)). The isotropic antenna is the model of an idealized antenna that does not exist in reality. However, it is of major importance for antenna design, as it serves as a reference for evaluating performance characteristics of real antennas. An

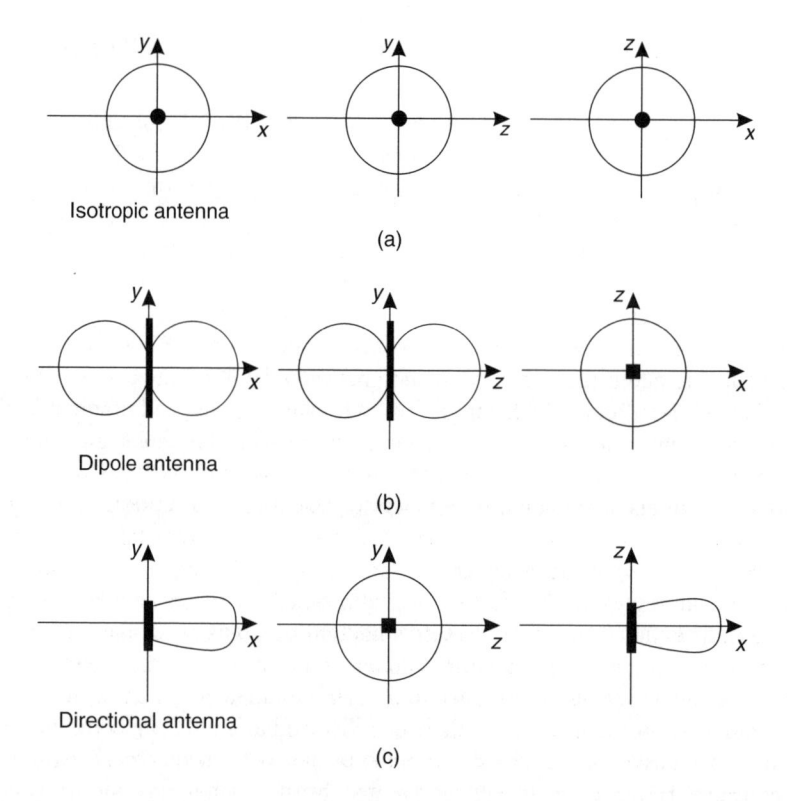

Figure 4.6 Radio patterns of isotropic, dipole, and directional antennas (Schiller 2000).

BASICS OF WIRELESS COMMUNICATIONS

essential parameter here is the *antenna gain*, which is a measure of the directivity of an antenna. It is defined as the power output in a certain direction, compared to that produced in any direction by an isotropic antenna (see (Stallings 2002a) for further details).

The simplest antenna one can build is the *dipole antenna*, which consists of two straight wires arranged along a line either horizontally or vertically and with a small feeding gap between them (see Figure 4.7). Both wires are of the same size and the entire dipole size must correspond to half of the wavelength of the signals to be transmitted, and hence this type of antenna is denoted also as *half-wave dipole*. Another version is the *vertical quarter-wave antenna*, which is usually installed on the roof of vehicles or used for portable radios. The relationship between antenna size and wavelength is essential for the efficient transmission of signals. Each signal is made up of several signal components of different frequencies, and the component whose half wavelength exactly matches with the antenna size is transmitted most efficiently.

Half-wave dipole Quarter-wave antenna
(a)   (b)

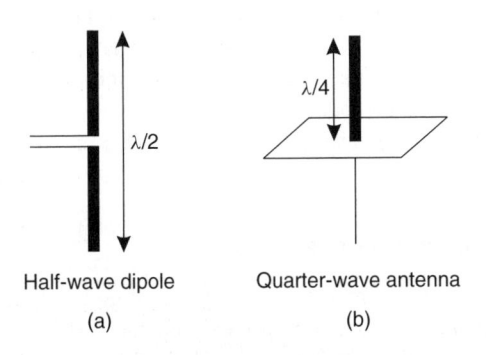

Figure 4.7 Half-wave dipole and quarter-wave antenna.

The dipole has an omnidirectional radio pattern with respect to the plane being perpendicular to the antenna. As demonstrated in Figure 4.6(b), the pattern corresponds to a circle when considered from the direction of this plane, while it looks like a horizontal eight when considered from the directions of the two other planes.

While the dipole with its omnidirectional propagation characteristics is the most rudimentary antenna, more complex antenna configurations aim at the concentration of signal strength in one direction as depicted in Figure 4.6(c). Most antennas deployed in today's mobile networks produce directional radio patterns with angles between 60 and 120°. The directional concentration of power is achieved by increasing the power in one direction at the expense of other directions (see also (Stallings 2002a)). Compared to omnidirectional antennas, directional ones provide better coverage and ease the cell planning in cellular systems, where different antennas using the same frequency bands need to be carefully aligned in order to avoid cochannel interferences.

In mobile cellular networks like GSM and UMTS, it is common practice to compose an antenna with a number of directional antenna elements, usually between three and six. The radio patterns of such *sectorized antennas* are depicted in Figure 4.8(a). In their most radical form, antennas can be made up of an array of multiple antenna elements that enable to accurately concentrate signal power in a narrow beam toward the receiver, thereby having

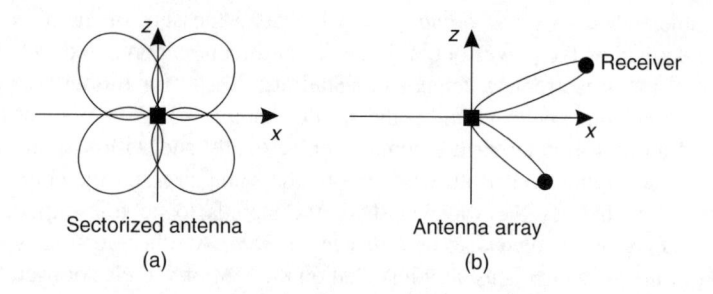

Figure 4.8 Radio patterns of sectorized antennas and antenna arrays.

a signal power of zero in any other direction (as shown in Figure 4.8(b)). In their initial form, antenna arrays were designed to be fixed and aimed at limiting interferences caused by multipath propagation. In recent years, research concentrates on adaptive antenna arrays (also referred to as smart antennas), which enable the beam to follow the movements of a mobile receiver. This is of particular importance for mobile networks, as this technology establishes another method for separating the transmissions of different users, which is known as *Space Division Multiplexing* (SDM) (see also Section 4.3.1). A comprehensive introduction to this emerging field is given by Lehne and Pettersen (1999).

4.2.3 Speed of Electromagnetic Waves

A characteristic that is of major importance for positioning rather than for transmission of data is the speed at which electromagnetic waves propagate. This speed was first derived by the Scottish physicist and mathematician James Clerk Maxwell in 1873 in his famous theoretical conclusions about the behavior of electric and magnetic fields. Maxwell discovered that the calculated speed of electromagnetic waves was equal to the speed of light, which, based on complex measurements, was predicted a few years earlier to be in the range of 3×10^8 m/s. From this observation, he further concluded that the phenomenon of light was actually an electromagnetic phenomenon, that is, visible light was an electromagnetic wave that adopted a small part in the electromagnetic spectrum. After numerous refinements in measurements, the exact speed of light in vacuum, and thus the speed of electromagnetic waves in vacuum, is now given by

$$c_0 = 299{,}792{,}458 \text{ m/s} \qquad (4.7)$$

Since the speed of light in vacuum is a constant quantity, it is convenient to use c_0 as a reference for measuring time and distance. This has lead to a redefinition of the meter in 1983, which nowadays corresponds to "the length of the path traveled by light in vacuum during a time interval of 1/299,792,458 of a second" (Bureau International des Poids et Mesures 1998). This, for example, is in contrast to an earlier version, where the meter was equal to 1/10,000,000 of the distance from the Equator to the North Pole.

Unfortunately, Equation 4.7 reflects the propagation velocity only in vacuum. If light travels through a medium such as air, water, or glass, its velocity is smaller than c_0 and depends on the density of the respective medium. The degree by which a wave is slowed down can be derived by the refraction index of the medium. Refraction occurs if a wave enters from one medium into another and is caused by the change of velocity when the wave crosses the boundary between the media. The wave is then bent either toward or away from the normal to the boundary, depending on whether it enters from a medium of lower into one of higher density or vice versa. The refraction index n is then the angle of incidence divided by the angle of refraction (both with regard to the normal) (see Figure 4.9). The refraction index n can also be derived by relating the different velocities as a wave travels in two media:

$$n = \frac{c_0}{c} \qquad (4.8)$$

where c_0 is the velocity of light in vacuum and c is the speed in some medium. Knowing the refraction index of a particular medium, one can hence calculate the speed of light in this medium. For example, the refraction index of air related to vacuum and with respect to light is 1.00029 and that of water is 1.33. However, it must be stressed that the refraction index of a particular medium is not the same for all electromagnetic waves but is, besides the density of the material, also a function of the wave's frequency (apart from vacuum, where all electromagnetic waves travel with the same velocity). This means that waves of different frequencies travel with different speeds through a given medium, a phenomenon that is called *dispersion*. For example, dispersion of visible light can be observed by means of a light beam that enters into a prism and is subdivided into separated beams of different colors when emerging from the prism.

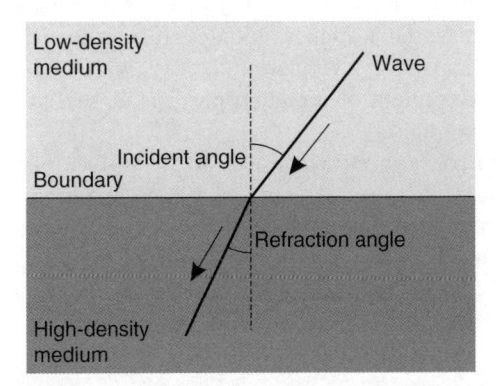

Figure 4.9 Refraction of electromagnetic waves.

Of particular importance, especially in the area of positioning, is the velocity of waves in the Earth's atmosphere. The atmosphere is subdivided into different layers that have different impacts on the propagation of signals. As illustrated in Figure 4.10, the lowest layer is the *troposphere*, which extends from the surface of the Earth to an altitude of approximately 20 km. The troposphere is subject to all weather phenomena and hence comprises liquid water and water vapor and also shows large variances in the temperature. It is also the layer where pollution is gathering. The next layer is the *stratosphere*, which extends to an altitude of approximately 50 km and which shows only little water vapor and a relatively constant temperature. The broadest layer is the *ionosphere*, which runs from the stratosphere up to an altitude of 1500 km. It contains a plasma of negatively charged

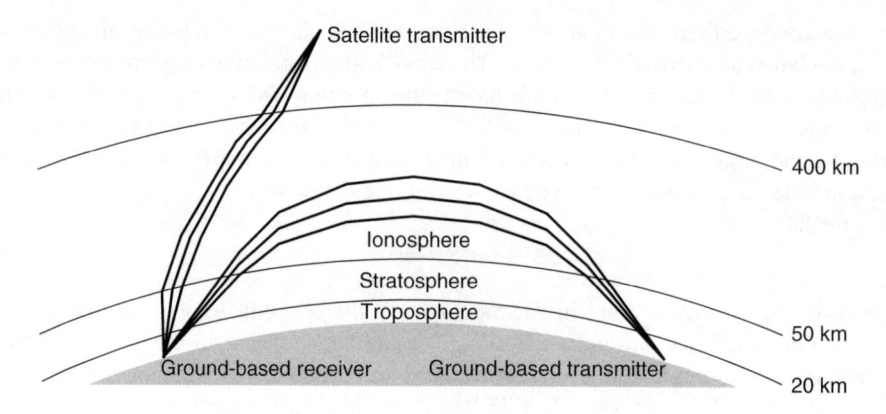

Figure 4.10 Signal propagation in the atmosphere.

electrons and positively charged atoms and molecules called *ions*, which is produced when ultraviolet light from the sun hits on gas molecules appearing in these regions of the atmosphere and separates them into electrons and ions. The degree of ionization is all but constant and depends on the density of atoms in the atmosphere and the intensity of ultraviolet light, which in turn varies with the time of day and the activity of the sun. Generally, it can be stated that the degree of ionization is the smallest in the lower regions of the ionosphere and increases at higher altitudes. In addition, the ionosphere is subject to high variances in the ionization level during the day, with the highest level occurring at noon when the respective portion of the atmosphere is closest to the sun and with small levels of ionization during night.

Generally, the atmosphere is most dense near the surface of the Earth and becomes thinner at higher altitudes. As a consequence, the atmosphere has a gradually decreasing refraction index, which is often specified in terms of the relative permittivity ϵ_r, which expresses the relationship between electric displacement and electric field intensity, and which changes with height. While the speed of waves in the troposphere is nearly independent of their frequency, the refraction of the ionosphere is a function of the wave's frequency and is given by

$$n = \sqrt{\epsilon_r} \approx \sqrt{1 - \frac{81 N_e}{f^2}} \qquad (4.9)$$

where f is the wave's frequency and N_e denotes the density of free electrons in terms of the number of electrons per cubic meter. Depending on the respective altitude, time of day, sun activity, and season, typical values for N_e are in the range of 10^{10} to 10^{12}. Given this magnitude for N_e, it is not possible for waves below a certain frequency band, which lies in the range of 30 MHz, to penetrate the ionosphere. Instead, waves below 30 MHz are reflected back to Earth. This reflection is actually a step-wise refraction, which gradually increases as the density of the ionosphere (and thus the refraction index) at higher altitudes decreases. This is illustrated in Figure 4.10.

The reflective properties of the ionosphere below 30 MHz are also the reason why communication between the ground and satellites takes place by using signals of much higher frequencies, usually in the range of several GHz. For most applications, the gradually

BASICS OF WIRELESS COMMUNICATIONS

changing refraction index of the atmosphere does not play a role, and the velocity of waves is commonly considered to be very close to the speed of light in vacuum. However, most positioning technologies such as GPS are based on distance measurements that exploit the speed of signal propagation, and small variances in this speed may cause significant errors in position data. For example, in GPS, tropospheric refraction is responsible for delays up to 240 cm, according to Schüler (2001), and delays or advances due to ionospheric refraction can amount to several tens of meters, according to Leick (2004). To compensate for these errors, it is therefore a major challenge to develop mathematical models for predicting the time-varying and altitude-dependent refractivity of the atmosphere. Leick (2004) provides a comprehensive overview of these issues.

4.2.4 Attenuation

When signals travel away from a transmitting antenna, they are exposed to a reduction in their strength, a phenomenon that is denoted as *attenuation* or *path loss*. The degree of attenuation primarily depends on the distance between the antenna and the point of measurement (e.g., the receiving antenna), the signal's wavelength, and the surrounding environment (e.g., indoor, outdoor, urban, rural, etc.).

Attenuation can be best explained by means of an isotropic antenna, which, as described in Section 4.2.2, radiates a signal in all directions equally and hence produces the radio pattern of a sphere. A signal emitted by such an antenna with some power P_t arrives attenuated at some distance d to the antenna, because the amount of P_t uniformly disperses over the surface of a sphere of radius d. The degree of attenuation is usually expressed by the ratio of the transmitted power P_t to the received power P_r. Assuming an isotropic antenna and no obstacles between transmitting and receiving antenna, the following relationship holds:

$$\frac{P_t}{P_r} = \frac{(4\pi d)^2}{\lambda^2} \qquad (4.10)$$

This formula is a simplification of the *Friis free space equation*, which in its original form describes the free space loss for real antennas, that is, the half-wave dipole, and also considers loss factors not related to propagation like filter and antenna losses (see (Friis 1946) and (Schelkunoff and Friis 1952)). Such additional loss factors usually appear in regions very close to the transmitting antenna, and hence the path loss at larger distances is usually not expressed with respect to P_t, but depends on the power that is measured at the so-called *far-field distance* d_f. The power Pr received at an antenna is then given by

$$P_r(d) = P_r(d_f) \left(\frac{d_0}{d}\right)^2 \qquad (4.11)$$

where $P_r(d_f)$ is either measured or predicted according to Equation 4.10. The far-field distance d_f is related to the signals' wavelength and the dimensions of the antenna. For mobile communication systems in the range of 1–2 GHz, it is typically chosen to be 1 m in indoor environments and 100 m or 1 km in outdoor environments (Rappaport 2002).

According to Equation 4.11, the signal strength falls as the square of the distance, and hence this equation is also known as the *inverse square law*. However, this law assumes a line of sight between transmitting and receiving antennas with no obstacles in between. However, in most cases the path from the transmitting to the receiving antenna in which a signal travels is obstructed by various obstacles like buildings, mountains, or smaller objects, which cause additional attenuation that is not related to the propagation itself. It is closely related to phenomena like reflection, diffraction, and scattering, which occur when a signal hits an obstacle. These phenomena are subsumed under the term multipath propagation and are explained in the following section. In order to consider the additional degree of attenuation, a more common version of the free space equation is given by

$$P_r(d) = P_r(d_f) \left(\frac{d_0}{d}\right)^\alpha \quad (4.12)$$

where α is the *path-loss gradient*, which reflects the degree of obstacles in the surrounding environment. Table 4.1 shows some typical values for the path-loss gradient. Usually, it varies to a larger extent for indoor environments than for outdoor areas. For large indoor open areas, it is even smaller than under outdoor line-of-sight conditions, while it rapidly increases up to $\alpha = 6$ approximately as soon as the signal path is obstructed (Pahlavan and Krishnamurthy 2002).

Table 4.1 Examples for path-loss gradients

α	Environment
2	Free space (vacuum)
2.5	Outdoor – rural areas
3–4	Outdoor – urban areas
4–5	Outdoor – dense urban areas
1.6–1.8	Indoor – large open areas and corridors
4–6	Indoor – non-line-of-sight environments

The Friis free space equation is obviously the simplest way to predict a signal's path loss, but the usage of different path-loss gradients is rather a very rudimentary and inaccurate way to cope with varying propagation conditions in different environments. A more accurate way is to use propagation models that are dedicated to special environments. For outdoor areas, the more prominent models are the Longley–Rice Model (Longley and Rice 1968), the Okumura Model (Okumura et al. 1968), and the Hata Model (Hata 1980), and for indoor environments the Log-distance Path Model and the Ericsson Multiple Breakpoint Model (Akerberg 1988). Rappaport (2002) as well as Pahlavan and Krishnamurthy (2002) present a comprehensive and detailed overview and comparison of these models.

4.2.5 Multipath Propagation

Multipath propagation means that a signal reaches the receiving antenna not only on a single path but on various paths of different lengths and with different degrees of attenuation. It basically results from the combination of *reflection*, *diffraction*, and *scattering* as illustrated in Figure 4.11. Which of these effects occur mainly depends on the wavelength of the signal in relation to the size of the respective obstacle.

Reflection occurs if the signal's wavelength is much shorter than the size of the obstacle. This condition is fulfilled for signals of typical mobile communication systems used today

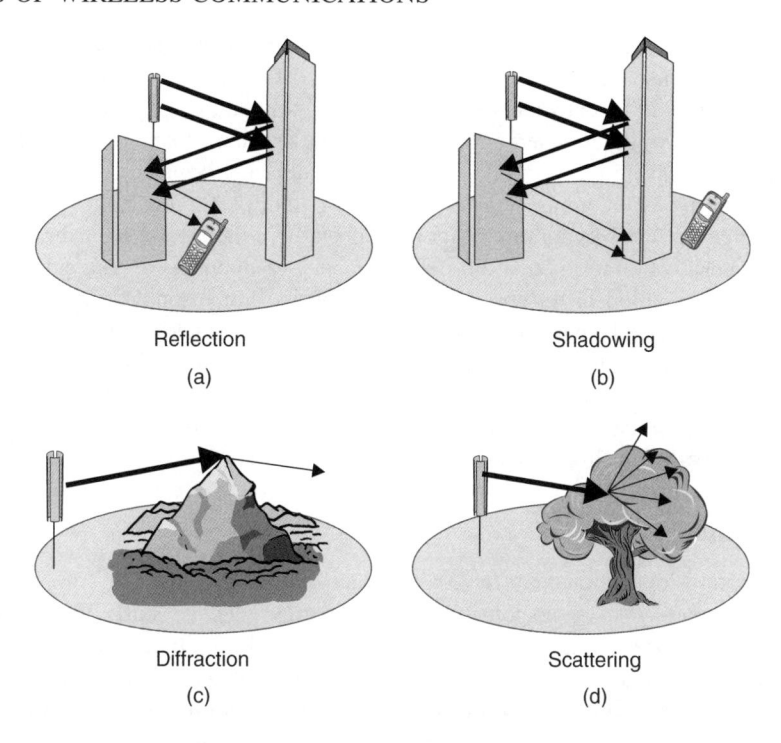

Figure 4.11 Multipath Propagation.

when they encounter buildings, walls, mountains, or simply the ground. Depending on the angle of incidence and the surface integrity of the obstacle, the reflected signal or parts of it can be much weaker than the arriving signal, that is, the reflected signal suffers from an additional attenuation (see Figure 4.11(a)). Closely related to reflection is *shadowing* (which actually does not belong to multipath propagation), which occurs if a reflecting obstacle is located between transmitting and receiving antennas (see Figure 4.11 (b)).

Diffraction describes the bending of a signal if it hits on irregularities of an obstacle, for example, edges or corners (see figure 4.11 (c)). Like reflection, this effect appears if the size of the obstacle is significantly larger than the wavelength of the signal. And, the signal is attenuated by diffraction to a certain extent. Another effect is scattering, which means that a signal is multiplied at obstacles whose size is of the same magnitude as the signal's wavelength or less. In mobile communications, this especially holds for obstacles like lampposts, traffic signs, or even the foliage of trees as illustrated in Figure 4.11(d). If a signal hits on such obstacles, several copies of it are generated, each of them being much weaker than the original and each of them further propagating in the surrounding environment.

Multipath propagation has both positive and negative effects on mobile communications. First, it must be stated that without reflection, diffraction, and scattering it would not be possible to transmit data if no line of sight between transmitting and receiving antenna exists. For example, a signal is typically reflected back and forth between obstacles several times before it finally encounters the receiving antenna. Phenomena like these enable that mobile cellular systems work in rural and indoor areas at all and accomplish a sufficient coverage with a reasonable number of base stations. Furthermore, efforts taken by the research community in recent years aim at exploiting multipath propagation in combination with antenna arrays in order to serve a user on a single frequency band over different signal paths, thereby increasing the data rate (Lehne and Pettersen 1999).

Simultaneously, overcoming negative effects of multipath propagation is one of the major challenges in wireless communications, especially in the area of receiver technology. One of these negative effects is that different frequency components of a signal are affected by multipath propagation in different degrees and ways. For example, some components may be more attenuated than others when hitting on an obstacle, or they may be even entirely erased, while others further propagate nearly unaffected. This causes significant distortion of the signal, which are denoted as *frequency-selective interferences*. Another drawback is that a signal may be heavily distorted when multiple copies of it (generated by scattering) overlay each other at the receiver.

4.2.6 Doppler Effect

While multipath propagation may have an impact on single frequency components of a signal, the *Doppler effect* is responsible for a general frequency shift. This shift occurs when the sender, the receiver, or both of them are moving during transmission as illustrated in Figure 4.12. As a result, the receiver experiences the frequency of the signal shifted by a certain amount. This shift is known as the *Doppler shift*, and it has been named after the Austrian physicist Christian Doppler, who discovered this phenomenon in 1842.

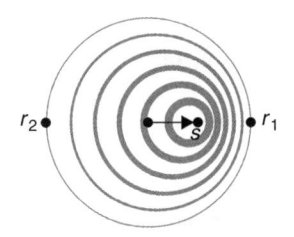

Figure 4.12 Doppler effect.

The Doppler shift results from the fact that a wavelength of a signal is compressed if sender and receiver are approaching each other during transmission (as can be seen from the point of view of receiver r_1 in Figure 4.12), and elongated correspondingly if they are departing from each other (as illustrated by means of receiver r_2). The amount of the Doppler shift depends on the velocity with which sender and receiver are moving toward or away from each other. It is expressed by the following equation:

$$f_d = \frac{v}{\lambda} \cos \alpha \qquad (4.13)$$

where v denotes the relative velocity between sender and receiver, λ is the signal's wavelength, and $\cos \alpha$ refers to the angle between the direction of motion and the direction of

BASICS OF WIRELESS COMMUNICATIONS 79

arrival of the incoming signal at the receiver. Note that Equation 4.13 returns a positive value if sender and receiver are moving toward each other, that is, the signal's frequency is increased, and a negative value if they are moving away from each other, that is, the signal's frequency is decreased.

The maximum Doppler shift at a given velocity is also referred to as *Doppler spread* and occurs if sender and receiver are moving toward or away from each other on a straight line, that is, when $\cos \alpha = 1$. In general, it can be stated that frequency shifts resulting from the Doppler effect can be neglected if the signal's bandwidth is much greater than the Doppler spread. This especially holds for terrestrial mobile systems where subscriber velocities are comparatively small. However, the Doppler effect is of concern in satellite communications with satellites being arranged on low Earth orbits or medium Earth orbits (LEO and MEO). At these orbits, the satellites circulate the Earth at very high velocities, resulting in Doppler shifts in the range of several kHz. Thus, the receiver of a signal emitted from or to a satellite must adjust this shift in order to properly receive the transmission. This adjustment causes some delays in the start-up phase of the receiver until the Doppler effect has been quantified.

4.3 Multiplexing and Multiple Access

A typical characteristic of wireless communications is that the medium, that is, the air interface, must be shared among different applications (e.g., radio and television, mobile cellular systems, LANs), and within a certain application between different users (radio and television stations, subscribers). Consequently, the electromagnetic spectrum is a very valuable source, where each application and user is assigned a limited amount of bandwidth. This is in contrast to wired infrastructures, which are often dedicated to a particular application (and in some cases even to a particular user), and where bandwidth can be comparatively easily scaled by laying out new trunks. This is of course not possible in wireless communications, as the air interface exists only once. An important challenge is therefore the division of the air interface into several channels and the (dynamic) assignment of these channels to applications and users, which is the essence of *multiplexing* and *multiple access*.

Multiplexing refers to the subdivision of a single medium into a number of channels that can be used independent of each other without interferences in data transmission. From the entire capacity of the medium, each channel is assigned a certain portion, which for digital data corresponds to the data rate the channel is able to realize and which must be carefully dimensioned in order to fulfill the requirements of the application using the channel. Multiple access, on the other hand, deals with the dynamic assignment of channels to users. It is necessary for all applications where users make use of channels only temporarily, as it is generally the case for all mobile communication systems. On the other hand, multiple access is not required for applications where channels are permanently assigned to users, like for radio and television broadcast stations.

Generally, a medium offers a number of resources each of which or a combination of which can be subdivided into several channels by the process of multiplexing. In wireless communications, the resources of the air interface are given by space, frequency, time, and code, and thus classified as *space*, *frequency*, *time*, and *code division multiplexing* (SDM, FDM, TDM, and CDM) and the corresponding multiple access schemes.

4.3.1 SDM and SDMA

SDM allows the establishment of several channels in the same frequency range for simultaneous use, assuming that they are sufficiently separated in space. This kind of multiplexing can be basically achieved in two ways: first, by exploiting the path loss of signals, and second, by propagating signals as highly directed beams instead of radiating them omnidirectionally. The first method is implicitly applied in all wireless communication systems. Signals of a given frequency range are attenuated as farther away they travel from the emitting antenna, and hence the same frequency range can be reused by another antenna that is located beyond the so-called *frequency reuse distance*. This distance must be chosen long enough to avoid mutual interferences between channels in the same frequency range, which are also referred to as *cochannel interferences*. The frequency reuse distance is mainly a function of radiating power and multipath conditions on the spot. The second method assumes a transmission using antenna arrays that are able to concentrate the signal power in a highly directed beam toward the receiver (see also Section 4.2.2). SDM is here possible without interferences only if the signal paths of different channels of the same frequency range do not cross during transmission. If multiple access between different users is realized by such beams of the same frequency range, this technique is referred to as *Space Division Multiple Access* (SDMA).

4.3.2 FDM and FDMA

In FDM (Frequency Division Multiplexing), the channels adopt different bands of the frequency range that has been assigned to an application (see Figure 4.13a). The channels are of a certain bandwidth that has to be selected in accordance with the data rate required by the application. Thus, a frequency channel is always characterized by its carrier frequency and the width of the frequency range around the carrier. As has been stated in Section 4.1.3, signals often have an infinite bandwidth, which can actually be narrowed down by the use of filters such that the resulting signal contains only the data-carrying frequency components above a certain minimum energy level. However, it is very hard to completely reject needless components, and hence, FDM generally suffers from *neighbor channel interferences*. They have the form of small side lobes below

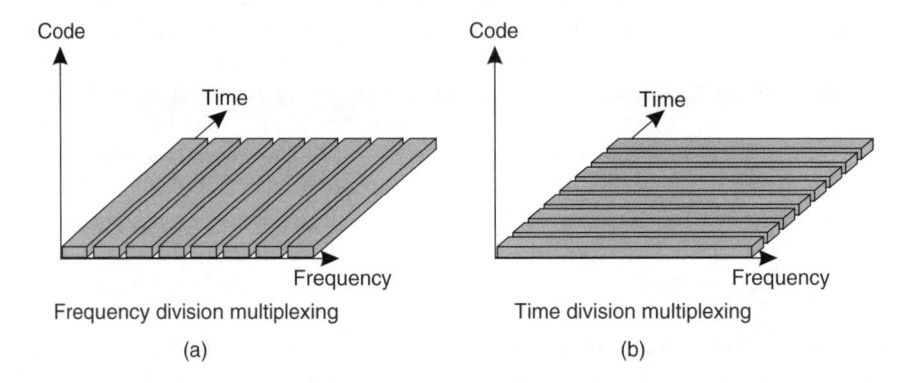

Figure 4.13 FDM and TDM.

and above a frequency channel and disturb the transmissions in neighboring channels more or less heavily. To avoid these impacts, neighboring frequency channels are separated by *guard bands*, which are not used for data transmission but only for absorbing needless side lobes. In its pure form, FDM is traditionally applied to radio and television and was also used by the earlier analog cellular systems. In today's cellular systems, it appears often in conjunction with TDM. The earlier analog cellular systems used pure FDM to separate different users, which was then called *Frequency Division Multiple Access* (FDMA).

FDM is also a common technique to separate *uplink* and *downlink channels* from each other. This is commonly referred to as *duplexing*. An uplink channel is typically defined as the channel that transfers data from a mobile terminal to a base station, while a downlink channel is used for transmission in the opposite direction. In satellite communications, the same terminology is used for transmission between a terrestrial station (which can be mobile or stationary) and a satellite or vice versa. If uplink and downlink channels are separated by frequency, this is referred to as *Frequency Division Duplexing* (FDD). Uplink and downlink frequency channels must be separated from each other by a reasonable frequency range to avoid interferences between them.

4.3.3 TDM and TDMA

In TDM (Time Division Multiplexing), different channels use a medium interchangeably in time, and therefore have to defer and interrupt their transmission as long as the medium is accessed by other channels. The amount of time that is assigned to a channel and the number of times the transmission of a channel is interrupted can be basically tailored to the special needs of the particular application, for example, depending on the required data rate or the maximum delay that can be tolerated. Telecommunication systems, whether they are wired or wireless, traditionally use a system where the time axis is subdivided into *frames*, and each frame in turn consists of a number of *time slots*. A channel is then represented by a certain time slot, and the transmission of a channel has to be interrupted until this time slot appears in the next frame (see Figure 4.13(b)).

Like FDM, TDM in wireless systems also suffers from a type of neighbor channel interference that occurs if the time channels have been assigned to different, spatially distributed senders. In this case, the senders have to be synchronized in order to have well-defined beginnings and endings of time slots. However, an exact synchronization is hard to achieve, since signals for synchronization are also subject to the finite speed of light and therefore always arrive with different delays at the senders. To compensate for this problem, neighboring time slots are separated by *guard times* during which data transmission is not allowed, and thus avoid the overlapping of transmissions on different channels. TDM is often used to coordinate multiple access and is then denoted as *Time Division Multiple Access* (TDMA)

Figure 4.14(a) illustrates the combination of FDM and TDM, where each frequency channel is subdivided into a number of time channels. Here, eight frequency channels are given, and a frame consists of two time slots, resulting in 16 independent channels. The combination of FDMA and TDMA also allows the deployment of a technique that is called *frequency hopping* and which is deployed in multipath environments in order to limit the impacts of frequency-selective interferences (see Section 4.2.5). Using frequency hopping,

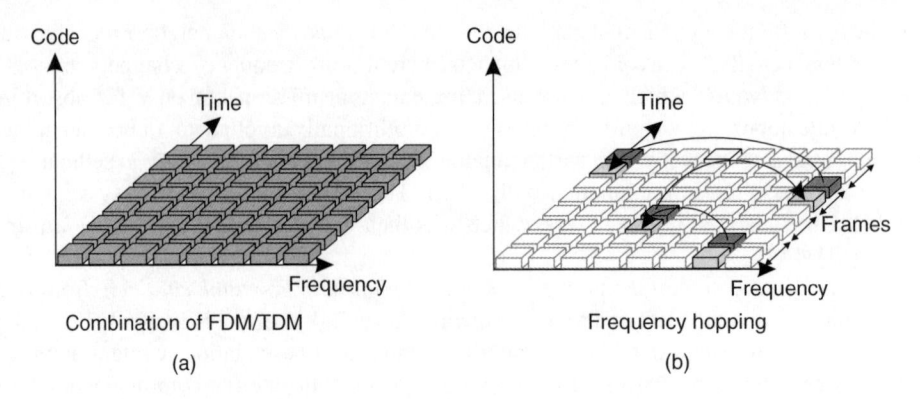

Figure 4.14 Combination of FDM/TDM and frequency hopping.

the frequency channel is rapidly changed, usually between subsequent frames, according to a well-defined *hopping sequence* that must be arranged between sender and receiver (see Figure 4.14(b)). If frequency-selective fading appears on a particular frequency range, a channel is thus not exposed permanently to it, but only for a very short time. While all applications are susceptible to permanent interferences, some can just tolerate very short ones. Frequency hopping is especially suited for speech telephony, where a single defective time slot arriving at the receiver is barely recognized by the user.

TDM can also be used for separating uplink and downlink channels, which is denoted as *Time Division Duplexing* (TDD). In this case, a number of time slots in a frame are reserved for transmission in the uplink, while the others are used in the downlink. Unlike FDD, TDD is only suited for local networks with a range of up to several hundred meters at the maximum. This is due to the fact that cell sizes of a few kilometers or more would require very large guard times between neighboring uplink and downlink time slots, which would waste valuable resources.

4.3.4 CDM and CDMA

Among all multiplexing methods, CDM (Code Division Multiplexing) is the most difficult to understand intuitively. The basic principle is that all channels transmit simultaneously in the same frequency range and in the same space, thereby interfering with each other to a large extent. This means that the signals of different channels are summed up during transmission and hence must be separated after reception at the receiver. To make this possible, different channels are separated by code (see Figure 4.15), that is, it is required to encode the data to be transmitted with a so-called *chipping sequence*, which is unambiguous for each channel. A chipping sequence is of a certain length and is continuously repeated. It consists of a number of binary symbols, which are termed *chips*. Following this definition, a chipping sequence is basically very similar to an ordinary sequence of bits. The major difference is that a chip length is much smaller than the length of a bit, or, at most, equal to the bit length. The chipping sequence used for a channel must of course be known to both sender and receiver. This principle can be utilized to coordinate the transmissions of different senders, where each sender is assigned a particular code channel. This method is then called *Code Division Multiple Access* (CDMA).

BASICS OF WIRELESS COMMUNICATIONS

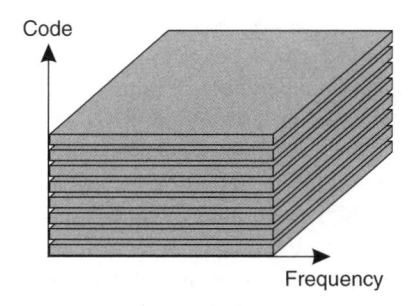

Figure 4.15 CDM.

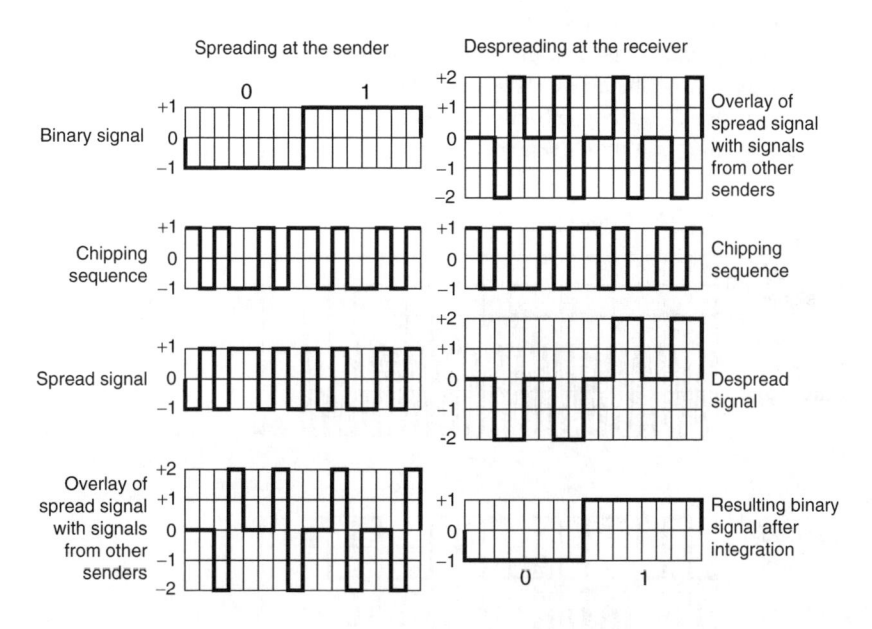

Figure 4.16 Encoding and decoding of data in CDMA. Adapted from (Roth 2002).

The encoding, transmission, and decoding of signals are illustrated in Figure 4.16. In CDMA, bits and chips are represented in bipolar notation, that is, by values of $+1$ and -1. At the sender, the data is encoded with the chipping sequence by multiplication, resulting in the output signal depicted in Figure 4.16. During transmission, the signals from different senders overlay and arrive in the form of a composed signal at the receiver. To reconstruct the data of different senders, the receiver has to apply the chipping sequence of the respective sender by simply multiplying it again with the composed signal. The resulting signal then just looks very similar to the original data, which is finally fully reconstructed by integration. The data from other senders can be reconstructed analogously. Note that if the receiver applies a wrong chipping sequence, multiplication and integration would produce ambiguous results, and a reconstruction would hence not be possible.

CDM and CDMA belong to a class of techniques, that is termed *spread spectrum*. The name is derived from the fact that a signal that originates from data encoded with a chipping sequence is spread over a much larger frequency range compared to a signal derived from pure data. Therefore, another common term for chipping sequence is *spreading code*, and the process of encoding data with a chipping sequence is also referred to as *spreading*. The relationship between the chip and the bit length is defined by the *spreading factor*, which in turn defines the number of chips to be used to encode a single bit.

Figure 4.17 compares the frequency ranges of a conventional despread signal with that of a spread signal. In this figure, the bit rate (of the despread signal) is denoted as C_b and the chip rate (of the spread signal) as C_c. Both signals have been modulated with binary PSK. Since the chip length is only a portion of a bit length, the chip rate is a multiple of the bit rate and thus results in a corresponding bandwidth required for transmission (which directly follows from the coherence between data rate and bandwidth as introduced in Section 4.1.3). Furthermore, the spread signal has the same amount of power as the despread signal, but this power is equally distributed over the much larger frequency band resulting in a smaller amount of power per frequency. Strictly speaking, the *power spectral density* of a spread signal is much smaller and may even seem like noise during transmission.

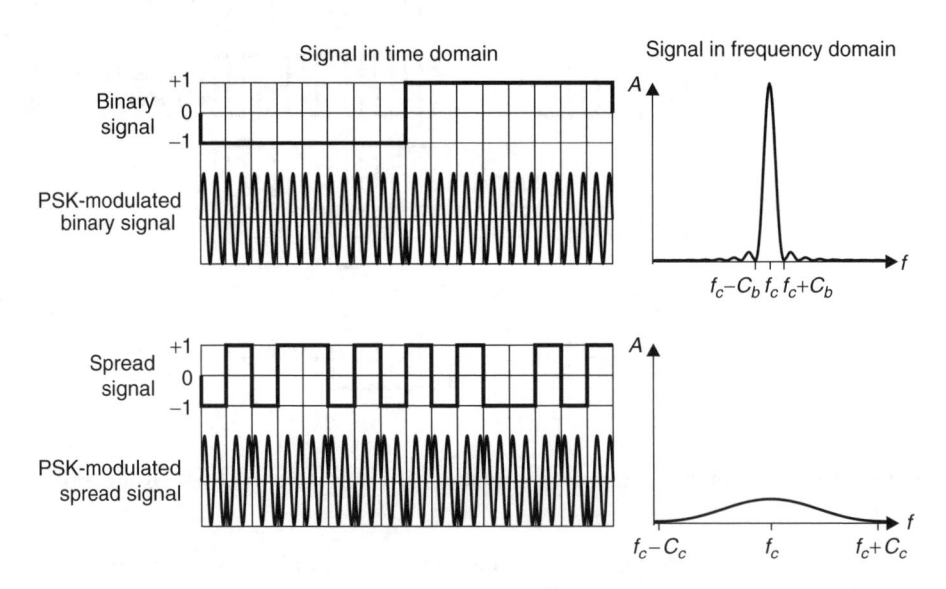

Figure 4.17 Spread and despread signals in time and frequency domain.

Spreading codes must have good *correlation* properties in order to be suitable for CDMA. Generally, correlation is a means for determining how much similarity one set of data has with another. In CDMA, two kinds of correlation are of relevance: *autocorrelation* and *cross correlation*. The former defines the correlation of a chip sequence c_i of N chips with regard to n shifts of itself:

$$\Phi_{ii}[n] = \frac{1}{N} \sum_{m=1}^{N} c_i[m]c_i[m+n] \qquad (4.14)$$

BASICS OF WIRELESS COMMUNICATIONS

with $n = 1, \ldots, N$. The cross correlation refers to the correlation of two chip sequences c_i and c_j, both consisting of N chips, with regard to n shifts:

$$\Phi_{ij}[n] = \frac{1}{N} \sum_{m=1}^{N} c_i[m] c_j[m+n] \qquad (4.15)$$

with $n = 1, \ldots, N$. Given these equations, values of autocorrelation and cross correlation can adopt values in the range of -1 and $+1$. Table 4.2 gives the meanings of values within this range that are of particular importance.

Table 4.2 Interpretation of correlation values

Correlation value	Interpretation
1	Both sequences match exactly.
0	There is no relation between the sequences at all.
-1	The two sequences are inverse to each other.

Autocorrelation is an important measure for achieving synchronization between sender and receiver. A chipping sequence has a good autocorrelation if

$$\Phi_{ii}[n] = \begin{cases} 1 & n = 0, N, 2N, \ldots \\ -\dfrac{1}{N} & \text{otherwise} \end{cases} \qquad (4.16)$$

This condition is also illustrated in Figure 4.18 for a chipping sequence of length $N = 8$. In order to synchronize with a sender, the receiver has to detect the beginning of a chipping sequence, which can be simply done by using this condition. The receiver monitors the incoming signal and compares it with the chipping sequence used by the respective sender by applying the autocorrelation function. If the result of this function shows a peak, that is, if $\Phi_{ii}[n] = 1$, it has detected the beginning of the code and is thus synchronized.

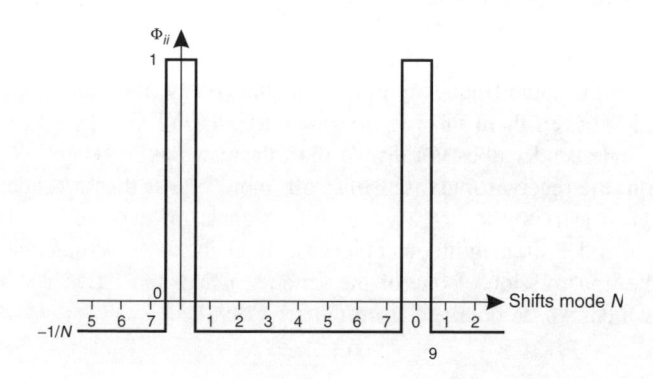

Figure 4.18 Autocorrelation of a chipping sequence.

A low cross correlation between different chipping sequences is essential in order to properly separate the different channels from the composed signal at the receiver. The more similar the two chipping sequences are, the more difficult it is to reconstruct the original data. If two chipping sequences show zero cross correlation, they are said to be *orthogonal*. Unfortunately, orthogonality usually exists only if unshifted versions of the chipping sequences are matched and is usually lost if they are shifted by any amount to each other. This is of major concern in CDMA, where different senders are usually not synchronized and any combination of shifted chipping sequences overlay during transmission. However, orthogonality is a very stringent condition and represents the best case for separating code channels. In most cases, it is also sufficient to merely have a low cross correlation (near zero), which is passably kept for all shifted versions of codes that may appear during transmission.

Thus, the challenge of CDM and CDMA is to generate chipping sequences with good correlation properties. Usually, these codes are generated by linear feedback shift registers that are arranged in a certain way in order to generate codes of the desired properties, for example, *m-sequences* and *Gold codes*. Other codes are composed by using matrices, for example, *Walsh codes*, or recursive tree structures, a method that is called *Orthogonal Variable Spreading Factor* (OVSF). It is far beyond the scope of this book to provide an introduction into these methods. The interested reader can find useful information on this issue in (Stallings 2002a) and (Dinan and Jabbari 1998).

CDM and CDMA have many advantages compared to other multiplexing and multiple access schemes. Spread spectrum is less susceptible to interferences and is much harder to intercept than conventional narrow-band signals. This is due to the fact that signals adopt a much broader frequency range and have a power spectral density in the range of noise, which makes it very hard for an interceptor to detect that there is an ongoing data transmission at all. Therefore, CDMA is very attractive for military applications and in fact was initially developed by and for the militaries. Unlike other schemes, CDMA does not suffer from cochannel and neighbor channel interferences (although the usage of codes with bad cross correlation may result in impairments that could be regarded as neighbor channel interferences). Furthermore, it is very flexible with regard to the data rate, which can be tailored to the special needs of an application. This is not possible with the rigid concept of fixed frequency and time channels in FDM and TDM respectively.

However, it must be stressed that CDMA is a very complex technique requiring sophisticated hardware in the senders and receivers. An important issue is power control. As all senders transmit in the same frequency range simultaneously, the radiating power must be carefully aligned between them in order to guarantee that all senders can be heard at the receiver and a single sender does not drown the transmissions of others. A sender located farther away from the receiver must transmit with more power than a sender located close by. However, when arriving at the receiver, both signals must be of equal strength or at least in the same order of magnitude; otherwise there is the risk that the receiver does not recognize the transmission of one of the senders. CDM and CDMA show many other particularities, which can be obtained from (Viterbi 1995).

4.4 Conclusion

This chapter has provided an overview of wireless communications. The fundamentals explained here are relevant to the transmission of any data for each kind of mobile service as well as for realizing positioning methods. Roughly speaking, it can be stated that wireless data transmission relies on the manipulation of radio signals as enabled by modulation and their coordinated transmission as enabled by multiplexing and multiple access. Positioning, on the other hand, is based on observing and measuring radio signals or their behavior, especially their traveling time or their attenuation. Nevertheless, positioning also requires the generation of special signals that are subject to measurements, the so-called *pilot signals*, which are generated in the same way as signals carrying user data. Unfortunately, these measurements are complicated by phenomena like multipath propagation, refraction in the atmosphere, or the Doppler effect. While these aspects are further discussed in Chapters 6 to 8, the following chapter first gives an introduction to cellular networks and location management.

5

Cellular Networks and Location Management

Initially, the development of LBSs progressed with the success of cellular networks, which achieved their high user penetration during the 1990s. At this time, operators were looking for new sophisticated services, which could provide a considerable added value to mobile subscribers in order to generate new revenues. As cellular networks track subscribers on a permanent basis for location management, offering services making use of a subscriber's "network location" was obvious. Recall from Chapter 2 that this kind of location refers to the topology of a (cellular) network, that is, to the subnetwork and base station the subscriber is currently attached to. The idea was to transfer a network location into a spatial or descriptive location using spatial databases and GIS and to exploit the resulting location information for LBSs.

However, very soon it turned out that network locations often do not meet the accuracy demands of many LBSs, and especially not those required for enhanced emergency services such as E-911 in the United States. This gave rise to the development of more accurate positioning methods and their integration into cellular networks. Nevertheless, location management is still of relevance in the field of LBSs, because it provides basic functions that these advanced positioning methods rely on. Also, there are some similarities between location management and location services with regard to the dissemination and maintenance of location information, which requires some knowledge about the architectures, topology, and limitations of cellular networks.

This chapter gives a general introduction to cellular networks and location management. Starting with an overview of different cellular systems, it focuses on the network architectures of GSM, GPRS, and UMTS. Subsequently, it identifies the common mechanisms of location management and illustrates their realization in these systems.

5.1 Overview of Cellular Systems

Mobile communications reached the mass market not before the 1980s. While the rudimentary transmission of analogous speech over a wireless link was well established at that time, the major challenge was to implement advanced mobility features such as handover, roaming, and the localization of subscribers, which required additional control channels between terminal and serving base station in addition to the channels carrying the speech. Since then, several systems went into operation, which are usually classified into *first-*, *second-*, and *third-generation systems* (1G, 2G, and 3G).

The development of 1G systems was inspired by the progress in digitization and the availability of the first low-cost microprocessors in the 1970s. These key technologies enabled the establishment of digital control channels for mobility management and call control, although speech transmission in these systems was still analogous. The most prominent 1G systems were the *Nordic Mobile Telephone* (NMT) system in Europe and the *Advanced Mobile Phone System* (AMPS) in North America, which is also known as *IS-88* and which is partly still in use today. Multiple access in these systems was organized by FDMA.

2G systems are characterized by the digital transmission of speech and low-rate data and consequently enable more sophisticated multiple access schemes such as a combined FDMA/TDMA or CDMA. GSM is not only the most prominent 2G system, but with its 1 billion subscribers today dominates mobile communications as no other system does. It is deployed by over 600 operators in 200 countries of the world. Other 2G systems are *CdmaOne* (also known as *IS-95*) and *DAMPS* (Digital AMPS, also known as *IS-136*) in North, Central, and South America and Asia and also the *Personal Digital Cellular* (PDC) system in Japan. GSM, DAMPS, and PDC use TDMA, while CdmaOne is based on CDMA. In recent years, some 2G systems were subject to extensions in order to prepare a smooth migration path from 2G to 3G systems. For GSM, extensions concern the introduction of a packet-switched mode, known as *General Packet Radio Service* (GPRS), and a new modulation scheme in the GSM frequency bands, which is referred to as *Enhanced Data Rates for GSM Evolution* (EDGE). GSM networks with GPRS/EDGE capabilities are also called *2.5G systems*.

At the time of writing this book, 3G systems are being installed in many countries of the world. Compared to 2G systems, they have been designed for higher data rates in the range of several hundreds of kbps and for packet-switched operation. In the long term, it is even envisaged to entirely replace all circuit-switched services, including speech telephony, by packet-switched services. Thus, 3G systems focus on many sophisticated and innovative applications like multimedia entertainment, video telephony, LBSs, or surfing the WWW, to name only a few. 3G systems are globally specified by the ITU under the term IMT-2000 and by a number of regional standardization bodies like 3GPP, 3GPP2, and TIA. All systems are based on CDMA, but slightly differ from each other in the parameters of their air interfaces (for example, bandwidth, spreading factors, frequency spectrum) and the protocols of the core network. For example, the European version of IMT-2000, UMTS, uses a variant of CDMA which is called *Wideband CDMA* (WCDMA), and which is also an option for 3G systems in Japan and South Korea. In North America, on the other hand, the preferred air interface standard is *Cdma2000*, because it can coexist with CdmaOne systems in the same frequency ranges. Besides WCDMA, Cdma2000 systems are also envisaged for Asia.

CELLULAR NETWORKS AND LOCATION MANAGEMENT

While the standardization of 3G systems has not been finished yet and their installation is still in progress, activities have been launched for a *fourth generation* (4G). These systems will be characterized by both new air interfaces for achieving even higher data raters and the coexistence of existing and future air interfaces, including WLAN and UMTS, under a common core network.

Table 5.1 provides an overview of 1G, 2G, and 3G networks. The following sections explain the common principles of these cellular networks and their location management mechanisms, and then demonstrates their realization by means of GSM, GPRS, and UMTS.

Table 5.1 Overview of cellular systems

System	Multiple access	Frequency range	Deployment
1st Generation			
NMT	FDMA	450/900 MHz	North Europe
AMPS/IS-88	FDMA	800 MHz	North America
2nd Generation			
GSM/GPRS	TDMA	850 MHz	North America
		900 MHz	Worldwide excl. North America
		1,800 MHz	Worldwide excl. North America
		1,900 MHz	North America
DAMPS/IS-136	TDMA	800 MHz	North America
CdmaOne/IS-95	CDMA	800/1,900 MHz	North/Central/South America and Asia
PDC	TDMA	800/1,500 MHz	Japan
3rd Generation			
UMTS/WCDMA	CDMA	around 2 GHz	Europe/Japan/South Korea
Cdma2000	CDMA	around 2 GHz	North America/Asia

5.2 Principles of Cellular Networks

A cellular network consists of a number of *radio cells*, where the term "cell" refers to the geographic coverage area of a base station. The size of this coverage area mainly depends on the base station's signal strength and the degree of attenuation, which, as explained in Section 4.2.4, also depends on the constitution and intensity of obstacles in the surrounding environment. The cell radii of today's cellular networks vary between 100 m in urban environments up to several tens of kilometers in rural areas. For example, GSM has been designed for cell radii of up to 35 km, which hence also represents the maximum distance between mobile and base station.

Each base station is assigned a certain number of channels for transmitting and receiving data, which is referred to as *cell allocation* (CA). An important characteristic of cellular networks based on FDMA is that the same cell allocation (or parts of it) can be used by several base stations simultaneously, assuming that they are separated by a fairly large distance from each other, which is also denoted as *channel reuse distance*. That is, to avoid interferences between cells (the so-called *cochannel interferences*), it must be guaranteed that neighboring base stations are assigned cell allocations of different channels. This is illustrated in Figure 5.1, where the cells of the same shade use the same cell allocation and are separated from each other by at least the channel reuse distance. Note that channel reuse is the key concept that is able to serve millions of subscribers with a comparatively small number of channels.

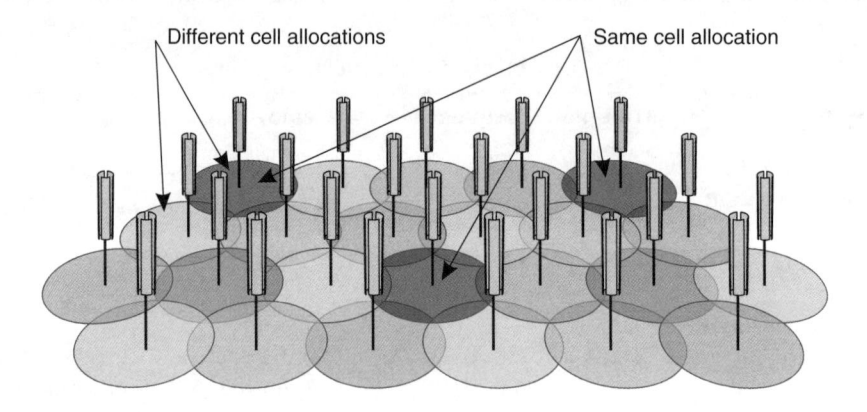

Figure 5.1 Cellular networks.

Basically, there exist two concepts for arranging a cell allocation, which is closely related to the kind of multiplexing and multiple access scheme applied in a particular system. In GSM, which realizes a combination of FDMA and TDMA, a cell allocation comprises a number of frequency channels, and neighboring cells must use different frequency channels in order to avoid cochannel interferences as far as possible. This separation by frequency is also the favorite concept used by most 1G and 2G systems mentioned earlier. In CDMA-based systems like UMTS, on the other hand, different terminals transmit in the same frequency range simultaneously and are separated by different code channels. This method can also be applied for cell separation. A cell allocation of a base station here consists of a set of spreading codes, and neighboring base stations usually transmit in the same frequency range, but always on different code channels.

As also indicated in Figure 5.1, there are no sharp borders between neighboring cells. Rather, they are overlapping each other to a greater or lesser extent, which means that a certain position is usually covered by several base stations simultaneously. In urban areas, a mobile device can typically hear a set of around 10 base stations simultaneously, and then selects from this set the base station with the strongest signal and the best signal quality for transmission. Cell overlapping is an important characteristic that is essential for advanced positioning methods that are based on lateration or angulation (see Chapters 6 and 8). Also note that, as opposed to Figure 5.1, the shape of a cell is not a perfect circle, but rather very irregular, which mainly goes back on the propagation conditions of the spot.

The number of cells a network is made up of is basically a function of the size of the area to be covered and the user penetration. As cellular networks are mostly operated on a national basis, the coverage area is given by the area of the respective country. However, when building up a new network, operators first concentrate on establishing coverage in congested urban areas before establishing base stations in rural areas, and hence coverage is sometimes incomplete. In this case, an operator may enter into roaming agreements with other operators in order to offer services over foreign networks in those regions that are not covered by himself. Furthermore, if a network runs the risk of becoming overloaded in a certain region, operators can increase the capacity by simply increasing the base station density. This goes along with an extension of the network by a number of new base stations and a complete or partial reorganization of CAs in the respective region. To gather CAs for the new base stations, the signal strength of existing ones is reduced, thereby also reducing the channel reuse distance. As a result, the same channels can be reused in a shorter distance than before, and the capacity of the network in the respective region is increased. This method is preferably applied in urban areas with a high user penetration, and hence cells in these regions are usually much smaller than in rural areas. Applying these principles, today's cellular networks consist of a high number of cells. For example, a typical medium-sized country with high user penetration, such as Germany, requires a number of base stations, in the order of tens of thousands, for nationwide coverage.

A cellular network, of course, not only consists of base stations, but also comprises a network infrastructure for interconnecting base stations, mobility support, service provisioning, and connection to other networks like ISDN or the Internet. The following sections introduce the infrastructures of GSM, GPRS, and UMTS in brief.

5.2.1 GSM Architecture

GSM was primarily designed to supply the mobile subscriber with *circuit-switched* telephony services as he knew from the *Plain Old Telephony System* (POTS) for many years. Telephone calls are made and received through a mobile terminal that, in official GSM terminology, is denoted as *Mobile Station* (MS). The terminal contains the technical equipment and a *Subscriber Identity Module* (SIM), which is a small chip card that stores subscriber-specific identifiers, addresses, as well as keys for authentication and encryption. It is composed by the network operator and delivered to each subscriber with a valid subscription.

A GSM network consists of several *access networks*, which include the radio equipment that is necessary to interconnect a terminal to the network. The access networks are interconnected by the core network, which also enables the interoperability with external telephone networks, for example, ISDN. In GSM, the access network is referred to as *Base Station Subsystem* (BSS), and the core network is denoted as *Mobile Switching and Management Subsystem* (SMSS). The resulting architecture is depicted in Figure 5.2.

The BSS is responsible for monitoring and controlling the air interface. It consists of two different components, which are called *Base Transceiver Station* (BTS) and *Base Station Controller* (BSC). BTS is the official GSM term for "base station" and thus contains transmitter and receiver equipment as well as an antenna. An important design goal was to keep the base stations as simple and as cheap as possible, and hence they are equipped with only very limited capabilities for signal and protocol processing. The bulk of the work, for example, allocation and release of channels at the air interface, is done by the BSC. Also, the BSC is responsible for control and execution of handover, a function which is needed to keep a circuit-switched connection if the subscriber moves between base stations. Each

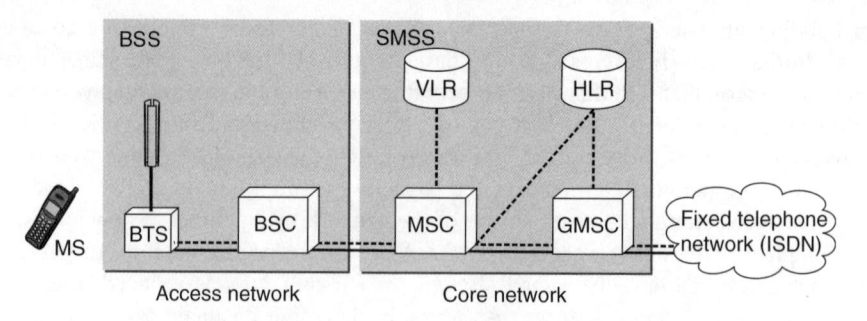

Figure 5.2 GSM architecture.

BSC controls several base stations, which are connected to the BSC via fixed lines or radio link systems.

The SMSS is a fixed network of switching nodes and databases for establishing connections from and to the mobile subscriber and maintaining them. The switching components are the *Mobile Switching Center* (MSC) and the *Gateway MSC* (GMSC), both differing from conventional fixed telephone switches in that they contain additional mechanisms for supporting the mobility of subscribers, for example, localization and handover. The MSC connects a number of BSCs to the network, while the GMSC is the interface between the GSM and an external fixed network like ISDN. Thus, the MSC is responsible for serving a limited geographic region, which is given by all base stations connected to the MSC over their BSCs. On the other hand, the GMSC is rather a central component that may serve an entire GSM network. However, the actual number of GMSCs is operator-specific and depends on network design criteria and the number and arrangement of access points between GSM and external networks.

When a connection is to be established, MSC and GMSC have to determine the next intermediary node, that is, another switch, to which to route this connection. In a mobile network, this decision depends on the current location of the mobile subscriber. To find out this location, GMSC and MSC are connected to a *Home Location Register* (HLR) and a *Visitor Location Register* (VLR), respectively, which store location information for each subscriber. The working of HLR and VLR will be explained later in this chapter in detail.

Finally, the management of the network is imposed on an *Operation and Maintenance Subsystem* (OMS) (not depicted in Figure 5.2), which supports the administration and commercial operation, security management, network configuration, and performance management, as well as maintenance tasks (Eberspächer et al. 2001). It consists of one or several *Operation and Maintenance Centers* (OMC), which control and coordinate the entire management, as well as databases for registering malfunctioning or stolen devices (*Equipment Identity Register*, EIR) and for generating the keys needed for authentication and ciphering (*Authentication Center*, AuC).

The fixed connections within the SMSS as well as between BSS and SMSS are realized by conventional transmission lines according to the *Synchronous Digital Hierarchy* (SDH) or *Synchronous Optical Network* (SONET) standards. Besides the transport of telephone calls, the network components must also exchange control information for establishing, maintaining, and releasing connections, for mobility management, advanced service

CELLULAR NETWORKS AND LOCATION MANAGEMENT 95

features, and also for controlling positioning. The exchange of such control data is subsumed under the term *signaling*. In the SMSS, signaling is realized by the protocols of the *Signaling System No.7* (SS7), which is a common protocol stack that has been standardized for ISDN (see (ITU–T Recommendation Q.700)). Since SS7 provides no functions for mobility support, an extension has been specified for GSM, which is called *Mobile Application Part* (MAP) and which enables the transfer of control data for mobility support (see (3GPP TS 29.002)). For signaling between SMSS and BSS as well as within a BSS, GSM-specific protocols have been specified.

5.2.2 GPRS Architecture

GSM was initially designed for offering voice services, which, by their very nature, are circuit switched. This means that a dedicated circuit of switching nodes and channels in between must be established before any data can be transferred. When establishing such a circuit, dedicated channels are allocated between switching nodes in the fixed network (part) as well as between terminal and base station at the air interface, which are reserved for exclusive use as long as the connection lasts, regardless of whether data is transferred at all. However, with the increasing popularity of the Internet in the 1990s, the introduction of *packet-switched* services was desired, and hence the classical GSM was extended by the *General Packet Radio Service* (GPRS). In a packet-switched network (PS network), data is segmented into several packets, either of fixed or variable size, which are passed through the network independently from each other. Channels between switching nodes are only allocated for the time it takes to transfer a packet and released immediately afterwards. Thus, packet-switched traffic by its very nature is discontinuous.

An important design goal of GPRS was to reuse the existing components of GSM networks as far as possible, especially the expensive and well-established infrastructure of base stations. Therefore, the BSS was extended by a number of new protocols for enabling the transmission of short data bursts over the air interface, thereby reusing the existing hardware to a large extent. However, for the fixed network part, it was necessary to build up a second network of new components that work parallel to the existing SMSS. These components are depicted in Figure 5.3.

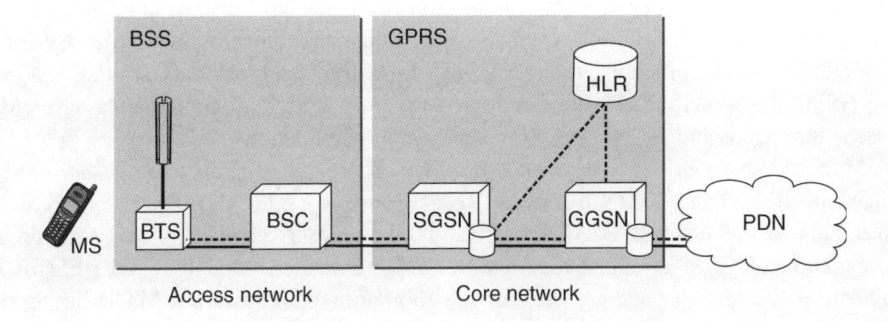

Figure 5.3 GPRS architecture.

Basically, the GPRS core network is an IP-based network composed of a number of routers for delivering packets between an external *packet data network* (PDN) and a BSS.

GPRS was specified for interoperability with X.25 networks and the Internet (although almost all operators today probably support only Internet access). The routers are termed *GPRS Support Nodes* (GSNs), and, in analogy to MSCs, differ from conventional routers as known from the Internet in that they provide enhanced mechanisms for mobility support. In addition, they are responsible for counting packets for volume-based charging. Two different types of GSNs exist. The *Gateway GSN* (GGSN) is the counterpart of the circuit-switched GMSC and is responsible for interconnecting the GPRS core network with an external PDN. The *Serving GSN* (SGSN) is the counterpart of the circuit-switched MSC and accordingly serves a limited geographical region defined by the set of base stations it is connected to. Depending on the respective configuration, a network may contain several GGSNs, which means that an SGSN has not assigned a fixed GGSN; it may deliver and receive packets to and from different GGSNs.

In contrast to MSCs and GMSCs, GSNs have integrated location registers for storing location data and user profiles for all subscribers being attached to the GPRS network. Thus, packet delivery can be accelerated, because delays that would occur when location data were to be requested from external location registers can be avoided. The location register of the SGSN replaces the VLR, while the location register of the GGSN adopts the role of the HLR. Nevertheless, HLR and VLR are connected to the GPRS network in order to support a common location management when subscribers are registered for both circuit-switched and packet-switched services. In addition, the HLR supports the GPRS location management if a subscriber moves between regions served by different SGSNs. Further details about GPRS can be found in (Sanders et al. 2003).

5.2.3 UMTS Architecture

During the design phase of UMTS, it was an important goal to determine a smooth migration path from 2G to 3G systems. Both types of networks should be integrated as far as possible, rather than operating two dedicated networks that are entirely separated from one another. This implies that UMTS terminals should also work in a GSM/GPRS mode and that interoperability, for example, authentication, handover, and location management, is enabled between both systems. Similar migration strategies have also been acquired for the replacement of other 2G by 3G systems, for example, for the crossover from CdmaOne to Cdma2000. Not surprisingly, today's UMTS infrastructure looks very similar to a combined GSM/GPRS network. Figure 5.4 depicts the most important components of a UMTS infrastructure according to Release 99 of the 3GPP specifications.

UMTS incorporates two access networks, which are called *GSM/EDGE Radio Access Network* (GERAN) and *UMTS Terrestrial Access Network* (UTRAN). The former basically corresponds to the original GSM access network as described earlier, and incorporates the conventional GSM air interface as well as the extended version EDGE. EDGE has been defined as an intermediary step in the migration path toward UMTS, but is only deployed by some operators (maybe as an alternative to UMTS). The counterpart of a BSS in UTRAN is termed *Radio Network Subsystem* (RNS). A base station is called *Node B* and it is controlled by a *Radio Network Controller* (RNC), which is the counterpart of GSM's BSC. Furthermore, a terminal is referred to as *User Equipment* (UE) in the official UMTS terminology. Although this topology looks very similar to that of the GSM BSS, there are

CELLULAR NETWORKS AND LOCATION MANAGEMENT

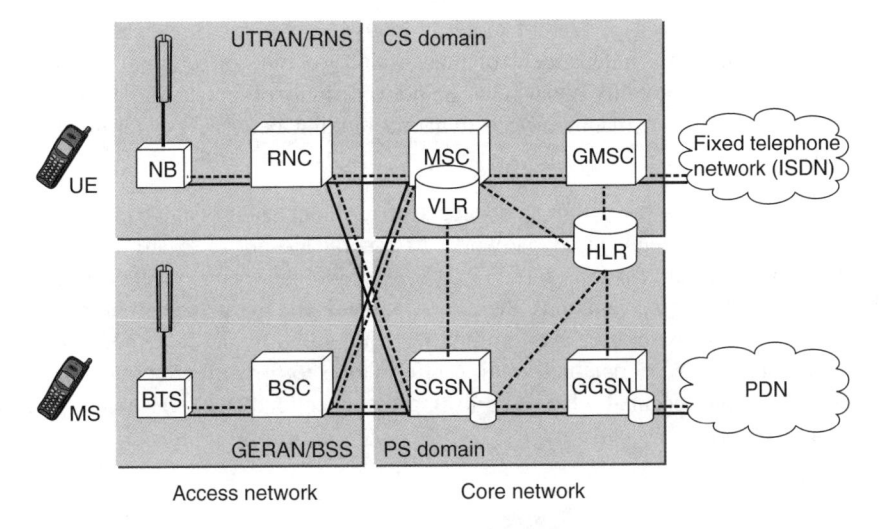

Figure 5.4 UMTS architecture.

fundamental differences that originate from the very different designs of the air interface in both systems. A detailed overview of this is given in (Kaaranen et al. 2001).

The former GSM and GPRS core networks are now represented by two domains, which are referred to as *circuit-switched* and *packet-switched domains* (CS and PS domains). They accommodate the components that have just been introduced in the previous sections. However, all of them have been exposed to extensions and modifications in order to cope with increased data rates, enhanced security mechanisms, new protocols and signaling for enhanced service features, and so on.

In the mid to long term, it is intended to completely replace the CS domain by services of the PS domain. The goal is to convert all network connections to IP, that is, not only within the access and core networks but also in between. Traditional services like voice as well as many new, more or less sophisticated multimedia services will then be realized by a new service environment referred to as *IP Multimedia Subsystem* (IMS) in the PS domain (see (3GPP TS 22.228)). Starting with Release 99, GSM, GPRS, and UMTS are further developed as an integrated system. The architectures presented in this and the two previous sections are specified in (3GPP TS 23.002).

5.3 Mobility Management

In all cellular networks, the support of mobility is a fundamental requirement. Mobility management comprises all operations that are needed to serve the user with services while on the move. In the area of cellular networks, there are two types of mobility a network has to support, which are as follows:

- **Terminal mobility.** This type refers to the spatial mobility of classical terminals like mobile phones, PDAs, and notebooks within the coverage area of a network. Supporting terminal mobility means that the network must offer appropriate mechanisms for establishing and maintaining a dynamic link between the network and the terminal.

- **Personal mobility.** This type of mobility is given if the subscriber can be identified toward the network independent of the terminal or type of terminal he uses. Supporting personal mobility means that the network must offer appropriate mechanisms for establishing and maintaining a dynamic binding between the terminal and the subscriber.

Figure 5.5 illustrates the different concepts of terminal and personal mobility. In the earlier days of telecommunications, networks in general did not offer any mechanisms at all for supporting mobility (see Figure 5.5 (a)). The terminal was connected over a wired trunk with a local switch, and thus there was a fixed and permanent binding between terminal and network. Also, user and terminal were identified by the same address (i.e., the telephone number), which contained routing information to forward telephone calls to the target switch and the terminal when establishing a connection. Thus, the binding between terminal and subscriber was also fixed.

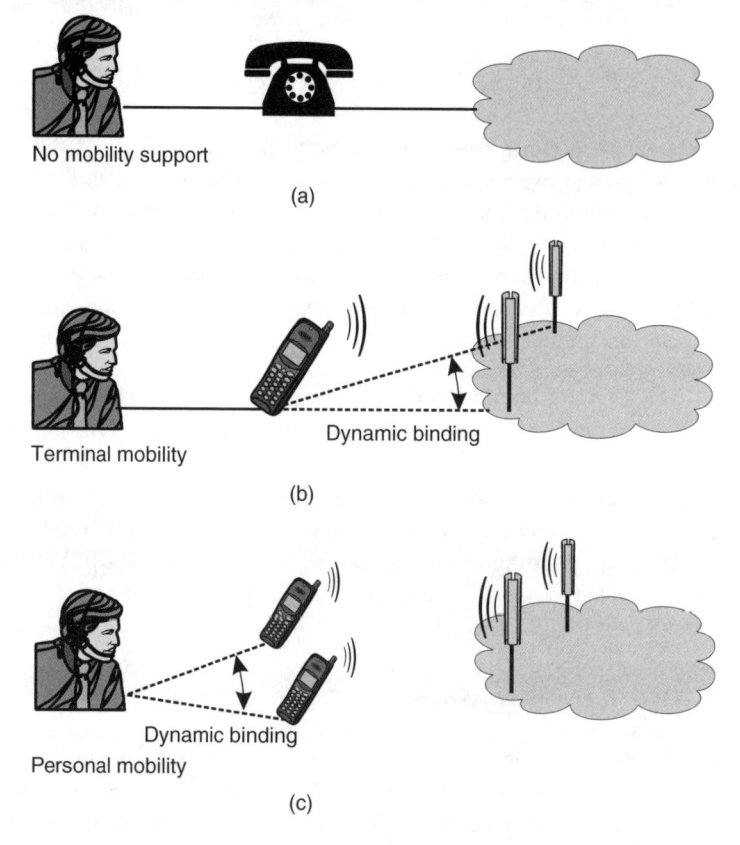

Figure 5.5 Mobility types. Adapted from (Roth 2002).

Today's cellular networks support both terminal and personal mobility. Terminal mobility is illustrated in Figure 5.5 (b). The main assumption for it is the capability to establish a wireless link between the terminal and a base station for carrying circuit or packet-switched traffic.

For example, if the mobile subscriber wants to make a telephone call (*mobile-originated call*) or if he wants to transfer data packets to the Internet (*mobile-originated packets*), he must attach to a nearby base station in order to get connected to the core network. On the other hand, if a telephone call (*mobile-terminated call*) or packets (*mobile-terminated packets*) arrive, the network must establish a wireless link between a base station and the terminal. In this case, it is first necessary to determine a nearby base station that is within the range of the subscriber. The localization must also be possible if the subscriber is roaming in a foreign network. Furthermore, full support of terminal mobility also requires that an ongoing data transfer, being circuit or packet-switched, is not interrupted if the subscriber moves to another cell. For circuit-switched traffic, the circuit must be reorganized and shifted to the new base station, a function that is called *handover*. For discontinuous packet-switched traffic, mobile-terminated packets must be dynamically rerouted to the new base station.

A dynamic binding between subscriber and terminal for personal mobility (see Figure 5.5 (c)) is basically achieved in that both of them are referenced independent of each other. A subscriber is referenced by personal addresses (e.g., telephone numbers), which are assigned on a permanent basis and which remain valid regardless of which terminal he is registered with in the network. Terminals, on the other hand, are referenced by temporary addresses, which are dynamically assigned if the subscriber registers with the network and which are updated during a registration. Closely related with personal mobility is the authentication of subscribers toward the network, which is required to verify his identity and to avoid that unauthorized individuals access services on his behalf. The address management, which comprises the assignment and update of addresses as well as address mapping mechanisms between them and the authentication procedures are controlled inside the network. On the terminal site, the addresses are stored in the SIM, which is also responsible to authenticate the subscriber upon request from the network site.

Although not clearly defined in literature, the term "mobility management" is related to all functions for supporting terminal and personal mobility, among other things, localization, handover, and security functions. Each of these functions is realized by numerous procedures and protocols between terminal, access, and core network. A subset of mobility management is location management, which is to be highlighted in the following sections.

5.4 Common Concepts of Location Management

Location management comprises all functions needed to localize the subscriber in the cellular network. In this chapter, the term localization is used differently from positioning (as considered in the context of LBSs) in that it describes the determination of a subscriber's position with regard to a base station in a cellular network or, more commonly speaking, with regard to the topology of a cellular network. Positioning, on the other hand, aims at determining a subscriber's position in terms of a descriptive or spatial location. To understand location management, it is necessary to consider the operations taking place between network and terminal as well as within the cellular network. The basic concepts of both will be explained in the following section.

5.4.1 Location Update and Paging

There are two fundamental approaches to how a network can track a mobile subscriber. In the first approach, the terminal reports its current location to the network. This function is called *location update*. Although there are numerous strategies concerning the point of time a location update can be initialized, consider at first a simple location update strategy where the terminal performs a location update whenever the subscriber enters a new cell. Upon arrival of such a location update in the network, the current cell of the user is stored in a database. In case of a mobile-terminated transmission, the current cell of the target terminal can be simply obtained by requesting this database.

In the second approach, the terminal does not report its current location to the network at all. Rather, the network searches the terminal in all cells, which is referred to as *paging*. If the target terminal recognizes that it is paged, it returns a paging response to the network. The base station the terminal has selected for passing this paging response is also the base station over which circuit or packet-switched data is then transferred.

An important goal of location management is to minimize the signaling traffic caused by location updates and paging respectively. For location updates, this overhead is a function of the cell size. The smaller this size, the higher is the frequency of location updates and vice versa. Especially, in urban areas with cell radii of 100 m or less and strong overlapping of cells, the signaling required for location updates may hence waste valuable resources at the air interface and of the fixed network part, which goes at the expense of channels for exchanging user data. In addition, each location update is associated with some power consumption and burdens the terminal's battery more or less heavily. In a similar manner, paging burdens the air interface. Following the approach introduced earlier, a terminal has to be paged in all cells of a network, which, considering the huge number of cells a network is made up of, is simply not practicable.

Owing to these reasons, a combined approach of location updates and paging is used, which is based on the concept of *location areas*. As illustrated in Figure 5.6, several cells are grouped to build a location area, which represents the smallest unit for which the network maintains the current location of a subscriber. Instead of reporting the current cell, the terminal now performs a location update only if the subscriber enters another location area and reports the identity of this location area to the network. In case of mobile-terminated traffic, the network then only pages the terminal in the location area the subscriber is registered with. Thus, the location-area structure represents a compromise between the pure approaches presented earlier, namely, either keeping track of a subscriber's location on cell basis or paging him in the entire network.

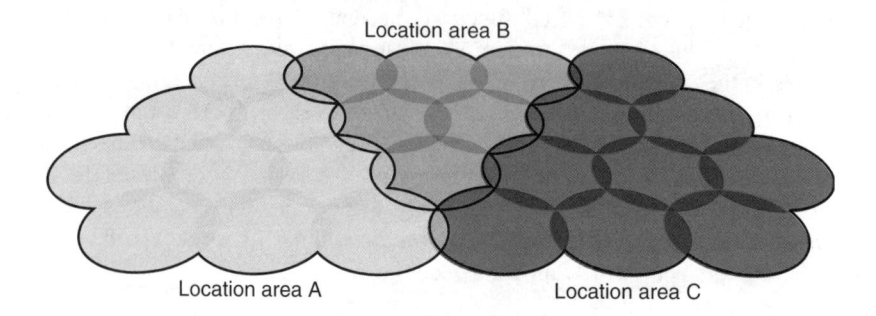

Figure 5.6 Location areas.

The shape of location areas and the number of cells they consist of are fixed after taking into consideration several parameters and the special conditions on the spot. From a theoretic point of view, the optimal size of a location area can be determined as a function of the signaling overhead caused by paging and location updates. This is illustrated in Figure 5.7[1]. The overhead of location updates decreases with an increasing number of cells per location, while the paging overhead increases. The margin where the paging overhead exceeds the location update overhead represents the optimal size of location areas. However, in reality it is not possible to determine these overheads very easily. It depends on the average call and mobility behavior of subscribers, where the call behavior has an impact on the paging frequency and the mobility behavior on the frequency of location updates. The call behavior is typically specified by the mean arrival rate of calls and packets respectively as well as the duration of calls. The mobility behavior is somewhat more complicated and comprises the velocity, the direction of motion, the duration of motion, the duration of residence at a particular location, and other parameters. Also, mobility behavior significantly depends on the infrastructure of the considered region, which is very different, for example, in the case of highways compared to that of cities.

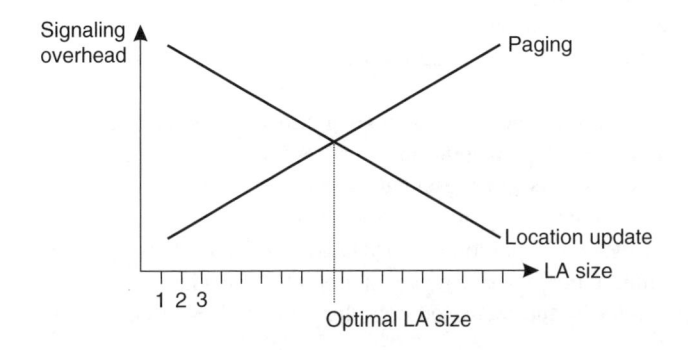

Figure 5.7 Overhead of paging and location updates depending on LA size.

In recent years, various improvements of the location-area concept have been suggested in the research community in order to further reduce the overhead associated with the related operations. Plassmann (1994) proposes an approach where a location area is subdivided into several *paging areas*, and each paging area in turn into several cells. As before, a terminal initializes a location update when it enters a new location area. However, the paging is not done in all cells of a location area simultaneously, but according to a sequence of *paging areas*. This sequence usually starts with the paging area where the terminal was last detected by the network. If the terminal does not respond to a paging request in a particular paging area, it is paged in the next one. Paging traffic can be reduced in this way, because on average paging is not performed in all cells of a location area. However,

---

[1] Note that the functions sketched in the figure are not necessarily linear ones. Their concrete appearance depends also on many circumstances, for example, call and mobility behavior on the spot, location-area shape, and the amount of data carried in update and paging messages. The intention of the figure is only to demonstrate the general impact the location-area size has on the overhead.

it must also be stressed that sequentially paging causes additional delay until the terminal is located.

Location update overhead can be reduced by the *two-Location algorithm*, which is suggested by Lin (1997). In this approach, the terminal stores the two most recently visited location areas. If the user enters a new location area, it is checked whether it is stored in memory. If not, the location area before the last one is replaced by the new one and a location update is performed. However, if the new location area is already in memory, a location update is not necessary. Paging is then performed according to a certain strategy, for example, in the most recently visited location area first or according to a random selection. Although (Lin 1997) has shown that his approach outperforms the classical location update strategy, this approach goes along with increased paging overhead and delay.

Other approaches deal with the structure of location areas. For example, a problem is caused by the fact that location update traffic appears only in those cells that are located at the borders of location areas. To equally distribute this traffic among all cells, Okasaka et al. (1991) propose to introduce so-called *multilayer location areas*. Each layer covers all cells of the network and fixes a certain structure of location areas, but the location areas of different layers are arranged in a way that their borders do not appear in the same cells. Terminals are assigned to different groups, where each group is associated with a certain layer and hence has to perform location updates according to the location-area structure given by this layer.

To draw a conclusion, there are a huge number of interesting scientific papers that deal with the optimization of paging and location update. Comprehensive overviews and comparisons of these works are presented in (Mukherjee et al. 2003), (Zhang 2002), and (Tabbane 1997). Although many advanced approaches have proven to significantly reduce the traffic caused by these operations, they have not found their way into operation. One reason for this might be the reluctance of operators to change their running networks. Another reason might be the increased complexity that goes along with some approaches.

5.4.2 Database Concepts

Upon arrival of a location update, the network stores the location area of the subscriber in databases, which are requested each time mobile-terminated data is to be transferred in order to find out in which cells to page the subscriber. In these databases, each subscriber is represented by an entry, which contains his personal addresses and numbers (e.g., his telephone number) as primary key together with other parameters, above all the location area, and also the subscriber's service profile and security data needed for authentication and encryption.

The simplest approach one can imagine for maintaining such subscriber data is that of a central database, which stores the data of all subscribers of a cellular network on a single node that is located anywhere in the network. Each location update and each location-area request is passed to this central database. While this approach would work very well for networks serving a few thousands of subscribers in a small region, it is not practicable for coping with millions of them and for networks with nationwide coverage. In this case, a central database would be faced with a huge amount of (nearly) simultaneous requests and would become overloaded very fast. In addition, many requests would have to cover large distances, that is, from the serving base station to the site where the database is located, which imposes a high load on the underlying signaling network as well. Thus, a central database approach for maintaining subscriber data does not scale for cellular networks very well.

Therefore, location management is supported by distributed databases. Nearly all systems, for example, GSM, use a two-level approach consisting of a central HLR and many distributed VLRs. A VLR represents a certain subregion of a network's coverage area. It is responsible for a number of associated location areas and holds data entries for all users currently visiting one of them. In particular, data entries in the VLR point to the current location area of a user. In GSM, each MSC has an associated VLR that stores the location areas of all subscribers staying within the MSC area. The HLR, on the other hand, holds data entries for all subscribers of the cellular network, like in the central database approach mentioned earlier. In particular, a data entry here points to the MSC in which area the respective subscriber is currently registered with.

The advantage of the two-level approach is that location updates affect only the local VLR if the new and the previous location area belong to the same VLR. In this case, it is not necessary to redirect a location update to the central HLR, thereby disburdening the HLR and avoiding long-distance signaling to a large extent. Only if previous and new location area are not served by the same VLR, it is necessary to establish an entry in the new VLR, delete the entry in the old one, and update the pointer in the HLR.

Like location update and paging strategies, database concepts were and are subject of intensive research aiming at the reduction of overhead caused by database queries, updates, and signaling. Jain et al. (1995) propose to establish so-called *forwarding pointers*. Instead of updating the remote HLR, a chain of forwarding pointers is established in the VLRs the subscriber has visited. The HLR points to the first VLR, this VLR to the VLR the subscriber has visited next, and so on. Although location updates to the HLR can be avoided this way, the entire chain of forwarding pointers must be passed every time the subscriber needs to be located. Thus, this approach is useful only if subscribers change their location much more frequently than they receive calls. If, on the other hand, the call rate is higher, caching and replication techniques may work more efficiently, as proposed, for example, by (Jain et al. 1994). Other authors propose to use multilevel databases instead of the two-level concept. One of the most famous approaches on this goes back to Wang (1993). However, the multitude of approaches in this area can be hardly covered here. An overview and a comparison can be found in (Mukherjee et al. 2003) and (Küpper 2001).

5.5 Location Management in CS Networks

This section introduces the location management in CS networks by taking GSM and the UMTS CS domains as examples. At first, it is helpful to have a closer look at the hierarchical topology location management is based on. This topology is illustrated in Figure 5.8. The first level is given by the *service area* of a GSM/UMTS network. As most networks operate on a nationwide basis, this area usually corresponds to a country. A service area is subdivided into a number of *MSC areas*, each of it having a responsible MSC for switching telephone calls originating from and terminating at this region. A network has at least one MSC area, but most of them are composed by several tens of them. An MSC area consists of a number of location areas, and a location area, in turn, is composed by a number of cells, as explained earlier.

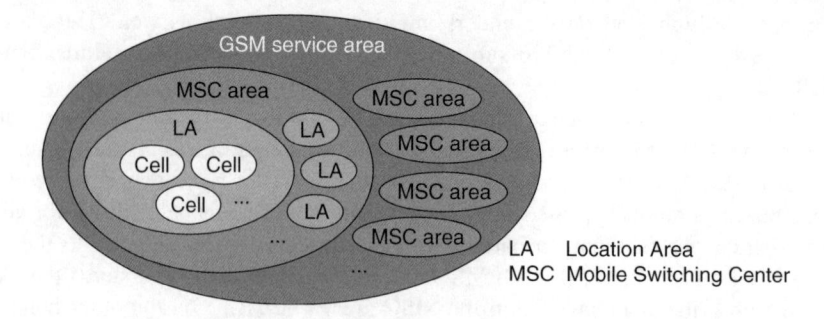

Figure 5.8 GSM/UMTS topology for location management.

5.5.1 Identifiers and Addresses

The support of terminal and personal mobility requires a number of addressing schemes for assigning either permanently or temporarily addresses to subscribers and terminals in order to enable a dynamic binding between subscriber and terminal as well as between terminal and network. In addition, location areas and cells need to be referenced as well. Figure 5.9 lists the different identifiers and addresses, which are used for both GSM and UMTS CS location management.

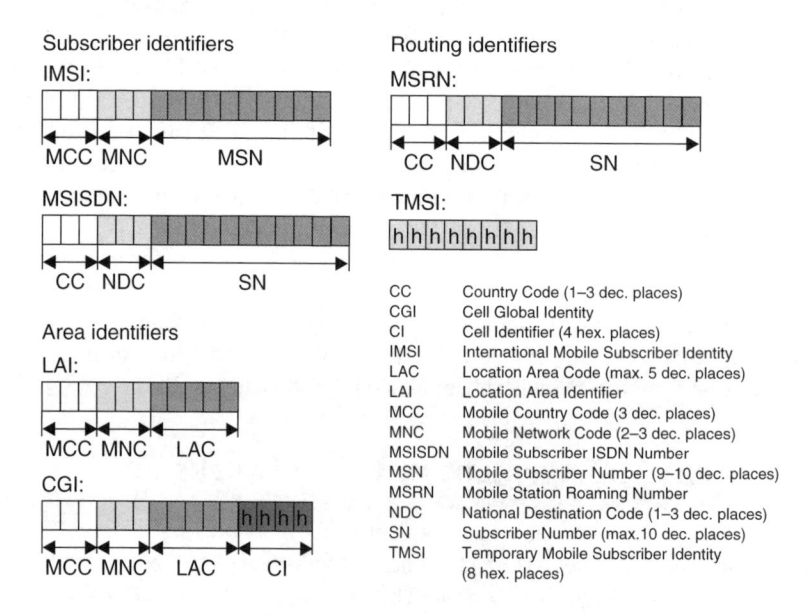

Figure 5.9 Identifiers used for CS location management.

Basically, there are different identifiers for referencing the subscriber, the terminal, and different types of areas of the topology of location management. The subscriber identifiers are:

- **International Mobile Subscriber Identity (IMSI).** Each subscriber can be uniquely identified by his IMSI. Its structure has been fixed by the ITU-T specification E.214, which defines numbering schemes for a worldwide unambiguous identification of fixed and mobile networks for signaling purposes (ITU–T Recommendation E.214). It is assembled from the *Mobile Country Code* (MCC), which reflects the country where the subscriber's home network is located, the *Mobile Network Code* (MNC), which denotes the network within a country, and the *Mobile Subscriber Number* (MSN), which uniquely identifies the subscriber within a network.

- **Mobile Subscriber ISDN Number (MSISDN).** The MSISDN is basically a subscriber's telephone number, which is dialed by another party when making a call. Thus, it is, in contrast to the IMSI, a public identifier. Its structure follows the ITU–T specification E.164, which is also used for assigning telephone numbers in fixed telephone network such as ISDN (ITU–T Recommendation E.164). The MSISDN consists of a *Country Code* (CC), a *National Destination Code* (NDC), and the *Subscriber Number* (SN). Although the meaning of CC, NDC, and SN are very similar to that of MCC, MNC, and MSN (as used for the IMSI), they are assigned according to different numbering schemes and hence are not equal.

IMSI and MSISDN allow for the separation of a subscriber's identity and his telephone number. The IMSI is of relevance for internal signaling procedures within a network, for example, authentication and accounting, or between networks, for example, roaming. In contrast to the MSISDN, it is not public and thus represents a key concept for protecting the subscriber's identity from misuse. Also, in contrast to other identifiers, the IMSI is not transmitted over the air interface (apart from a few exceptions). The association between the IMSI and the MSISDN of a subscriber is maintained in the HLR. Both identifiers are also stored on the SIM and in the VLR. Each subscriber can be assigned several MSISDNs, for example, in order to distinguish between different circuit-switched services such as voice, fax, and data.

The following identifiers are used to route an incoming telephone call through the topology of a cellular network to the target terminal:

- **Mobile Station Roaming Number (MSRN).** The MSRN represents a certain terminal with regard to the MSC area it is currently located in. It can be seen as a temporary location-dependent telephone number that is only valid for the duration a subscriber remains in the MSC area and that is changed when he enters another one. Also, it has the structure of a telephone number (i.e., an MSISDN), that is, it consists of CC, NDC, and SN. The MSRN is for internal routing purposes only. It is neither visible for subscribers or third parties nor is it possible to make a call by dialing this number. The MSRN is generated by the responsible VLR and passed to the HLR, which maintains the association between MSISDN and MSRN for each subscriber.

- **Temporary Mobile Subscriber Identity (TMSI).** The TMSI identifies a subscriber in a particular location area and is hence used for paging a terminal. It is assigned by the responsible VLR each time a terminal registers with the network, and it is renewed if he enters another location area. The TMSI consists of eight hexadecimal places. In contrast to the MSRN, it is only of local relevance and not passed to the HLR.

Finally, it is necessary to assign identifiers to location areas and cells:

- **Location Area Identifier (LAI).** Each location area is assigned an LAI, which consists of MCC, MNC, and a *Location Area Code* (LAC) and which is thus unambiguous. The LAI is broadcast in all cells that belong to the corresponding location area and is observed by the terminals in order to detect whether the subscriber has entered a new location area. If so, the terminal initializes a location update. For each subscriber in a particular MSC area, the VLR holds the association between TMSI and LAI.

- **Cell Global Identity (CGI).** Each cell is assigned a CGI, which is assembled by the LAI and a *cell identifier* (CI).

The following sections demonstrate the interaction of these identifiers during the routing of incoming telephone calls through the cellular network topology and for location updating.

5.5.2 Localization and Routing

The localization of subscribers is necessary for mobile-terminated calls. It is organized as a stepwise process, where each step corresponds to one layer of the hierarchical network topology. Figure 5.10 illustrates the localization procedure and the routing of a call, which arrives at the GMSC from an external network, for example, ISDN.

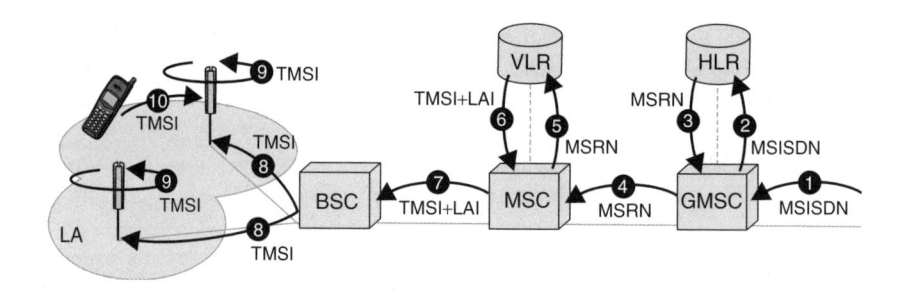

Figure 5.10 Localization and routing of incoming calls.

The external network passes the call together with the MSISDN of the target subscriber to the GMSC (1). In a first step, the GMSC must determine the current MSC area of this subscriber. It therefore passes the MSISDN to the HLR (2). In today's networks, the amount of subscriber entries is usually shared between several HLRs, and each HLR is responsible for a certain subrange of MSISDNs. The HLR searches the database and returns the MSRN (3), which points to the MSC area where the subscriber currently resides. The call is then routed to this MSC (4). In order to prepare for the paging procedure, the MSC passes the MSRN to the local VLR (5) in order to obtain the subscriber's TMSI and LAI (6). The former is included in the broadcast message to page the subscriber, while the LAI must be known to decide in which location area to page him. Knowing the LAI, the MSC can ask the responsible BSC or BSCs for paging (7). The BSC forwards the TMSI to all base stations of the particular location area (8) and initializes the paging (9). Each base station then broadcasts the TMSI on a broadcast channel that is permanently monitored by all terminals. If the target terminal recognizes that it is paged, it sends a paging response to the network (10). This response is directed over the base station the terminal has determined as that with the best signal quality, and thus the subscriber's cell is determined. In the last step, a dedicated channel is established at the air interface and the call is passed to the terminal (not shown in the figure).

For calls between different cellular networks or for internal calls between subscribers of the same network, this procedure is basically the same. In the latter case, the request to the HLR can be avoided if calling and called subscriber belong to the same VLR region. Also, localization for the delivery of short messages is similar.

5.5.3 Location Updates

In GSM and the UMTS CS domains, there are two kinds of location updates:

- **Location update on location-area crossing.** As already mentioned, this type of location update is executed whenever the subscriber enters a new location area.

- **Periodic location update.** In addition, the terminal performs location updates periodically regardless of whether a new location area has been entered.

The periodic location update is only optional, and the period according to which it appears can be configured by the operator by simply broadcasting a timer value to the terminals. It is advantageous in cases of inactive terminals (no calls and no location-area change for a longer period of time) in order to indicate the terminal's presence to the network. Also, it eases recovery arrangements in the case of database failures.

Whether a terminal performs location updates at all depends on its state. The different states and transitions between them are defined by the state model for CS location management depicted in Figure 5.11. If the state of a terminal is DETACHED, it is usually switched off, has no SIM card inserted, or, especially in UMTS, is not registered for services of the CS domain. A terminal is transferred from the DETACHED to the IDLE state if it initializes an *IMSI attach* or location update procedure. The IMSI attach procedure is performed if the LAI received first after switching the terminal on is the same as it had received last before it was switched off (i.e., the subscriber still stays in the same location area). However, if the terminal receives another LAI after it is switched on, a location update is performed in order to update the VLR and to assign a new TMSI. Analogously, the terminal returns to the DETACHED state if it performs an *IMSI detach*, for example, when the subscriber switches off his terminal. Finally, if the subscriber establishes or receives a call, the terminal is transferred from the IDLE to the BUSY state (an alternative term is CONNECTED), where it remains until the call is finished.

Location updates, either on location-area crossing or periodic ones, are only performed if the terminal is in IDLE state. They are not required when it is in BUSY state. In this case, the network implicitly keeps track of the subscriber, because there is a radio link between the terminal and the network, and this link is kept even if the subscriber moves to another cell (handover). Thus, during this time the network knows the subscriber's position even on a cell basis, which consequently includes also his current location area. If, on the other hand, the terminal is detached, it is of course not possible to perform location updates, that is, the terminal's location is not known in the network and it is not reachable.

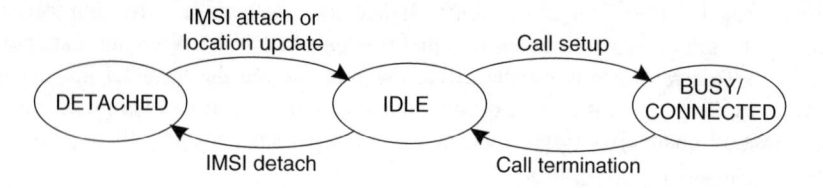

Figure 5.11 CS location management state model.

Before explaining the location update procedure in detail, it is important to know that a terminal in IDLE state permanently monitors and compares the signal strength and the signal quality from a number of base stations in its surrounding environment. From the set of monitored base stations, it selects the one with the strongest signal and listens to signaling messages transmitted by that base station. These messages are sent on special broadcast channels, which are used to inform the terminal about parameters of the air interface (e.g., number of frequency channels and arrangement of signaling and user channels) and for paging the terminal.

Usually, the received signal strength and quality vary, which is due to path loss, for example, if the subscriber moves closer to another base station, or due to effects of multipath propagation. Thus, if the received signal of another base station becomes stronger, the terminal switches to the broadcast channels of that base station. Note that monitoring these broadcast channels does not require a dedicated connection between terminal and base station; the terminal is idle and just listens to them (comparable to television broadcast). Thus, the switching from one base station to another in IDLE state should not be confused with handover (which assumes an ongoing connection and is hence executed in BUSY state only).

The location update procedure is illustrated in Figure 5.12. In this example, BTS_1 is assigned to location area LA_1, and BTS_2 to LA_2. Furthermore, VLR_1 attends to location area LA_1, and VLR_2 to LA_2. At the beginning, the terminal listens to the broadcast of BTS_1 (1), which regularly transfers the LAI of that cell. If the received signal quality of BTS_2 becomes better than that of BTS_1 (2), the terminal starts to listen to the broadcast of BTS_2 and recognizes that the LAI has changed. This event represents the trigger for performing a "location update on location-area crossing". The terminal initiates a location update request with the expired $TMSI_1$ and LAI_1 as parameters (3). This request is directed to VLR_2, which is associated with the new location area. Since the real identity of the subscriber is so far unknown (VLR_2 cannot derive the IMSI from $TMSI_1$), VLR_2 requests the IMSI from VLR_1 (4) (which can be found out by means of LAI_1). After having received the IMSI, VLR_2 generates a new MSRN for the subscriber and sends it together with the IMSI to the HLR (5). The HLR updates the subscriber's database entry and asks the previous VLR_1 to delete the subscriber's entry (6). After the successful execution of these procedures has been acknowledged to VLR_2, it generates a new $TMSI_2$ and returns it to the terminal (7). $TMSI_2$ is used from now on to page the terminal.

This example is special in that the location areas LA_1 and LA_2 belong to different MSC areas and thus affect different VLRs. However, in most cases, a location update takes place within a single MSC area, and steps 4, 5, and 6 can be skipped. Also, VLR and HLR do not necessarily belong to the same network. In order to enable the localization of roaming subscribers, the database entry in the HLR, strictly speaking the MSRN, points to the

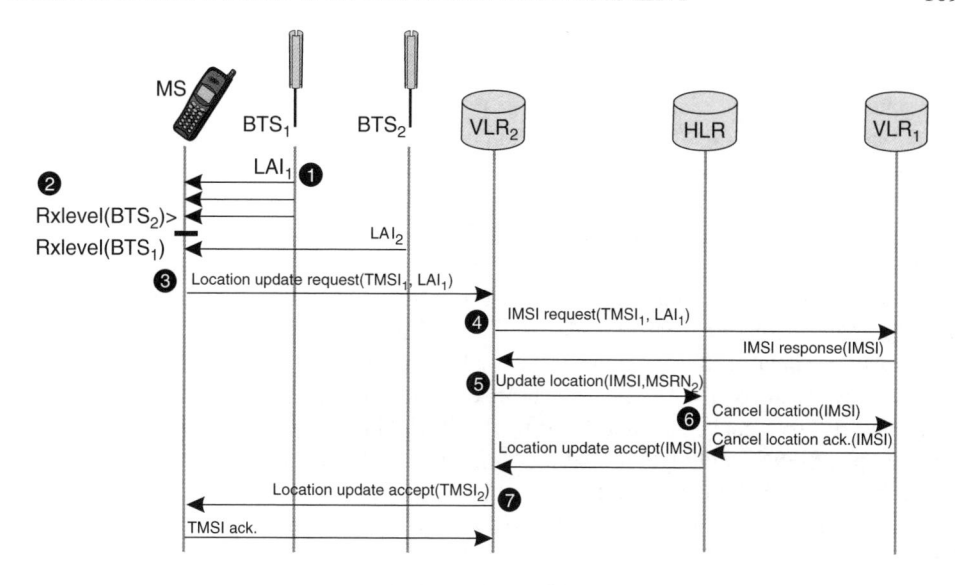

Figure 5.12 Location update procedure in CS networks (simplified).

VLR of a foreign network. Note also that the sequence diagram in Figure 5.12 is merely a simplified version. Actually, a location update is closely related to security issues and entails the generation and distribution of security keys needed for authentication and encryption. Because of simplicity, these operations have been disregarded here. The interested reader can find additional information on this in (Eberspächer et al. 2001).

5.6 Location Management in PS Networks

Location management for PS networks is somewhat more complicated and requires an extended hierarchical organization of geographical areas. This is primarily constituted by the traffic characteristics of packet-switched data transmission, which significantly differs from that of circuit-switched telephone calls. The following section demonstrates the routing of packets through the PS network, followed by an introduction of the mechanisms of packet-switched location management.

5.6.1 Localization and Routing

Recall from Section 5.5.2 that localization and routing in the CS domain are realized by a mapping of different identifiers: the MSISDN is mapped onto the MSRN, and the MSRN is mapped onto the LAI and the TMSI. These mapping procedures are supported by the MAP, which is a signaling protocol within the SS7 network. Signaling is a fundamental concept of the telecommunications domain for controlling the establishment, maintenance, and release of telephone calls. It is not available for PS networks like the Internet. Instead of signaling, each packet carries control information in its header, above all, source and destination addresses of the packet. Thus, the question arises how to forward packets to mobile subscribers, which move between different base stations and SGSNs. The fundamental idea is to use *packet encapsulation* and *tunneling* for routing packets to the target terminal and, where necessary, to combine it with signaling protocols, for example, for controlling medium access, localization, and paging in the access network. It is assumed in the following text that the GPRS/UMTS PS domain carries IP-based traffic only, that is, the X.25 variant of GPRS is not considered here.

Like any host in the Internet, a terminal is referenced by an IP address. This IP address can be of type IPv4 or IPv6, and it is officially denoted as *Packet Data Protocol address* (PDP address) in the GPRS/UMTS terminology. Traditionally, an IP address identifies a network connected to the Internet and a host attached to this network. In the special case of a PDP address, the network part identifies a GPRS/UMTS PS network, and the host part addresses a terminal. From the point of view of the Internet, the GPRS/UMTS PS network is visible by means of the GGSN, which appears from there as a conventional router that sends and receives packets to and from the GPRS/UMTS PS domain. Other components, for example, SGSNs and also the terminals, are not visible from outside.

At first, the following explanations are related to GPRS only. The GPRS core network is an IP-based network of SGSNs and GGSNs, each of it represented by internal IP addresses (*SGSN addresses* and *GGSN addresses*). Mobile-originated packets are routed from an SGSN to the responsible GGSN, and mobile-terminated packets are routed from the GGSN to the SGSN to which the target subscriber is currently attached. For this internal routing, it is not possible to simply replace the destination addresses of packets (i.e., either that of an Internet host or the terminal's PDP address) with the internal GGSN and SGSN addresses respectively, because the actual destination terminal would be lost and packets could never be delivered. Therefore, a technique is applied between SGSNs and GGSNs, which is known as *tunneling*.

Tunneling allows the transfer of packets between two networks over a third one, whereby this intermediate network is completely transparent to the networks interconnected in this way. It is realized by encapsulating IP packets into packets of a tunneling protocol, which is here denoted as *GPRS Tunneling Protocol* (GTP) (see (3GPP TS 29.060)). For each terminal that is attached to the GPRS network, a tunnel is arranged between GGSN and SGSN and vice versa. It is identified by the IP address of the target component (i.e., GGSN or SGSN) together with a tunnel identifier, which is used to discriminate packets from and to different terminals (and maybe from and to different applications of the same terminal). Thus, the combination of SGSN address and tunnel identifier can be regarded as the counterpart of the MSRN of the CS domain. A tunnel can be seen as a logical link through the core network that is maintained regardless of whether or not data is transferred. If the subscriber moves to an area served by another SGSN, it is required to shift the tunnel to the new SGSN. Note that tunneling is also used for Mobile IP in the Internet (see (Perkins 1997)). Basically, the approach applied by Mobile IP for redirecting packets between home and visited network is very similar to the concept applied in GPRS. Tunneling represents also the core technology for establishing *Virtual Private Networks* (VPNs) that interconnect private networks over a public network. In this case, tunneling goes along with encryption.

GPRS tunneling now works as follows (see Figure 5.13): if a mobile-terminated IP packet arrives at the GGSN (1), the GGSN obtains the address of the responsible SGSN and the tunnel identifier by means of the PDP address (contained in the packet header) from its internal location register. It then inserts the IP packets into a *Packet Data Unit*

(PDU) of the tunneling protocol (2) and sends them to the target SGSN (3). There, the IP packet is removed from the tunneling PDU (4) and further processed in order to forward it to the terminal. The latter procedure may involve a paging request if the cell of the target terminal is not known, which is initialized and controlled by the SGSN (5). The terminal is paged by its *Packet-TMSI* (P-TMSI), which is different from the TMSI in the CS domain. After the cell has been obtained, the IP packet is transferred from the SGSN to the terminal for which the *Subnetwork Dependent Convergence Protocol* (SNDCP) is used (6 + 7) (see (3GPP TS 44.065)). In the opposite direction, that is, for mobile-originated traffic, the packets are transferred from the terminal to the SGSN via SNDCP, and then tunneled from the SGSN to the GGSN.

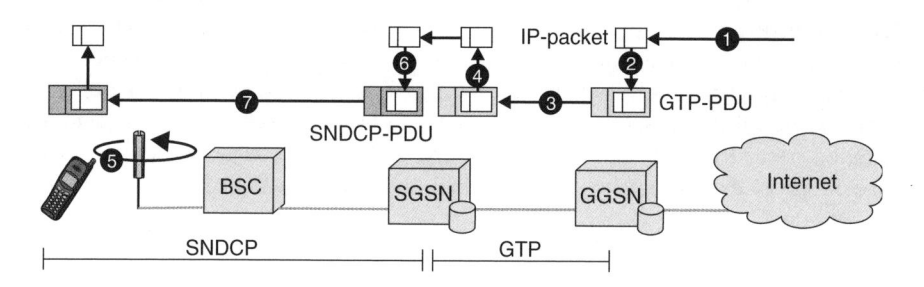

Figure 5.13 Routing of incoming packets.

For the UMTS PS domain, the tunnel is terminated at the RNC, and not the SGSN (see (3GPP TS 29.060)). Therefore, the SNDCP for interconnecting core network and terminal is not needed here. The advantage of this approach is that signaling, for example, for paging the target terminal, can be handled within the access network and does not need to be exchanged between SGSN and access network, which in GPRS heavily burdens the interface in between and causes additional delays. For the exchange of packets between RNC and terminal, standard protocols of the access network are used.

Finally, it is necessary to have a look at how subscription data of the GPRS/UMTS PS domain is managed. The subscriber registers for PS services via a *GPRS attach* procedure (*PS attach* in UMTS), whereupon a so-called *PDP context* is negotiated between terminal and network. A PDP context consists of the following elements:

- **PDP type.** The PDP type fixes the type of PDP used, for example, IPv4 or IPv6.

- **PDP address.** The PDP address is used to reference the terminal within the external PDP. It is included in the source address field for mobile-originated packets and in the destination field of mobile-terminated packets. Its format depends on the PDP type.

- **QoS class.** Depending on the needs of the respective applications, different *Quality of Service* (QoS) parameters can be negotiated between terminal and network. QoS parameters are related to service precedence, reliability, delay, and throughput.

- **GGSN address.** This is the address of the GGSN that serves as the gateway to the external PDN.

112   CELLULAR NETWORKS AND LOCATION MANAGEMENT

The PDP address can be assigned permanently or temporarily. In the former case, the subscriber has obtained a fixed IP address from his operator for exclusive use. In the latter case, he gets another IP address each time he attaches to the network. In most of today's networks, IP addresses are predominantly assigned on a temporary basis. However, it is expected that with the introduction of IPv6, which has a much larger address space, addresses will also be assigned permanently. The PDP context is stored in the terminal, the serving SGSN, as well as in the GGSN.

5.6.2 Characteristics of CS and PS Traffic

Recall from Section 5.3 that circuit-switched services assume the establishment of a circuit between two parties before any data transmission can take place. At the air interface, such a circuit is represented by a channel that is exclusively used for data transmission as long as the call lasts and regardless of whether or not data is transferred at all. On the other hand, for packet-switched services, channels at the air interface are assigned on demand only for the duration needed for transmitting a data burst and then released for other subscribers. Figure 5.14 gives examples of circuit-switched and packet-switched traffic patterns in a certain period of time.

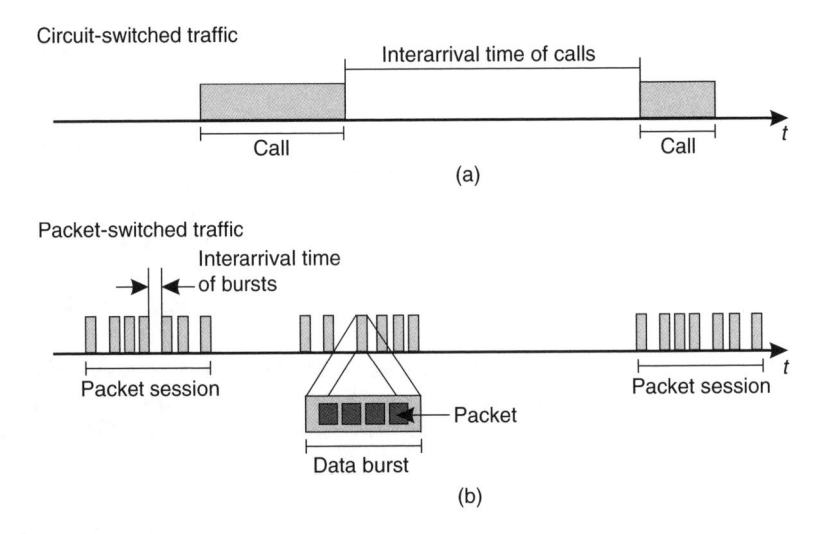

Figure 5.14 Characteristics of circuit and packet-switched traffic.

Figure 5.14a shows the temporal distribution of a subscriber's telephone calls over a certain period of time as well the duration of each call. The time between the end of a call and the beginning of the next one is called *interarrival time*. Figure 5.14b shows the temporal distribution of data bursts, their lengths, and interarrival times. The counterpart of a call is denoted as *packet session* here and may represent a subscriber surfing the WWW, for example, in order to read newspaper or to make home banking. It is very obvious that the duration of a data burst is much shorter than that of a telephone call and, in most cases, the average interarrival time of bursts is also very short compared to that of calls.

Accordingly, the arrival rate of data bursts (i.e., the number of bursts arriving in a certain period of time) is in most cases much higher than the arrival rate of telephone calls. To better understand this, consider just a user navigating a WWW site, a procedure that is characterized by a fast sequence of clicking links and receiving content.

To understand the problem of packet-switched location management, imagine that location updates are still performed on location-area crossing. In this case, the subscriber would have to be paged several times in the course of a packet session in all cells belonging to his location area. Strictly speaking, the network would have to page the subscriber for each data burst that is to be transmitted on the downlink. This is due to the fact that the subscriber does not necessarily remain in the same cell during the interarrival time of data bursts arriving on the downlink. Also, it is possible that, for example, a WWW request on the uplink and the corresponding WWW response on the downlink are transmitted in different cells if the subscriber has changed the cell in the intermediary time. Note that for circuit-switched traffic this problem does not exist in this manner, because during an ongoing connection cell changes are supported by the handover procedure. To conclude, packet-switched location management would suffer from increased paging costs if location updates were performed on location-area crossing. Figure 5.15[2] compares the paging costs for circuit-switched and packet-switched traffic.

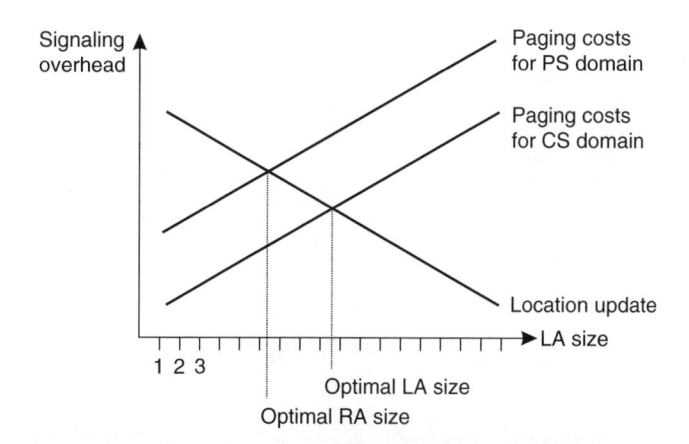

Figure 5.15 Comparison of location management overhead in CS and PS domains. Adapted from (Sanders et al. 2003).

The figure also highlights how to reduce the increased paging costs. It shows that the margin where the paging costs exceed the location update costs is much lower (in terms of number of cells per location area) for packet-switched than for circuit-switched traffic. Consequently, the solution lies in a reduced size of location areas in order to balance paging against location update overhead more appropriately for PS networks.

---

[2] Note that the functions sketched in the figure are not necessarily linear ones. Their concrete appearance depends also on many circumstances, for example, call and mobility behavior on the spot, LA shape, and the amount of data carried in update and paging messages. The intention of the figure is only to demonstrate the general impact the LA size has.

114 CELLULAR NETWORKS AND LOCATION MANAGEMENT

Therefore, for GPRS an additional level has been introduced into the geographical topology of GSM networks, which is given by *routing areas*. Figure 5.16 shows the resulting topology of a combined GSM/GPRS network. A routing area also comprises a number of cells, but is much smaller than a location area. A location area contains an integer number of routing areas, that is, routing areas do not overlap location-area borders.

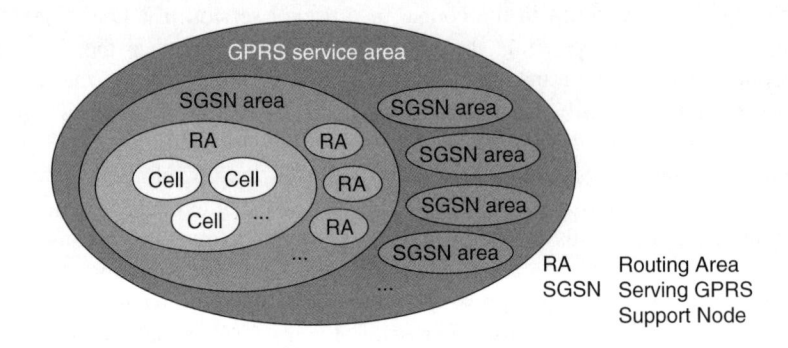

Figure 5.16 GPRS topology for location management.

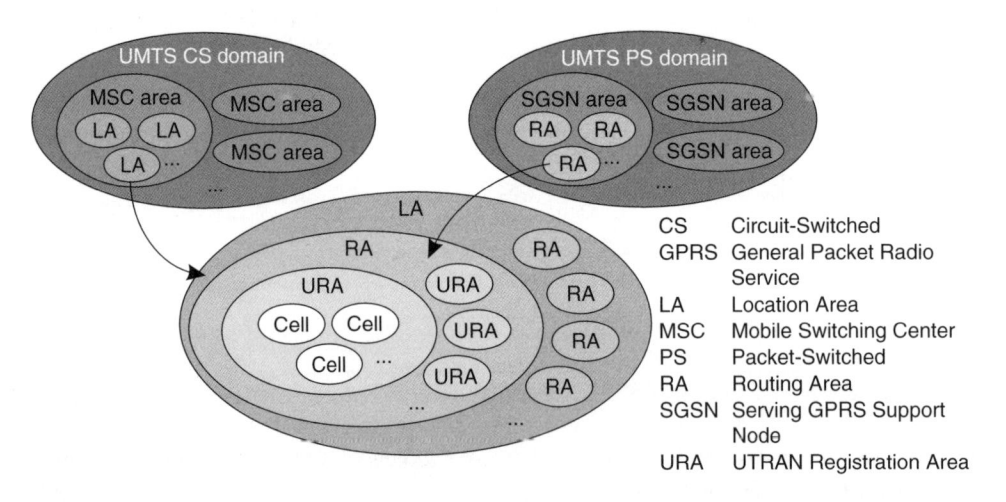

Figure 5.17 UMTS topology for location management.

This concept has been further refined for UMTS, which contains another level of so-called *UTRAN Registration Areas* (URAs). Figure 5.17 shows the different types of areas for both PS and CS location management. A location area contains several routing areas, a routing area comprises several URAs, and an URA consists of several cells. Again, there is no overlapping between areas of different types. In the next section, the location update procedure is explained with regard to this enhanced topology.

5.6.3 Location Updates

In GPRS and UMTS PS domains, it is differentiated between the following types of location updates, which reflect the geographical topology introduced earlier:

- **Location update on routing-area crossing.** This type is executed whenever the subscriber enters a new routing area. It is hence called *routing-area update*.

- **Location update on URA crossing.** This type is executed whenever the subscriber enters a new URA. It is called *URA update* and is only available in the UMTS PS domain (i.e., not for GPRS).

- **Location update on cell crossing.** This is the most fine-grained type of location update. It is executed whenever the subscriber enters a new cell, and it is referred to as *cell update*.

- **Periodic location update.** Like in CS networks, location updates are also performed periodically. Different versions exist, which periodically report the current RA, URA, or cell.

On the basis of experiences gained with the first GPRS implementations, the location management of the UMTS PS domain has been further extended and optimized with regard to the overhead caused by location updates and paging. Both approaches are introduced in the following sections.

5.6.3.1 Location Updates in GPRS

Similar to the CS domain, the type of update a terminal performs in GPRS mode depends on its state. The state model for GPRS location management is depicted in Figure 5.18. The terminal is in IDLE state if the subscriber has not registered for GPRS. In this state, no valid PDP context exists between network and terminal, and the subscriber can neither send nor receive any packets. Updates are only performed on location-area crossing, assuming that the terminal is switched on at all and the subscriber has registered for circuit-switched services.

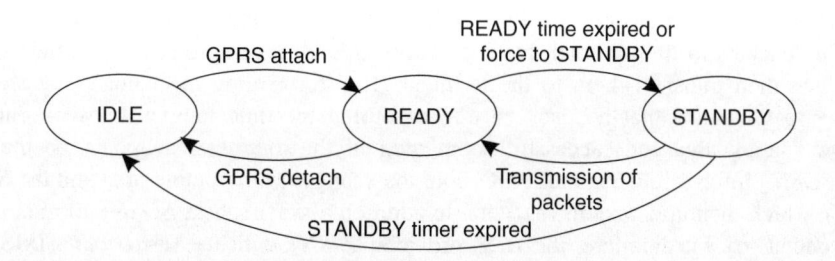

Figure 5.18 GPRS location management state model.

A subscriber registers for GPRS via the GPRS attach procedure, whereupon the network generates a PDP context including an IP address for the terminal, and the terminal is transferred from the IDLE to the READY state. Simply speaking, the READY state represents an ongoing packet session where data bursts are exchanged at short time intervals. In this state, the terminal performs cell updates, that is, it initializes a cell update procedure whenever it starts to listen to another base station. Thus, in the network the subscriber's current cell is known and it is hence not necessary to page him in case of incoming packets.

However, although paging is entirely omitted in the READY state, cell updates would unnecessarily burden the air interface if no packet transmissions take place at all. Therefore, a terminal changes from READY to STANDBY state upon expiry of the READY timer (or if it is explicitly forced by the network to do so). The READY timer is restarted whenever a packet transfer occurs, and it expires when the terminal has neither sent nor received any data within the period of time defined by the expiry value of this timer. This value is chosen by the network and transferred to the terminal on a broadcast channel. Typical values are in the range of tens of seconds. In STANDBY state, the terminal then performs only routing-area updates, and returns to READY state as soon as a data transfer takes place. Similar to the READY timer, a STANDBY timer determines the period of time a terminal remains in STANDBY state before it changes into the IDLE state. Thus, in this case the terminal is automatically detached from the GPRS network.

Given this state model, it is possible to balance location update and paging overhead depending on a terminal's activity. The location update overhead is reduced during idle periods, while paging overhead is set to zero for terminals being in a packet session.

Basically, cell and routing-area updates work very similar to location updates in the CS domain, compared to Figure 5.12. One of the main differences is that a subscriber's location is not maintained by the VLR but by the internal location register of the SGSN. In READY state, the terminal's current cell is reported to the SGSN, while in STANDBY state the current routing area is reported. As an example, consider Figure 5.19, which illustrates the routing-area update procedure.

In this example, routing area RA_1 is assigned to $SGSN_1$, and RA_2 to $SGSN_2$. Like the LAI, the RAI (Routing Area Identifier) is broadcast in each cell and received by all terminals. If the terminal recognizes that the RAI has changed, it initializes a routing-area update, which contains the P-TMSI and the old RAI (1). If the routing-area update arrives at the new $SGSN_2$, it requests the PDP context from $SGSN_1$ (2). Upon arrival of this request, $SGSN_1$ transfers the PDP context to $SGSN_2$ and stops the transmission of incoming packets to the terminal. Instead, it forwards all incoming packets to the $SGSN_2$ (3), which then transfers them to the terminal. This forwarding mechanism is necessary in order to guarantee that no packets are lost during the time interval between entering the new routing area and successful completion of the routing-area update. In the next steps, $SGSN_2$ informs GGSN and HLR about the change of the routing area and the SGSN (4 + 5), which then update their registers. In addition, the HLR asks $SGSN_1$ to remove the PDP context (6). Furthermore, the HLR provides $SGSN_2$ with the subscriber's IMSI and GPRS subscription data (7). In the last step, $SGSN_2$ informs the terminal about successful completion of the routing-area update and transfers a new P-TMSI to it (8), which is used from now on to page the terminal when incoming data packets arrive.

In most cases, however, the change of routing areas takes place within an SGSN area. In this case, the only components involved in a routing-area update are the terminal and the SGSN, and steps 2–7 in Figure 5.19 can be omitted. The sequence diagram given here is again a simplified version, which does not consider security procedures that are

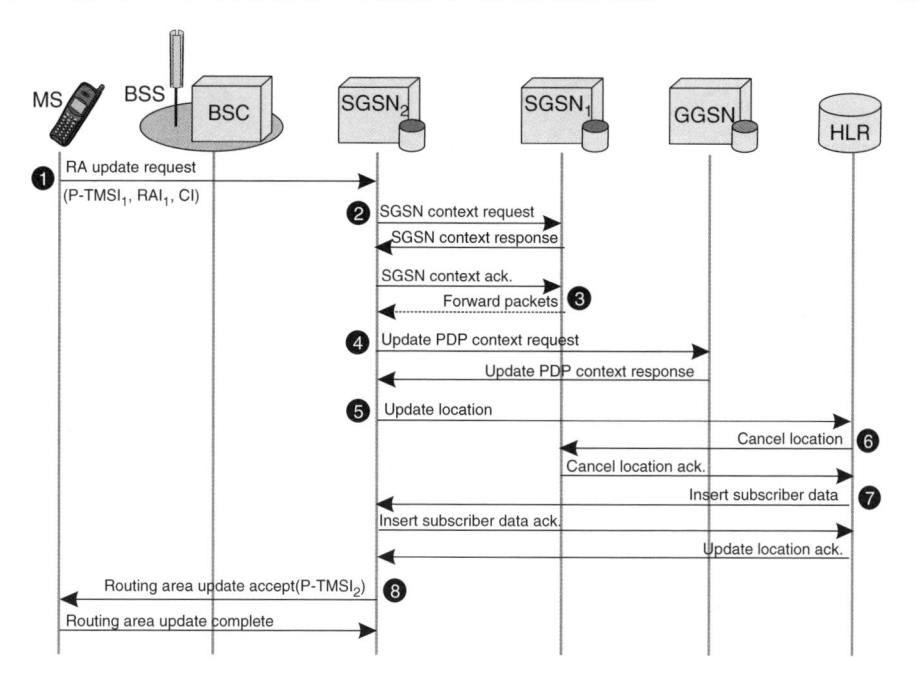

CELLULAR NETWORKS AND LOCATION MANAGEMENT   117

Figure 5.19 RA update procedure in GPRS.

associated with an update. Also, it is possible to perform a combined routing-area/location-area update if the new routing area belongs to another location area. The different versions of routing-area and cell updates can be obtained from (3GPP TS 23.060).

5.6.3.2 Location Updates in the UMTS PS Domain

Compared to GPRS, location management has been significantly refined for the UMTS PS domain. A drawback of GPRS is that, like in the CS domain, location management is exclusively controlled by the core network, that is, by the SGSN. As a result of this, all procedures of location management pass the interface between access and core network. For example, especially cell updates, which occur at high intervals, burden this interface. To cope with this problem, UMTS envisages location management functions in the access network as well. The main idea is to track the subscriber on the basis of routing areas in the core network, and on the basis of cells and URAs in the access network.

Consequently, two state models are defined for the UMTS PS location management, one for the core and another for the access network (see Figure 5.20). The state model for the core network is very similar to the GPRS state model (consider Figure 5.18 again). In PMM DETACHED state, the terminal is not registered for the PS domain, and, if at all, location updates are performed on location-area crossing. The PMM CONNECTED state represents an ongoing packet session. In contrast to the GPRS READY state, the SGSN does not track the subscriber on cell level, but only follows him on routing-area level. In this state, the serving RNC of the respective access network is responsible for a more fine-grained tracking, which is explained below. The PMM IDLE state corresponds to the GPRS

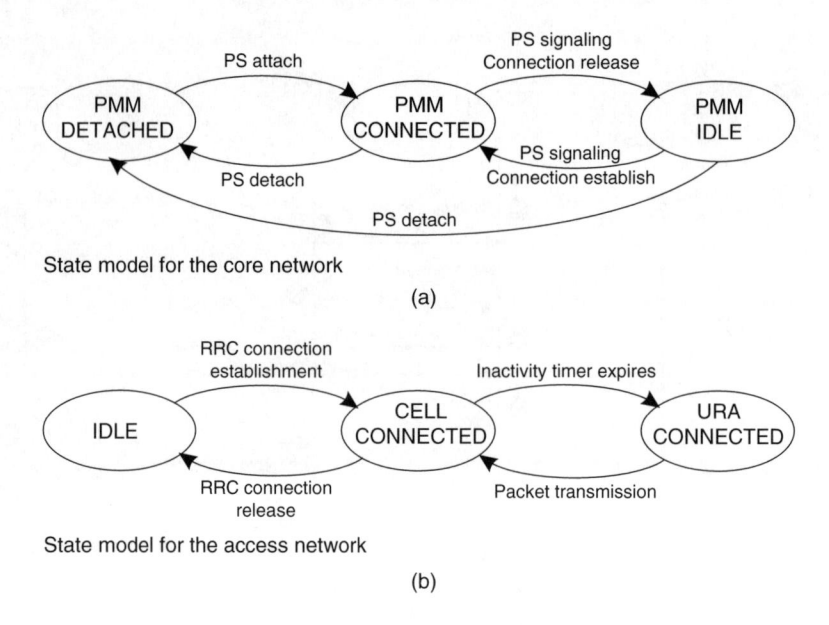

Figure 5.20 UMTS PS location management state models.

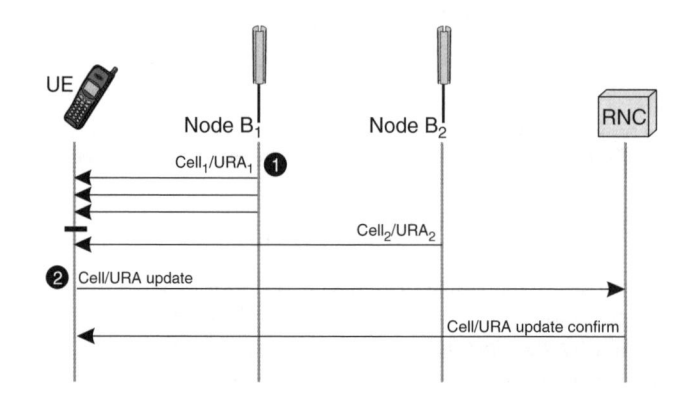

Figure 5.21 Location update procedure in UTRAN.

IDLE state, and updates are performed in this state on routing-area crossing, too. Like for GPRS, transitions between these states are controlled by attach and detach procedures and by timers.

In the access network, cell or URA updates (see Figure 5.21) are established if a signaling connection between terminal and RNC exists. This signaling connection is organized and controlled by the *Radio Resource Control* (RRC) protocol, which is defined in (3GPP TS 25.331). The RRC is used, for example, to set up, maintain, or release channels for exchanging user data. If the terminal is in IDLE state, which corresponds to the PMM IDLE state of the access network, no signaling connection exists and the subscriber's cell and URA are not known to the RNC. In the CELL CONNECTED state, the terminal sends cell updates to the RNC whenever switching to a new cell. In the URA CONNECTED state, the terminal only reports URA changes. In case of incoming data, the RNC must then page the terminal in all cells of the registered URA. The transitions between both states are again controlled by timers with regard to the last packet transfer. However, a state transition can also be forced by the network, for example, depending on the mobility behavior of subscribers (Kaaranen et al. 2001).

5.7 Conclusion

This chapter has introduced the network architectures and topologies of cellular networks and explained their implications on location management. Generally, it can be concluded that the network permanently tracks a subscriber for the delivery of mobile-terminated calls or data and that the granularity of tracking, that is, the accuracy of "network locations" maintained in the network, strongly depends on the terminal's state. If there is a call or a data transfer in progress or if a data transfer has recently taken place, the network is kept informed about the cell where the subscriber currently resides. If, however, a data transfer has not taken place for a longer period of time, the location is only known in terms of URAs and routing areas respectively. In the worst case, the terminal is only registered with the CS domain, where the network only tracks the location areas of idle terminals. Table 5.2 gives an overview of the different levels of user tracking in GSM, GPRS, and UMTS, and summarizes the network components storing location information.

LBSs can be supported by the operations of location management in that they are used to obtain a target subscriber's current cell, which is a positioning method that is commonly known under the term *Cell-Id* and which is meanwhile supported by most cellular networks today. It is specifically suited for reactive LBSs, where the LBS user initializes a service session, for example, in order to receive some location-dependent content. When initializing this session, the terminal has to connect to a base station first, and in this way the current cell can be made available to the LBS provider (a procedure which will be explained in detail in Chapter 8). If LBS user and target are not the same person, it may be required to page the target in his current location or routing area in order to derive his current cell. Proactive LBSs, on the other hand, are much more difficult to realize. They are automatically initialized (or perform some other actions) in a push fashion as soon as the target enters a predefined location. In contrast to reactive services, they require to track the subscriber on a permanent basis, even if the terminal is in idle mode. Unfortunately, location management in this mode only delivers the target's location or routing area, which is much too coarse grained for almost all LBSs. Permanently keeping a connection to the target terminal or

Table 5.2 Areas tracked by network nodes (Lin et al. 2001)

| | MSC/VLR | | | SGSN | | UTRAN |
	GSM	GPRS	UMTS	GPRS	UMTS	UMTS
Cell	No	No	No	Yes	No	Yes
URA	–	–	No	–	No	Yes
Routing area	–	No	No	Yes	Yes	No
Location area	Yes	Yes	Yes	No	No	No

paging would cause too much signaling overhead and is therefore all but a practicable approach.

However, it must also be stressed that knowledge about a target's current cell in many cases does not meet the accuracy demands of LBSs, especially not those claimed for enhanced emergency services such as E-911 in the United States. This is due to the fact that cell sizes vary between several 100 m in urban and up to 35 km in rural areas. To compensate for this weakness considerable efforts were launched for developing and standardizing more accurate positioning methods, which are based on lateration and angulation, and for their integration into cellular networks. The following chapter provides the fundamentals of these methods.

Part II

Positioning

6

Fundamentals of Positioning

Positioning is a process to obtain the spatial position of a target. There are various methods to do so, which differ from each other in a number of parameters such as quality, overhead, and so on. In general, positioning is determined by the following elements:

- one or several parameters observed by measurement methods,
- a positioning method for position calculation,
- a descriptive or spatial reference system,
- an infrastructure, and
- protocols for coordinating the positioning process.

The core function of any positioning is the measurement of one or several *observables*, for example, angles, ranges, range differences, or velocity. Such an observable usually reflects the spatial relation of a target relative to a single or a number of *fixed points* in the surrounding environment, where a fixed point denotes a point of well-known coordinates. They are often measured by utilizing the physical fundamentals of radio, infrared or ultrasound signals, such as their velocity or attenuation. These signals when used for positioning measurements are also referred to as *pilot signals* or simply *pilots*. Furthermore, measurements are sometimes classified into *radiolocation* and *non radiolocation methods*. In the former category, observables are directly or indirectly measured by radio signals, while the latter category falls back on other physical quantities, for example, of optical or acoustic nature.

After the required observables have been determined, the target's position must be derived after taking into consideration the measurement results and the coordinates of the fixed points. This determination is usually based on a certain method that strongly depends on the types of observables used. Examples are *circular* and *hyperbolic lateration* or *angulation*. Table 6.1 shows an overview of the basic positioning methods derived from (Hightower and Borriello 2001a) and the associated observables and measurement methods.

Location-based Services: Fundamentals and Operation Axel Küpper
© 2005 John Wiley & Sons, Ltd

Table 6.1 Overview of positioning methods

Positioning method	Observable	Measured by
Proximity sensing	Cell-Id, coordinates	Sensing for pilot signals
Lateration	Range or	Traveling time of pilot signals
		Path loss of pilot signals
	Range difference	Traveling time difference of pilot signals
		Path-loss difference of pilot signals
Angulation	Angle	Antenna arrays
Dead reckoning	Position and	Any other positioning method
	Direction of motion and	Gyroscope
	Velocity and	Accelerometer
	Distance	Odometer
Pattern matching	Visual images or	Camera
	Fingerprint	Received signal strength

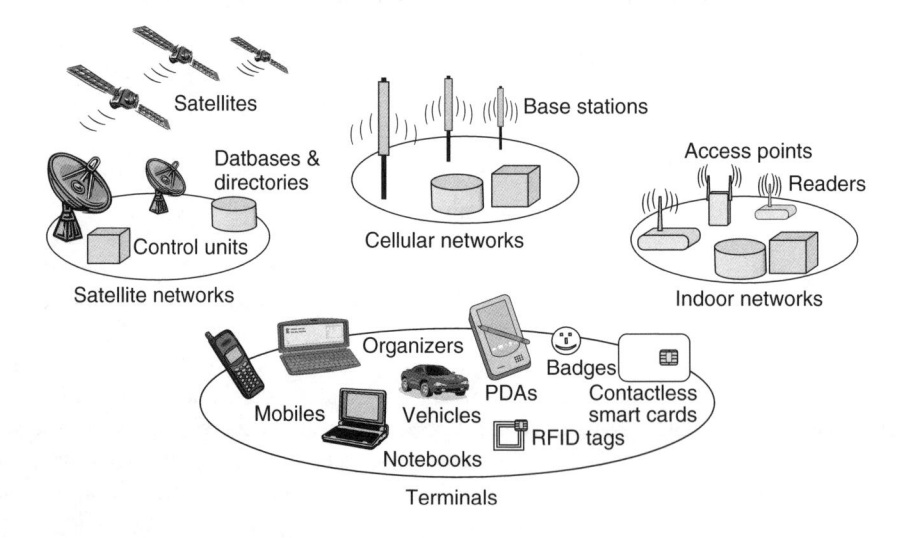

Figure 6.1 Positioning infrastructures.

A positioning method delivers a target's position with regard to a descriptive or spatial reference system as introduced in Chapter 2. The selection of this reference system strongly depends on the positioning method used. Some positioning systems deliver descriptive locations in terms of cell identifiers, room numbers, and floors, while others provide spatial coordinates based on WGS-84, UTM, or other reference systems.

Because a target is not able to derive its position autonomously, a distributed infrastructure that implements positioning is required. Figure 6.1 depicts the different components such an infrastructure may consist of. Each target to be located must be equipped with a terminal, that is, a terminal is always the entity whose position is, at first, unknown. Furthermore, base stations are required for most positioning methods either to perform measurements or to assist the terminal in doing so. Apart from a few exceptions (e.g., fingerprinting), the base stations are the aforementioned fixed points of well-known coordinates against which the spatial relation of a terminal is measured. As will be explained later in detail, one can distinguish between *satellite*, *cellular*, and *indoor systems*, as well as between *integrated* and *stand-alone infrastructures*. Depending on the type of infrastructure used, satellites, GSM/UMTS base stations, WLAN access points, or tag readers serve as base stations, while mobile phones, organizers, notebooks, badges, tags, and sensors may serve as terminals during positioning. At the network site, additional components may be required, for example, GIS databases, servers, or control units for coordinating positioning as well as for processing and distributing the measurement results and position data.

Furthermore, positioning needs to be coordinated and controlled by protocols between these infrastructure components. The kind of protocols as well as their complexity and appearance significantly differ in the various networks and systems. Generally, they are applied between a central control unit and the base stations as well as between base stations and the terminal. For example, they are used to transfer base station coordinates and measurement instructions to the terminal (so-called *assistance data*), for transferring measurement data to the network, or for exchanging the calculated position data, which is also called *position fix*.

Finally, there are several criteria for evaluating the quality of a certain positioning method and of a position fix respectively:

- **Accuracy and precision.** The most important quality parameters are *accuracy* and *precision*, which are often misleadingly mixed and used interchangeably. Accuracy refers to the closeness of several position fixes to the true but unknown position of a target. The farther away a position fix is from the true position, the lesser is the degree of accuracy and vice versa. Precision, on the other hand, refers to the closeness of a number of position fixes to their mean value. Section 6.4 covers accuracy and precision in detail.

- **Yield and consistency.** For many applications, it is desired to obtain position data independent of the environment (indoor, rural, urban) where the target person currently resides. However, some positioning methods are not available in certain environments or their accuracy strongly depends on the environment. This is indicated by *yield* and *consistency*. The yield of a positioning method refers to its ability to obtain position fixes in all environments, while consistency is a measure for the stability of accuracy in different environments.

- **Overhead.** Positioning is inevitably associated with a certain amount of *overhead* that occurs at the terminal, within the positioning infrastructure, as well as at the air interface in between. It can be classified as *signaling* and *computational overhead*. The former reflects the amount of messages that are exchanged between terminal and infrastructure as well as within the network to control positioning. Computational overhead refers to the waste of processing power that occurs in the control units or databases of the network and at the terminal. The overhead of positioning must always be evaluated under consideration of accuracy and precision. Generally, a high degree of accuracy or precision of position data causes a high overhead and vice versa. Therefore, the decision for an appropriate positioning method should always be carefully weighted between these quantities.

- **Power consumption.** *Power consumption* can be seen as an additional category of overhead. It is only of importance at the terminal since mobile devices in general are short of battery resources, and a high power consumption is at the expense of standby and talk times. Power consumption at the terminal resulting from positioning is in most cases strongly correlated with signaling and computational overhead.

- **Latency.** *Latency* refers to the time period between a position request and the subsequent delivery of a position fix. During this time, one or several base stations must be selected, the entire positioning process must be coordinated between the involved components by signaling, resources need to be allocated, measurements have to be done, and finally the position has to be calculated from the measurement results. If a target is tracked over a longer period of time by several consecutive position fixes, most of these steps need only to be executed once at the beginning of tracking. Therefore, an important indicator is the latency of the first position fix, which is called the *Time To First Fix* (TTFF). In today's implementations, the TTFF is in the order of magnitude of seconds. The size of the TTFF that can be tolerated heavily depends on the type of LBS. Especially in the case of reactive LBSs, which usually imply much interactivity with the user, a high TTFF is negatively experienced and may result in less attractiveness and acceptance. However, in the context of proactive services, higher TTFF values usually do not appear to the user and can be tolerated.

- **Roll-out and operating costs.** *Roll-out costs* are required either to install the infrastructure of base stations, databases, and control units, or to extend an existing infrastructure, for example, of a cellular network, to implement positioning there, which is an important design criteria in the introduction of a particular method. *Operating costs* are strongly related to the complexity of the positioning infrastructure. Indoor networks cause less or even no operating costs, whereas satellite networks generally suffer from exhaustive running charges.

This chapter provides an introduction to the fundamentals of positioning. It starts with a scheme for classifying positioning infrastructures and methods in the next section. After this, the working and mathematical foundation of the basic methods identified in Table 6.1 are explained, followed by an introduction to the particularities and problems of range measurements. Finally, error sources having an impact on the quality of positioning are identified and different accuracy measures are presented. Thus, this chapter serves as a basis for understanding the implementation of positioning in satellite, cellular, and indoor systems, which will be covered in subsequent chapters.

6.1 Classification of Positioning Infrastructures

Positioning and positioning infrastructures can be classified with respect to different criteria. The most common distinctions are made between *integrated* and *stand-alone positioning infrastructures*, *terminal* and *network-based positioning*, as well as *satellite*, *cellular*, and

FUNDAMENTALS OF POSITIONING

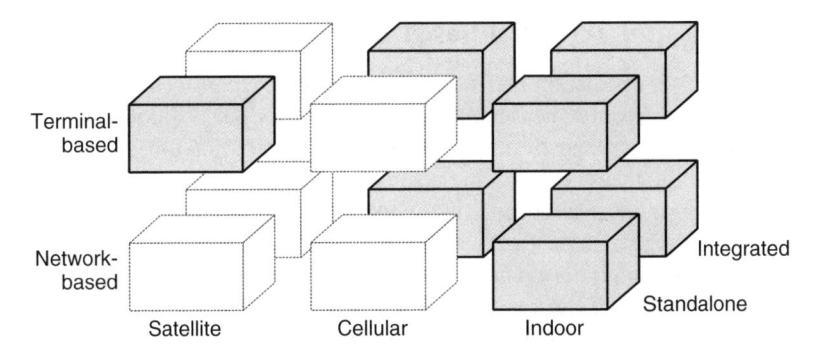

Figure 6.2 Classification scheme for positioning infrastructures.

indoor infrastructures. Figure 6.2 gives an overview of constellations of these classes and indicates the constellations being applied today (indicated by the gray-shaded boxes).

6.1.1 Integrated and Stand-alone Infrastructures

An integrated infrastructure refers to a wireless network that is used for both communication and positioning purposes. Typically, these networks have initially been designed for communication only and are now experiencing extensions for localizing their users through standard mobile devices, which especially holds for cellular networks. Components that can be reused are base stations and mobile devices as well as protocols of location and mobility management. Extensions concern the installation of new software releases in network components for executing and controlling positioning and the installation of new hardware components for clock synchronization (see Section 6.2), for collecting measurement data, and for other purposes.

An integrated approach has the advantage that the network does not need to be built from scratch and that roll-out and operating costs are manageable. On the other hand, positioning additionally burdens the network, which is at the expense of capacities that would be available for user traffic otherwise. Measurements in most cases must be done on the existing air interface, whose design has not been optimized for positioning but for communication, and hence the resulting implementations seem to be somewhat complicated and cumbersome in some cases.

A stand-alone infrastructure works independently of the communication network the user is attached to. In contrast to an integrated system, the infrastructure and the air interface are exclusively intended for positioning and are very straightforward in their designs. The most prominent example of a stand-alone infrastructure is GPS, but especially for indoor environments, for example, offices or airports, a number of proprietary solutions have been developed as well. Disadvantages are high roll-out and operating costs and the fact that targets cannot be located via a standard mobile device, but require additional, sometimes proprietary equipment. Furthermore, interoperability mechanisms have to be implemented between positioning and the communication infrastructure if position data is to be processed for an LBS.

6.1.2 Network and Terminal-based Positioning

Furthermore, there is a distinction between *network-* and *terminal-based positioning*, which refer to the site that performs measurements and calculates the position fix. For network-based positioning, this is done by the network, while for terminal-based positioning the terminal carries out the measurements and position calculations. A hybrid approach is given if the terminal performs measurements only and then transmits the results to the network, where the position is calculated from these results. This is called *terminal-assisted network-based positioning* (or simply *terminal-assisted*). The reverse configuration is referred to as *network-assisted terminal-based positioning*, but is applied quite rarely.

The decision which class of methods to use primarily depends on the characteristics of the respective LBS, for example, on the site where position data is needed and further processed, and on the capabilities of the respective mobile device. Network operators upgrading their networks for positioning are usually concerned about what is called a *smooth migration*. Usually, subscribers refuse to exchange or to modify their devices solely in order to use the new network features that go along with the upgrade. Network-based methods have the advantage that even legacy handsets can be located without any modifications on the handset. This is the reason why operators of cellular networks emphasize the implementation of at least one network-based method.

6.1.3 Satellites, Cellular, and Indoor Infrastructures

Another criterion to classify positioning is to consider the type of network in which it is implemented and operated.

6.1.3.1 Satellite Infrastructures

By their very nature, satellites cover huge geographical areas, and thus satellite positioning can pinpoint the location of a target on a whole continent or even the entire world. The most prominent example is GPS, which operates with at least 24 satellites and which thereby enables worldwide coverage. Similar systems are the Russian *GLONASS* (Globalnaya Navigationnaya Sputnikovaya Sistema) and the European *Galileo*. Because satellite systems for mobile communications, like Iridium, have not been designed for positioning (and also never achieved a significant market penetration), satellite positioning is always achieved by a stand-alone infrastructure of several satellites that is not used for communication. Also, satellite positioning is always terminal based (apart from military systems used for espionage).

The undisputable advantages of satellite positioning are the aforementioned global availability and the comparatively high accuracy of position fixes. However, with regard to LBSs it suffers from certain drawbacks that are typical for satellite-based systems in general. First, the signals radiated from satellites are usually very susceptible to shadowing effects and are easily absorbed by buildings, walls, and mountains. As a consequence, positioning only works if a direct line of sight between the satellites and the receiver is given, which usually is not the case when staying indoor. Second, satellite positioning of today suffers from a comparatively high power consumption at the satellite receiver. OBUs as used in vehicles for navigation may cope with this problem since they are directly supplied by the vehicle's electric circuit. However, standby and talk times of battery-operated mobile devices with

FUNDAMENTALS OF POSITIONING

integrated satellite receiver are negatively affected by this. Third, the roll out and operation of a satellite system come along with enormous capital investments. For example, it is estimated that the US DoD has approximately raised $12 billion for GPS till date and will spend further hundreds of millions of dollars within the next few years. One reason for these huge amounts is the short life span of satellites, which is typically in the range of five to seven years, and the associated necessity to permanently replace them.

6.1.3.2 Cellular Infrastructures

Cellular positioning refers to mechanisms in cellular networks like GSM or UMTS for obtaining the position of a subscriber. Basically, the procedures of location management can be exploited for this. Actually, this is the principal way LBSs work in today's cellular networks. However, as pointed out in the previous chapter, using only location management for supporting LBSs has some deficiencies, for example, insufficient accuracy and inappropriate location formats like cell and location-area identifiers. Thus, network operators have considerably extended their networks (or are planning to do so) by introducing auxiliary components and protocols in order to perform positioning in a more appropriate, accurate, and efficient manner.

Usually, operators of cellular networks tend to implement and run several positioning methods in parallel in order to improve availability and to deploy them adaptively depending on the user and service requirements, current network utilization, migration aspects, and other criteria. The different positioning methods are specified by the responsible standardization groups (e.g., 3GPP), which also recommend the set of mandatory positioning methods to be implemented at a certain stage of network expansion. Some methods are network based, and hence legacy devices without positioning capabilities can be smoothly and easily integrated into the system. As cellular networks usually operate nationwide, availability in terms of yield is quite good and, unlike satellite positioning, almost always work indoor. However, the roaming subscriber may experience the unavailability of his default positioning method if the foreign network is situated in an earlier stage of expansion than his home network.

Another concern is the waste of resources at the air interface and the fixed network site. When designing cellular networks two decades ago, a lot of efforts were made to develop an efficient location management to save these resources. Although a substantial progress with regard to computational power and bandwidth has been achieved since then, the air interface is still a valuable resource that is hard to scale. Cellular positioning can be very expensive with regard to signaling overhead, especially if high accuracy is required, and capacities used for it are not available for the transmission of voice or data. Although this is of minor concern in today's networks, where only a very small fraction of subscribers use LBSs, this might become a serious problem in future, when millions of subscribers use sophisticated and complex tracking services, for example, for navigation and community applications.

6.1.3.3 Indoor Infrastructures

Indoor positioning focuses on a deployment in buildings, on university campuses, and company premises. From its very beginning, it has been formed by activities in the area of ubiquitous computing and is often denoted by alternative terms like *location sensing*. It is based on radio, infrared, or ultrasound technologies with a very limited communication range. Indoor positioning is realized either by a stand-alone infrastructure, for example, as done in the ActiveBadge system, or in conjunction with Wireless Local Area Networks (WLANs) like IEEE 802.11. In the former case, targets are equipped with proprietary tags or badges, which operate on the basis of infrared, ultrasound, or *Radio Frequency Identification* (RFID) technology, and which receive pilot signals from or transmit them to a nearby base station. If positioning is integrated into WLANs, standard components like PCMCIA cards and access points can be used.

The benefits of indoor positioning are the low power consumption of the involved devices and the comparatively high accuracy because of the short range of underlying radio or ultrasound technologies. Unfortunately, there have been only proprietary systems so far, in most cases developed in the context of research activities or focusing on dedicated applications, and standardization issues have not advanced or even started yet. Thus, there is a lack of open, well-specified systems that can be universally applied like GPS or cellular positioning.

6.2 Basic Positioning Methods

This section explains the basic positioning methods that have been identified in Table 6.1. It does not deal with concrete implementation issues (which are treated in Chapters 7, 8, and 9), but only gives an introduction to the fundamentals.

6.2.1 Proximity Sensing

The easiest and most widespread method to obtain the position of a target relies on the limited range of coverage of radio, infrared, or ultrasound signals. The position of a target is derived from the coordinates of the base station that either receives the pilot signals from a terminal on the uplink or whose pilot signals are received by the terminal on the downlink. In the following discussion, this principle will be denoted as *proximity sensing*. It is illustrated in Figure 6.3. Figure 6.3 (a) shows a configuration with omnidirectional antennas, while 6.3 (b) shows proximity sensing with a sectorized antenna. The known position of the base station that either sends or receives the pilot signals is then simply assumed to be the position of the target.

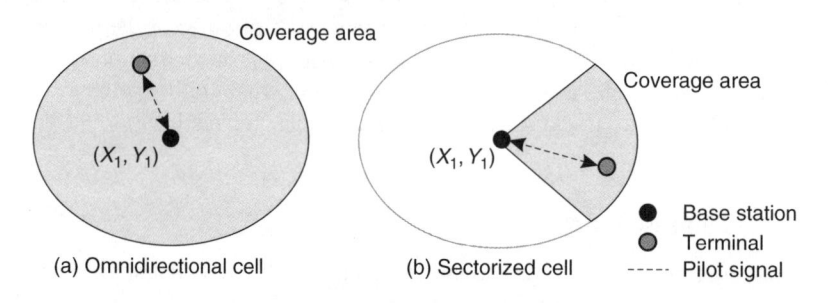

Figure 6.3 Proximity sensing.

The implementation of proximity sensing may be done in several ways, some of them being standardized by official standardization authorities and others being operator- or vendor-specific solutions. Its concrete appearance may also depend on whether it is realized in an integrated or stand-alone infrastructure.

In cellular systems, proximity sensing is known by the terms *Cell of Origin* (CoO), *Cell Global Identity* (CGI), or simply *Cell-ID*. It may be applied during an ongoing connection or if the terminal is idle. In the former case, the network simply adopts the coordinates of the base station that currently serves the terminal. There are no dedicated pilot signals or observables. Thus, this variant is a network-based approach. If the terminal is idle, proximity sensing can be realized as terminal-based positioning, where the terminal listens to the broadcast transmissions of nearby base stations. From the cell identifier received in this way, it can derive a position fix by contacting a remote database that performs a mapping from cell identifiers to base station coordinates. This database may be part of a more sophisticated GIS. It may be maintained by the network operator or by an external actor (see Huber et al. (2003)). Alternatively, some operators include the base station coordinates in the broadcast messages, which is very convenient and avoids any overhead. The network-based approach is subject to standardization, while there exist various proprietary solutions for terminal-based proximity sensing.

In indoor systems, proximity sensing has been implemented in various research projects, but has so far not been standardized. Examples for research prototypes are the *ActiveBadge* or the *Wireless Indoor Positioning System* (WIPS) (see (Want et al. 1992) and (Royal Institute of Technology, Sweden 2000)). These and similar prototypes work on the basis of infrared or ultrasound beacons that are emitted for detecting proximity. Proximity sensing is also the underlying principle of RFID tags. A more detailed overview is given in Chapter 9.

In cellular systems, proximity sensing has become very popular, because it requires only minor modifications on existing infrastructure and causes less overhead. However, the major drawback here is obviously the limited degree of accuracy it provides, which is strongly related with the cell radii, and may vary between 100 m in urban areas and several tens of kilometers in rural areas. Whether this degree of accuracy is tolerable depends on the respective LBS. In indoor systems, accuracy is much better, which is due to the limited coverage range of the radio, infrared, and ultrasound technologies used here.

6.2.2 Lateration

For *lateration*, the observable is either the range or the range difference between the target and a number of at least three base stations. Both are used to compile a system of n nonlinear equations for calculating the target's position, where n denotes the number of base stations. For $n = 3$, lateration is also known as *trilateration*. If positioning is based on range measurements, the position fix is calculated by circular lateration, while range differences form the basis for hyperbolic lateration. Both methods will be explained in the following text. It is of course important to deal with the question of how to measure range and range differences respectively. This will be explained in Section 6.3 in detail. Regardless of which method is used, these measurements are always subject to errors, and hence the resulting values are often denoted as *pseudoranges*, which differ from the true range according to a certain error potential.

6.2.2.1 Circular Lateration

For circular lateration, assume at first that the ranges r_i between the target and a number of base stations $i = 1, \ldots, n$ are known. The 2-D solution is then demonstrated in Figure 6.4. Knowing the range between a terminal and a single base station limits the target's position to a circle around the base station, with the range being the radius of the circle (see Figure 6.4 (a)). If the range to an additional base station is taken into account, then the target's position can be further reduced to the two points where both circles intersect (see Figure 6.4 (b)). The range to a third base station then finally leads to an unambiguous position of the target (see Figure 6.4 (c)).

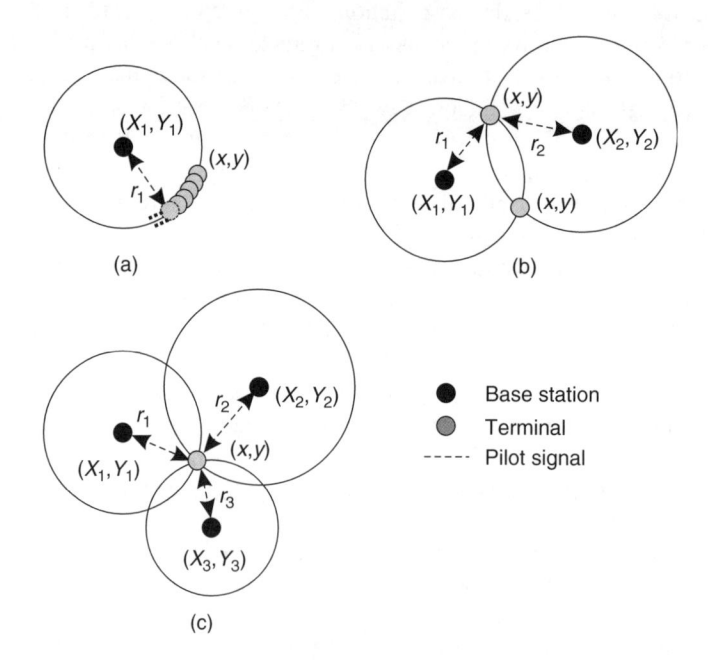

Figure 6.4 Circular lateration in 2D.

Calculating a target's position is then based on Pythagoras theorem. If (X_i, Y_i) are the known coordinates of the ith base station with respect to a Cartesian coordinate system (e.g., ECEF), and (x, y) are the target's unknown coordinates to be derived, then the range r_i between the ith base station and the target can be expressed by the following equation:

$$r_i = \sqrt{(X_i - x)^2 + (Y_i - y)^2} \tag{6.1}$$

If the coordinates of base stations are given by latitude and longitude, or if the target's position is to be expressed in latitude and longitude, the elliptical coordinates must be transferred into Cartesian ones and vice versa respectively in order to apply Equation 6.1.

The circular multilateration for 3D is shown in Figure 6.5. Instead of a circle, each range now defines a sphere around each base station where the target may be located. The intersection between two spheres results in a circle (see Figure 6.5 (a)) and the intersection

FUNDAMENTALS OF POSITIONING

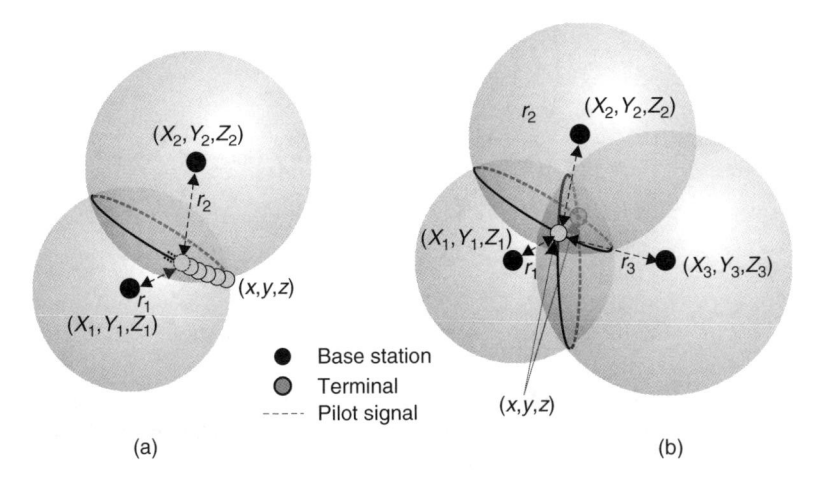

Figure 6.5 Circular lateration in 3D.

between three spheres restricts the target's position to two points (see Figure 6.5 (b)). In most cases, one of these points can be canceled, as its coordinates usually represent a rather curious or unrealistic place, for example, somewhere in outer space. Alternatively, the range to a fourth base station can be included to get an unambiguous position. In some systems, for example, GPS, ranges to at least four base stations (i.e., satellites) are needed for clock synchronization. In analogy to Equation 6.1, a target's 3-D position is given by the following equations:

$$r_i = \sqrt{(X_i - x)^2 + (Y_i - y)^2 + (Z_i - z)^2} \tag{6.2}$$

where the parameters z and Z_i represent the altitude of the target and the ith base station respectively.

As mentioned earlier, the measured pseudorange p_i deviates from the actual range r_i by an error ϵ due to inaccurate clock synchronization, refraction, and multipath propagation:

$$r_i = p_i + \epsilon \tag{6.3}$$

As a consequence, the three circles in Figure 6.4 do not intersect at a certain point, but span an area of some size, which depends on the degree of accuracy of range measurements. Figure 6.6 shows the error potential that results from inaccurate range measurements and limits the accuracy of the position fix. The same holds for the intersection of spheres in Figure 6.5. Therefore, the system of equations specified in 6.1 and 6.2 is in most cases inconsistent and has no unique solution. In order to achieve a result yet, different mathematical approaches can be used.

A well-established method is to apply an iterative process of least square fits in order to approximate the solution in a stepwise manner. At first, it is necessary for this to transfer the nonlinear equations into a system of linear ones. According to Foy (1976) and Torrieri (1984), this can be implemented with the Taylor series expansion. In general, the Taylor series is used to describe a function at a certain point by a power series. Let I denote an

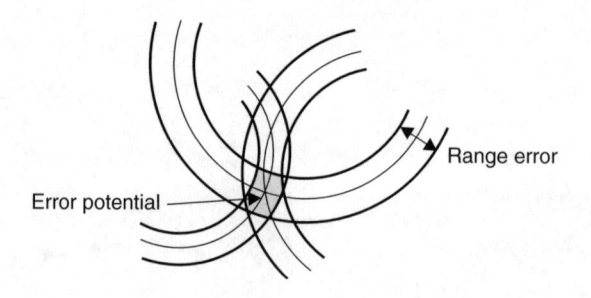

Figure 6.6 Error potential of circular lateration.

interval, a an inner point of I, and f a function with derivatives up to the order of $(n+1)$ on I. Then, for $x \in I$ the function f can be expressed in the following manner:

$$f(x) = \sum_{i=0}^{n} \frac{f^{(i)}(a)}{i!}(x-a)^i + R_n(x,a) \tag{6.4}$$

where $R_n(x,a)$ represents the remainder after $n+1$ terms. In the following text, the linearization is explained for the 3-D case. At first, an estimate of the target's position is determined, which is denoted by $(\tilde{x}, \tilde{y}, \tilde{z})$. The goal is now to calculate a correction vector $[\Delta x, \Delta y, \Delta z]$ that can be applied to the estimated position. The function of the pseudorange to the ith base station p_i, which is to be expressed by a Taylor series, then appears as follows:

$$p_i(x,y,z) = \sqrt{(X_i-x)^2 + (Y_i-y)^2 + (Z_i-z)^2} \tag{6.5}$$
$$= p_i(\tilde{x}+\Delta x, \tilde{y}+\Delta y, \tilde{z}+\Delta z)$$

For the position fix, only the first order expansion is determined, which is given by

$$p_i(\tilde{x}+\Delta x, \tilde{y}+\Delta y, \tilde{z}+\Delta z)$$
$$= p_i(\tilde{x}, \tilde{y}, \tilde{z}) + \frac{\partial p_i}{\partial \tilde{x}}\Delta x + \frac{\partial p_i}{\partial \tilde{y}}\Delta y + \frac{\partial p_i}{\partial \tilde{z}}\Delta z \tag{6.6}$$

Solving the partial differentials in this equation yields

$$\frac{\partial p_i}{\partial \tilde{x}} = \frac{-X_i + \tilde{x}}{\sqrt{(X_i-\tilde{x})^2 + (Y_i-\tilde{y})^2 + (Z_i-\tilde{z})^2}} = a_i$$

$$\frac{\partial p_i}{\partial \tilde{y}} = \frac{-Y_i + \tilde{y}}{\sqrt{(X_i-\tilde{x})^2 + (Y_i-\tilde{y})^2 + (Z_i-\tilde{z})^2}} = b_i \tag{6.7}$$

$$\frac{\partial p_i}{\partial \tilde{z}} = \frac{Z_i + \tilde{z}}{\sqrt{(X_i-\tilde{x})^2 + (Y_i-\tilde{y})^2 + (Z_i-\tilde{z})^2}} = c_i$$

Because the coordinates (X_i, Y_i, Z_i) of the ith base station as well as the target's estimated position $(\tilde{x}, \tilde{y}, \tilde{z})$ are known, the coefficients a_i, b_i, and c_i are numerically known too. Thus, we obtain

$$p_i(\tilde{x}+\Delta x, \tilde{y}+\Delta y, \tilde{z}+\Delta z)$$
$$= p_i(\tilde{x}, \tilde{y}, \tilde{z}) + a_i\Delta x + b_i\Delta y + c_i\Delta z \tag{6.8}$$

FUNDAMENTALS OF POSITIONING

In this equation, the value of $p_i(\tilde{x}, \tilde{y}, \tilde{z})$ corresponds to the pseudorange between the estimated position and the position of the respective base station and is therefore numerically known. Let Δp_i denote the difference between this pseudorange and the observed pseudorange. Then, for n observed base stations with $i = 1, \ldots, n$ we obtain a system of n linear equations given by

$$\Delta p_i = a_i \Delta x + b_i \Delta y + c_i \Delta z \tag{6.9}$$

Converting this system to matrix notation yields

$$\mathbf{b} = \mathbf{A}\mathbf{x} \tag{6.10}$$

with

$$\mathbf{b} = \begin{bmatrix} \Delta p_1 \\ \Delta p_2 \\ \ldots \\ \Delta p_n \end{bmatrix}, \mathbf{A} = \begin{bmatrix} a_1 & b_1 & c_1 \\ a_2 & b_2 & c_2 \\ \ldots & \ldots & \ldots \\ a_n & b_n & c_n \end{bmatrix}, \mathbf{x} = \begin{bmatrix} \Delta x \\ \Delta y \\ \Delta z \end{bmatrix} \tag{6.11}$$

The matrix $\mathbf{A}$ is composed from the partial differentials of Equation 6.8 and contains the coefficients of the system of equations. It is called *design matrix*. The vector $\mathbf{b}$ contains for each base station the deviation of the observed pseudorange from the range that is based on the estimated position and, as mentioned earlier, is numerically known. It is also referred to as *misclosure vector*. Finally, $\mathbf{x}$ represents the searched correction vector of the estimated position. Under ideal conditions, such a system has a unique solution for the correction vector $\mathbf{x}$ that can be obtained by calculating the inverse of $\mathbf{A}$ and rearranging Equation 6.10 as follows:

$$\mathbf{x} = \mathbf{A}^{-1}\mathbf{b} \tag{6.12}$$

However, in this case the system of equations is overdetermined, that is, it contains more equations than unknowns. In addition, it is based on estimations and inaccurate observations, and hence in most cases a solution does not exist. In order to obtain a result yet, one has to approximate a solution $\tilde{\mathbf{x}} = [\Delta \tilde{x}, \Delta \tilde{y}, \Delta \tilde{z}]^T$ by applying a least square fit. In the first step, the squared Euclidean distance of the residual vector $r = \mathbf{b} - \mathbf{A}\tilde{\mathbf{x}}$ must be built:

$$\|r\|_2^2 = r^T r = (\mathbf{b} - \mathbf{A}\tilde{\mathbf{x}})^T (\mathbf{b} - \mathbf{A}\tilde{\mathbf{x}}) \tag{6.13}$$
$$= \mathbf{b}^T\mathbf{b} - 2\tilde{\mathbf{x}}^T\mathbf{A}^T\mathbf{b} + \tilde{\mathbf{x}}^T\mathbf{A}^T\mathbf{A}\tilde{\mathbf{x}}$$

The condition of the least square fit is given by minimizing the squared Euclidean distance of the residual vector:

$$\min \|\mathbf{b} - \mathbf{A}\tilde{\mathbf{x}}\|_2^2 \tag{6.14}$$

which is achieved by building the derivative of Equation 6.14 and setting it to zero:

$$-2\mathbf{A}^T\mathbf{b} + 2\mathbf{A}^T\mathbf{A}\tilde{\mathbf{x}} = 0 \tag{6.15}$$

This leads to the following set of normal equations for which a unique solution exists:

$$\mathbf{A}^T\mathbf{A}\tilde{\mathbf{x}} = \mathbf{A}^T\tilde{\mathbf{x}}$$
$$\Leftrightarrow \quad \tilde{\mathbf{x}} = (\mathbf{A}^T\mathbf{A})^{-1}\mathbf{A}^T\mathbf{b} \tag{6.16}$$

The system of linear equations can now simply be solved for the vector $\tilde{\mathbf{x}} = [\Delta\tilde{x}, \Delta\tilde{y}, \Delta\tilde{z}]^T$, which yields a correction for the previously estimated position. The correction vector is applied to the estimated position, for which steps 6.8–6.16 are then repeated. This iteration process stops if the calculated correction values fall below a predefined threshold. More mathematical details on position calculation with Taylor series expansion and the least square fit method can be obtained from (Foy 1976), (Klukas 1997), and (Krakiwsky 1990).

Positioning methods based on circular lateration in combination with timing measurements are usually subsumed under the term *Time of Arrival* (ToA). GPS is the most prominent example where this method is applied. A GPS receiver determines pseudoranges to at least three satellites and calculates its position from that. For clock synchronization between receiver and satellite, it is usually required to observe the signals from a fourth satellite. This will be explained in Chapter 7 in detail.

6.2.2.2 Hyperbolic Lateration

Instead of determining the absolute range, hyperbolic lateration is based on range differences. A hyperbola is defined to be the set of all points for which the difference in the range to two fixed points is constant. These fixed points are denoted as *foci*, and, in the special case of positioning, are given by two base stations. For the 2-D case, consider at first the range difference between a target and two base stations as depicted in Figure 6.7. The range to the first base station is given by r_1, and the range to the second one by r_2. The difference $r_2 - r_1$ now limits the target's position to a hyperbola as shown in Figure 6.7 (a). If the range difference is determined for another pair of base stations, for example, to the second and third one, an additional hyperbola is created, and the intersection of both then delivers the target's position (see Figure 6.7 (b)). In 3D, the constant range difference between a target and two base stations is given by the surface of a hyperboloid, and it is necessary to determine a minimum of three hyperboloids to unambiguously fix the target's position in 3D.

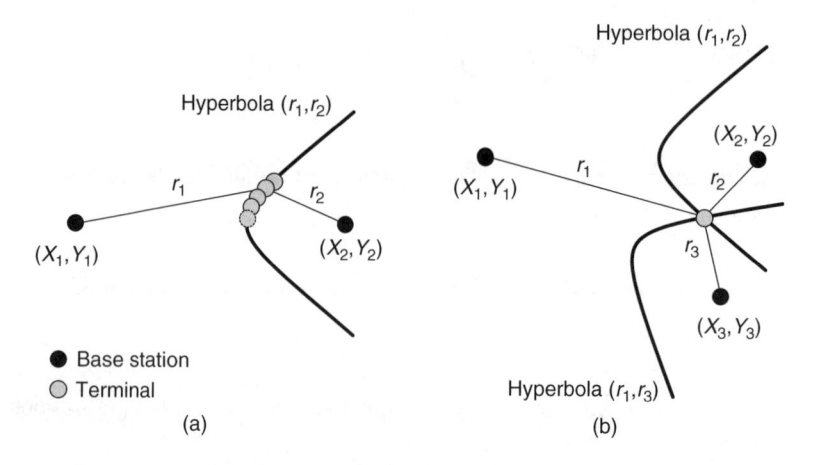

Figure 6.7 Hyperbolic lateration.

FUNDAMENTALS OF POSITIONING

Like for circular lateration, a system of equations can be arranged, where each equation expresses the range difference with regard to a particular pair of base stations. For the 3-D case, this system is given by

$$d_{ij} = r_i - r_j \qquad (6.17)$$
$$= \sqrt{(X_i - x)^2 + (Y_i - y)^2 + (Z_i - z)^2} - \sqrt{(X_j - x)^2 + (Y_j - y)^2 + (Z_j - z)^2}$$

where d_{ij} is the difference between the ranges r_i and r_j of the ith and jth base station and $i \neq j$.

It is common practice to measure all range differences with regard to a selected reference base station. In the following text, this reference is indicated by $i = 1$. The method for solving the system of nonlinear equations in 6.18 is basically the same as for circular lateration. On the basis of an estimation fix of the target's position, a Taylor series expansion is built, which yields the following coefficients for the design matrix **A** (compare with Equations 6.8 and 6.10):

$$\frac{\partial d_{1j}}{\partial \tilde{x}} = \frac{-X_1 + \tilde{x}}{\tilde{r}_1} - \frac{-X_j + \tilde{x}}{\tilde{r}_j} = a_i$$
$$\frac{\partial d_{1j}}{\partial \tilde{y}} = \frac{-Y_1 + \tilde{y}}{\tilde{r}_1} - \frac{-Y_j + \tilde{y}}{\tilde{r}_j} = b_i \qquad (6.18)$$
$$\frac{\partial d_{1j}}{\partial \tilde{z}} = \frac{-Z_1 + \tilde{z}}{\tilde{r}_1} - \frac{-Z_j + \tilde{z}}{\tilde{r}_j} = c_i$$

with

$$\tilde{r}_i = \sqrt{(X_i - \tilde{x})^2 + (Y_i - \tilde{y})^2 + (Z_i - \tilde{z})^2} \qquad (6.19)$$

being the range between target and ith base station with regard to the estimated position.

Like circular lateration, hyperbolic lateration has an error potential because of inaccuracies of range difference measurements. The error potential of two intersecting hyperbolas is depicted in Figure 6.8. Therefore, a least square fit can be used again to approximate a solution. In analogy to Equation 6.8, one obtains a system of equations that expresses the deviation Δd_{1j} between the range difference of the reference and jth base station with regard to the target's observed position on the one hand and the range difference based

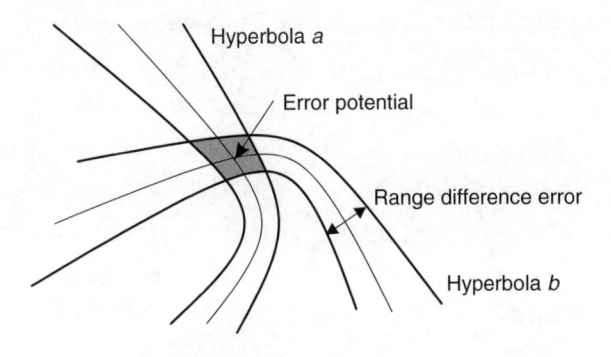

Figure 6.8 Error potential of hyperbolic lateration.

on its estimated position on the other. The deviations Δd_{1j} for each pair of base stations involved in positioning form the misclosure vector **b** that is used to perform the least square fit in analogy to Equations 6.14–6.16. Again, Equation 6.16 delivers a correction vector that is added to the initial position estimate, and the result is then used for the next iteration.

A number of extensions or alternatives to the approach presented here have been proposed, which aim at reducing the number of iterations or providing a better accuracy. One method, for example, is to compose a covariance matrix **Q** of the observations, which is applied to Equation 6.16 as follows:

$$\tilde{\mathbf{x}} = (\mathbf{A}^T \mathbf{Q}^{-1} \mathbf{A})^{-1} \mathbf{A}^T \mathbf{Q}^{-1} \mathbf{b} \qquad (6.20)$$

Details on this can be found in (Krakiwsky 1990), (Chan and Ho 1994), and (Klukas 1997). Other methods have been developed by (Fang 1990) and (Friedlander 1987), to name only a few. In addition, numerous works deal with performance and accuracy aspects of hyperbolic lateration systems, partially in comparison with circular lateration, for example, (Klukas 1997), (Mizusawa 1996), and (Aatique 1997). Hyperbolic lateration is preferably used in cellular networks and is termed *Time Difference of Arrival* (TDoA).

6.2.3 Angulation

Angulation is another method to estimate a target's position from the known coordinates of several base stations. In contrast to lateration, the observables here are the angles between the target and a number of base stations. Angulation is also termed *Angle of Arrival* (AoA) or *Direction of Arrival* (DoA).

For deriving these angles, it is required that either base stations or terminals are equipped with antenna arrays, depending on whether positioning is terminal or network-based (see also Section 4.2.2). In most systems where angulation is applied today, however, these arrays are usually arranged at the base stations because, for reasons of economy and complexity, a terminal-based solution is not practicable (at least not with the technologies available today).

The basic principle behind angulation is illustrated in Figure 6.9. The angle of an incoming pilot signal is measured at the base station and thus restricts the target's position along a line that intersects both the target's and the base station's position. If the angle to a second base station is taken into account, another line is defined and the intersection of both lines then represents the target's position. Thus, from a theoretical point of view, it is sufficient to make angle measurements at two base stations in order to obtain a position in 2D.

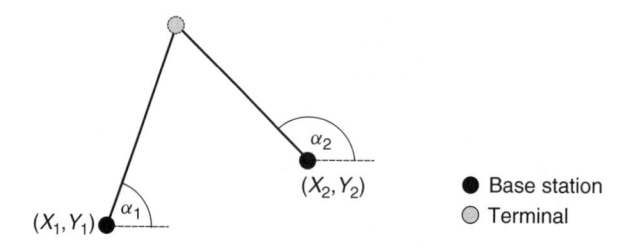

Figure 6.9 Angulation.

FUNDAMENTALS OF POSITIONING

However, angulation may suffer from a bad resolution of antenna arrays, and hence the observed angle is rather a rough approximation of the actual angle. This approximation is more accurate if the target is located closer to the base station and vice versa. Furthermore, like for lateration, multipath propagation is a major problem if no line of sight exists between target and base station. In this case, the pilot signal is subject to one or multiple reflections at obstacles and may arrive from any direction at the base station. Therefore, measurements at three or more base stations are recommended in practice in order to compensate for these errors. The error potential resulting from these effects is illustrated in Figure 6.10.

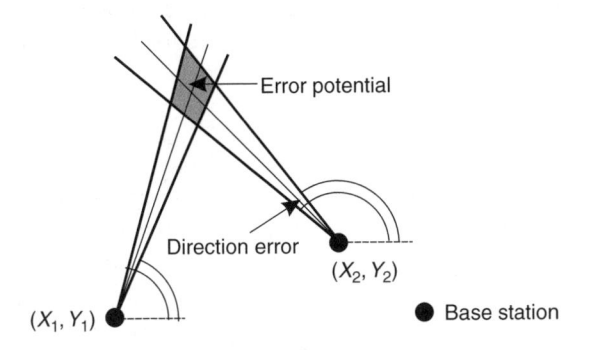

Figure 6.10 Error potential of angulation.

Calculating a target's position from several angle observations works very similar to the approaches presented earlier. In analogy to lateration, the observed angle ϕ will be denoted as *pseudoangle* here. It differs from the actual angle α in the following way:

$$\alpha = \phi + \epsilon \tag{6.21}$$

where ϵ represents the error imposed on the angle observations because of the effects mentioned earlier. The starting point is to establish a system of equations where the angle α_i of the arriving pilot signal at the ith base station is expressed as a function of opposite and adjacent leg (see also Figure 6.9):

$$\alpha_i = \arctan\left(\frac{Y_i - y}{X_i - x}\right) \tag{6.22}$$

where (X_i, Y_i) and (x, y) are again the coordinates of the ith base station and the target respectively. Note that the angles of arriving pilot signals must be measured against the same axis at all base stations. Since these angles are not exactly known, a solution for 6.22 must be approximated by a least square fit that starts with a linearization using the Taylor series expansion. This step is in analogy to Equations 6.6 to 6.9, that is, the observed angle is expressed as a function ϕ_i of an estimated position $(\tilde{x}, \tilde{y})$ and a correction vector $[\Delta x, \Delta y]$:

$$\phi_i(x, y, z) = \arctan\left(\frac{Y_i - y}{X_i - x}\right) \tag{6.23}$$
$$= \phi_i(\tilde{x} + \Delta x, \tilde{y} + \Delta y)$$

The partial differentials of the Taylor series expansion are then given by:

$$\frac{\partial \phi_i}{\tilde{x}} = \frac{\sin \phi_i}{r_i} = a_i$$
$$\frac{\partial \phi_i}{\tilde{y}} = -\frac{\cos \phi_i}{r_i} = b_i \qquad (6.24)$$

where r_i represents the range between the target and base station i. The coefficients a_i and b_i are used to compose the design matrix **A** of the resulting system of linear equations, and the misclosure vector **b** is given by the angle difference between the observed angle and the angle derived from the estimated position. The solution of the system is then achieved in analogy to Equations 6.14–6.16.

6.2.4 Dead Reckoning

Dead reckoning is obviously one of the earliest positioning methods that had been applied. It has its origins in seafaring and is also known to be the method that was used by Christopher Columbus in his voyages to discover the New World. In spite of its long history and owing to the development of advanced sensor technology, dead reckoning has gained major importance in recent years and is used today for the navigation of aircraft, ships, and automobiles. The term "dead reckoning" has been established as an abbreviation for *deduced reckoning* and is sometimes also referred to as *inertial navigation*.

Dead reckoning means that the current position of a target can be deduced or extrapolated from the last known position, assuming that the direction of motion and either the velocity of the target or the traveled distance are known. In order to distinguish the position deduced in this way from a position fix (which basically is the last known position derived by an alternative method such as GPS), it is referred to as *position estimate* here. The principle behind dead reckoning is demonstrated in Figure 6.11 (see also (Nehmzow and McGonigle 1993)). Let (x_0, y_0) denote the known position fix of a target with regard to a

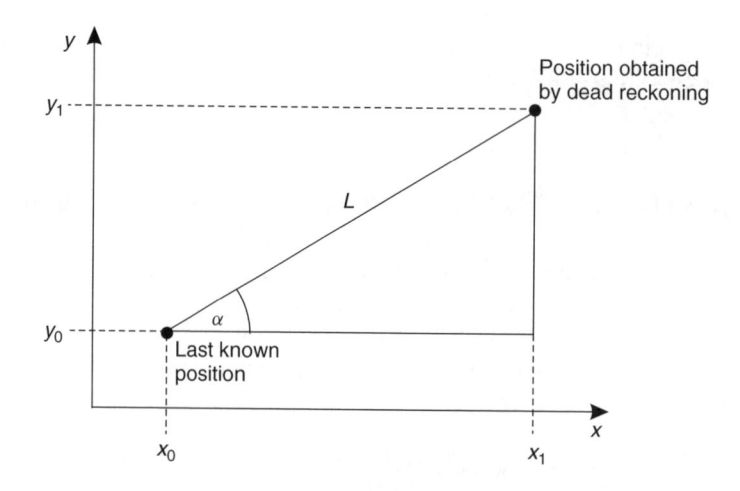

Figure 6.11 Dead reckoning.

2-D Cartesian coordinate system, α the direction of motion, and L the distance covered in a certain amount of time. The position estimate (x_1, y_1) derived via dead reckoning is then

$$x_1 = x_0 + L\cos\alpha, \quad y_1 = y_0 + L\sin\alpha \tag{6.25}$$

If the distance L is not known, it can be obtained from the velocity v of the target and the travel time Δt:

$$L = v\Delta t \tag{6.26}$$

Obviously, the crucial point in dead reckoning is to obtain the starting position, the direction of motion as well as the distance and speed respectively. The starting position must be determined by another positioning method, preferably one with increased accuracy such as lateration or angulation, for example, GPS. That is, dead reckoning is not a stand-alone technique but always used in combination with another method (in the earlier days of navigation, when technologies for positioning were not available, one used well-known control points registered on a map). Direction, distance, and speed of motion can be basically obtained in two ways: either by deducing them from two or more position fixes that have been measured in the past, or by additional sensors the target is equipped with. The types of sensors deployed for dead reckoning are:

- accelerometers for measuring the acceleration of a target (from which the velocity can be derived),

- odometers for obtaining the traveled distance by counting wheel turns, and

- gyroscopes for deriving the direction of motion of the target.

A detailed introduction to the working of these sensor technologies can be obtained from (Allen et al. 2001) and (Grejner-Brzezinska 2004).

Regardless of whether direction and distance or speed may be derived from a number of past position fixes or by sensors, the accuracy of dead reckoning is degraded with an increasing Δt. Inaccuracies in the absence of sensors may result from the fact that changes in the target's direction or speed between the position fix and the position estimate remain undetected, which becomes more probable if more time is passed between the position fix and the estimate. If, however, sensors are deployed, these changes can be taken into account, and hence accuracy is generally much higher. Nevertheless, the precise working of sensors is affected by many circumstances and this is why accuracy degradations cannot be entirely eliminated. For example, in case of an odometer, a possible unevenness of the surface the distance is measured on may cause errors in the range between 0.1 and 0.01 percent (Da and Dedes 1995).

Today, dead reckoning is deployed in OBUs for car navigation in order to refine position fixes provided by a GPS receiver or to maintain navigation instructions for the driver in case a sufficient number of GPS satellites is temporarily not visible (which happens, for example, if the car passes tunnels). For these purposes, dead reckoning adopts and processes floating car data delivered by sensors inside the car. In the context of LBSs, dead reckoning can be used to optimize the frequency of position updates. These updates are executed, for example, when position data obtained by a terminal-based method needs to be processed on a remote server. For example, Walther and Fischer (2002) present a tourist guide application that adopts dead reckoning for optimizing the frequency of position updates between the terminal and the server.

6.2.5 Pattern Matching

Positioning can also be realized by pattern matching. The main principle here is to observe the site (or scene) where positioning is to be applied and to draw conclusions about the position of a target from these observations. It is classified as *optical* and *nonoptical pattern matching*.

In the optical version, also known as *scene analysis*, visual images of a site are generated by a camera, which are compared with each other. Scene analysis can be further subdivided into *static* and *dynamic* (or *differential*) *scene analysis*. In a static scene analysis, the position of a target can be derived by comparing a snapshot of a scene generated by an observer with a number of preobserved simplified images of the scene (taken from different positions and perspectives), where the target is either the observer itself, or it is an object located in the scene. In case of dynamic scene analysis, the position is derived from differences between images taken successively from a scene. Scene analysis so far is not a topic in the area of LBSs, but is predominantly applied in the context of mobile robotics and augmented reality.

In nonoptical pattern matching, other physical quantities are taken into account. A popular method is to detect a position from the propagation characteristics of radio signals which the target's terminal experiences at a certain position on a site. In this case, pattern matching is also known as *fingerprinting*, and it is executed in two phases (Kaemarungsi and Krishnamurthy 2004). In an off-line phase, the site of interest is covered by a grid, and for each point of the grid the observer collects the *received signal strength* (RSS) from multiple base stations, resulting in a vector of RSS values for each point. This vector is called a *fingerprint*. In the on-line phase, the target's terminal composes a sample vector of RSS values received at its current position and reports it to a server. The server then tries to match the sample with the fingerprints generated in the off-line phase and estimates the target's position. The most common algorithm is to compute the Euclidean distance between the RSS sample and each fingerprint (Kaemarungsi and Krishnamurthy 2004). Alternative methods are to use neural networks (Small et al. 2000) or Bayesian modeling (Castro et al. 2001).

Pattern recognition is in most cases realized as terminal-assisted approach, where the terminal performs measurements and transfers the results to the network for position calculation. The advantage is that in some systems the terminals require no or only minor modifications. For example, in GSM, each terminal in busy mode transmits permanently RSS values of neighboring base stations to the network for handover decision, which could also be utilized for fingerprinting (see (Laitinen et al. 2001)). Nevertheless, GSM fingerprinting has only been evaluated in some prototype implementations and is currently not subject to standardization. However, it is an option for indoor systems and is preferably used in conjunction with WLAN, for example, in the RADAR system from Microsoft Research (Bahl and Padmanabhan 2000). The realization of WLAN fingerprinting is introduced in Chapter 9.

6.2.6 Hybrid Approaches

Basically, it is possible to implement any combination of the positioning methods presented earlier and consider several different observables for a position fix. The main motivation to do so is to provide increased accuracy, which is, for example, either necessary for meeting accuracy standards defined by national authorities in the context of enhanced emergency

FUNDAMENTALS OF POSITIONING

services or desired if available channel parameters, for example, those collected for handover decision, can be utilized for improving accuracy.

An example for the latter motivation is proximity sensing in combination with range or angle measurements as depicted in Figure 6.12. In cellular networks, the access network determines the approximate range for each terminal that is connected to a base station, which is a facility of medium access (not positioning) and required for synchronization. Although this range is very inaccurate and coarse grained compared to the range values needed for lateration (it is measured by round trip times), it can restrict the terminal's position to a ring around the base station, or to the part of a ring if sectorized antennas are used. Furthermore, if the base station is equipped with an antenna array, it is possible to further restrict the position. Nevertheless, it must be stressed that methods that combine proximity sensing with other observables are much more inaccurate than lateration or angulation.

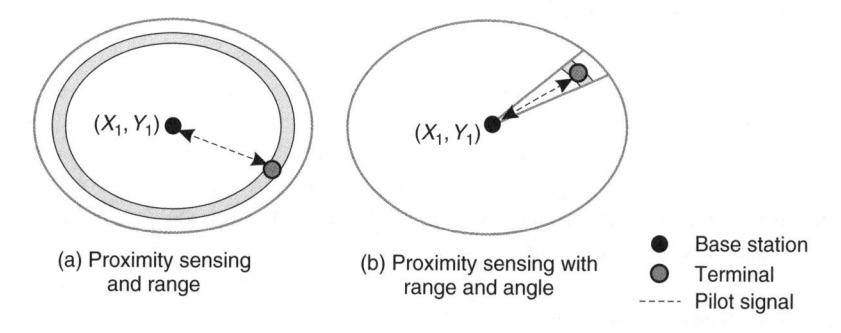

Figure 6.12 Combination of proximity sensing with range and angle.

However, both pure lateration and pure angulation also provide insufficient results if the geometry of base stations involved in measurements is poor. This typically occurs if coverage is to be restricted to highways between major cities and therefore base stations are arranged along the highways, as explained by (Proietti 2002). For pure hyperbolic lateration, the hyperbolas intersect the highway, which makes it very difficult to determine whether the target is located at the highway or somewhere away from it. Angulation, on the other hand, restricts the target's position to the highway, but it is hard to fix the actual point between the two base stations. An obvious method is therefore to combine lateration and angulation, which provides an intersection of the hyperbola and the direction of the radio beam (see Figure 6.13).

These are only two prominent examples of hybrid approaches. Other configurations are less common so far, but nevertheless there is a high potential for future research in this area.

6.3 Range Measurements

Ranges and range differences as used for circular and hyperbolic lateration can be obtained either by performing time measurements or by observing the *received signal strength* (RSS). Both approaches will be explained in the following section.

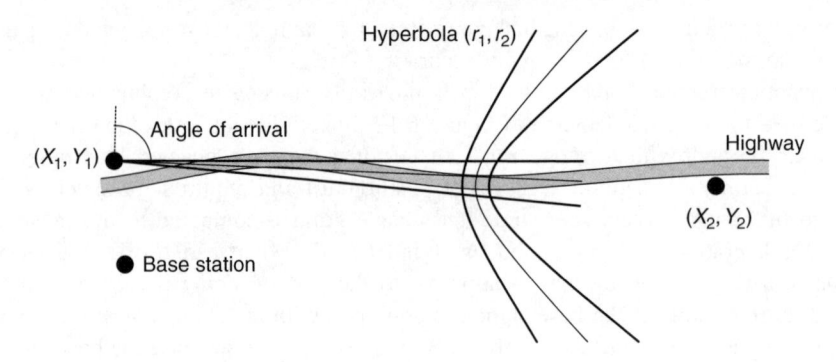

Figure 6.13 Combination of angulation and hyperbolic lateration.

6.3.1 Time Measurements

Time-based measurements for circular lateration (or ToA), and hyperbolic lateration (or TDoA) can be made in the downlink or uplink for either realizing terminal or network-based positioning. In the downlink, several pilots, generated and emitted by a number of base stations, are evaluated at the terminal. For uplink methods, a single pilot is emitted by the terminal, which is then received and observed at different base stations. Basically, radio, infrared, or ultrasound signals can be used as pilots, but the preferred method in nearly all systems is based on radio signals.

6.3.1.1 Measurement Methods

According to (Caffery 2000), there are three different measurement methods for radio signals:

- **Pulse ranging.** A pilot is represented by a single pulse, whose ToA is measured at the receiver (see Figure 6.14a). The range is then derived from the traveling time, for which the time of signal emission must also be known. For TDoA this is not necessary, but the pilots must be transmitted at the same time here in order to derive the time differences and, in a next step, the range differences. Pulse ranging is applied, for example, in GSM where the pulses are given by the beginning of time slots of the underlying TDMA scheme.

- **Carrier phase ranging.** The arriving pilot signal is compared with a reference signal that is generated at the receiver (see Figure 6.14b). At the time of pilot emission, reference and pilot must be in phase. The range can then be obtained from the phase shift between the delayed pilot arriving at the receiver and the reference signal. For TDoA, the phase shifts of different pilots are observed. Carrier phase ranging suffers from a problem known as *integer ambiguity*, which refers to the unknown number of phase cycles between sender and receiver. This number depends on the pilot's wavelength and the covered range. The easiest method for resolving it is to choose pilots of very low frequencies, where the wavelength is in the order of kilometers. However, this is not practicable in most cases (due to antenna design), and hence pilots have wavelengths in the range of some centimeters. In these cases, the number

FUNDAMENTALS OF POSITIONING   145

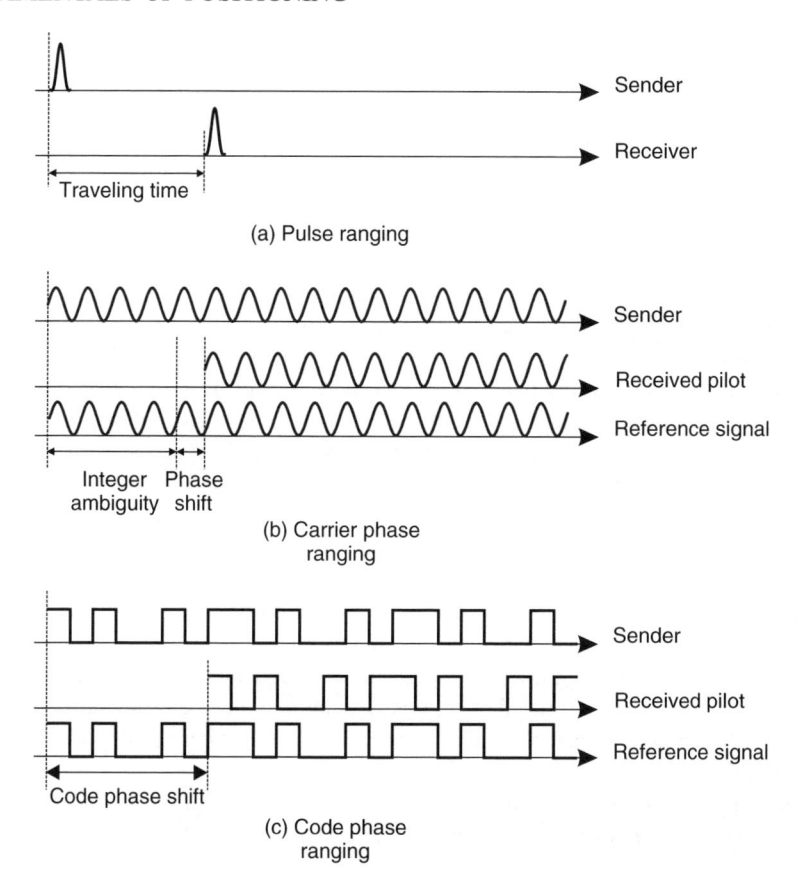

Figure 6.14 Time-based ranging methods.

of cycles is counted by advanced methods, for example, frequency differencing, time of arrival information, or, in satellite systems, by observing the Doppler shift (see (Caffery 2000)). Carrier phase ranging is often applied in GPS in order to achieve very high accuracy position fixes, for example, as required for surveying and mapping.

- **Code phase ranging.** Code phase ranging is based on spread spectrum technique as described in Section 4.3.4. The pilot is given by a signal that is modulated with a known spreading code (see Figure 6.14 (c)). The procedure is then very similar to carrier phase ranging. Sender and receiver start generating the spreading code simultaneously. The pilot arrives time delayed at the receiver, and the traveling time is represented by the shift in terms of chips between the received pilot and the reference signal. This shift is derived by computing the autocorrelation function for the received pilot and a number of shifted versions of the reference code until a shift is found for which both codes correlate. For TDoA, the shifts of different pilots are compared with each other. Integer ambiguity can be avoided by choosing spreading codes of sufficient length. Code phase ranging is the standard measurement method applied in GPS.

As can be derived from these explanations, all measurements utilize the speed of electromagnetic waves and require exact clock synchronization either between terminal and base stations (ToA) or only between the different base stations (TDoA). Clock synchronization is a significant problem for all time-based methods, which is easy to understand when imagining that a time offset of only 1 µs between sender and receiver causes a range error of 300 m (based on the speed of electromagnetic waves).

6.3.1.2 Clocks

The main components of a clock are an oscillator and a counter. The oscillator generates a consistent frequency, and the counter translates the number of the frequency's cycles into a common time unit. The need for clock synchronization results from the fact that no oscillator matches exactly the desired frequency and, even worse, is subject to fluctuations more or less heavily. The former aspect refers to the clock's accuracy, while the latter is related to its stability. The difference between both quantities is illustrated in Figure 6.15, which shows frequency variations of an oscillator over time. The dotted line represents the desired frequency, and the solid one the oscillator frequency.

Figure 6.15 (a) shows a perfect oscillator that matches the desired frequency exactly and does not change over time. A stable but inaccurate oscillator is shown in Figure 6.15 (b). Here, the frequency is constant but does not match the desired frequency. The opposite case is shown in Figure 6.15 (c), where the generated frequency centers around the desired frequency but changes over time, that is, this oscillator is accurate but unstable. Finally, Figure 6.15 (d) depicts an unstable and inaccurate oscillator, where both phenomena occur in combination.

When comparing two clocks, the difference in the absolute time produced by them is called *time offset*. If this offset remains constant over an infinite period of time, it can

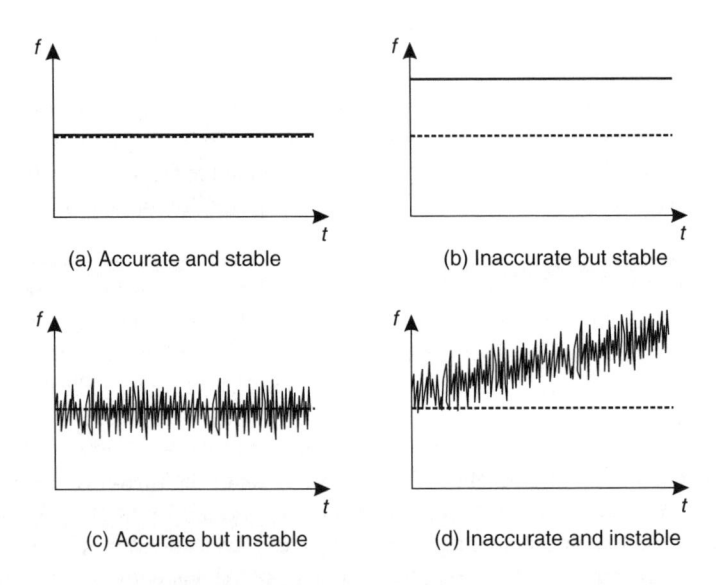

Figure 6.15 Accuracy and stability of clocks.

FUNDAMENTALS OF POSITIONING

be assumed that the oscillators of both clocks are perfectly accurate and stable. However, generally the time offset changes over a period of time (e.g., one second per day), which is referred to as *clock drift* and which results from inaccurate oscillators. If this clock drift were constant, its amount could be described by a linear function, and the time offset expected for a certain point of time in future could be predicted. Unfortunately, the oscillators' instability causes variations in the drift (*drift rate*), and hence such predictions are only possible for short time intervals. The instability is caused by a number of time-varying parameters such as temperature, aging, vibration, and capacity load, while the degree of accuracy primarily depends on the quality of the oscillators used.

In today's equipment, different versions of quartz-crystal oscillators are deployed. The common principle behind them is that the frequency is derived from the vibrations of a quartz crystal, which result from applying a voltage to it. The maximum amplitude of vibrations is achieved if the changing voltage matches the crystal resonance frequency. Unfortunately, this frequency changes with environmental parameters like temperature, humidity, and pressure, which makes the oscillator inaccurate and instable. One method to compensate for this is to enclose the crystal in a temperature-controlled chamber called an *oven*. Accordingly, oscillators using this method are called *Oven-Controlled Crystal Oscillators* (OCXOs). They have a frequency offset between 10^{-8} and 10^{-10}, which in the worst case approximately corresponds to a phase deviation of 1 ms per day. An alternative method is to measure the temperature by a sensor and to generate a correction voltage from that. This is used by *Temperature-Compensated Crystal Oscillators* (TCXOs), which have a frequency offset of 10^{-6} or a phase deviation of 1 µs per second. OCXOs are much more expensive than TCXOs. They are deployed in the base stations of cellular networks, while TCXOs are used in mobile devices.

GPS satellites are equipped with very expensive atomic clocks, where the frequency is derived from the motion of an electron and nucleus within the caesium atom. One second exactly corresponds to 9,192,631,770 cycles of this motion, which is the resonance frequency of the cesium atom. The frequency offset is typically in the order of 10^{-14}, which approximately corresponds to a phase deviation of 1 ns per day. Alternatively, rubidium oscillators are used, which show less accuracy. An overview of the different oscillator types used for positioning is given in Table 6.2.

Table 6.2 Overview of oscillators used in positioning equipment

Oscillator type	Used in	Frequency offset
OCXO	GSM and UMTS base stations	10^{-8} to 10^{-10}
TCXO	Mobile devices	10^{-6}
Caesium	GPS satellites	5×10^{-12} to 1×10^{-14}
Rubidium	GPS satellites	5×10^{-10} to 5×10^{-12}

A major challenge for positioning is to eliminate the time offsets between the clocks used for timing measurements. For keeping base stations synchronized among each other, time offset, drift, and drift rate are measured either against a universally valid time source (e.g., GPS time), or against the clock of a selected reference base station. Both are applied, for example, for positioning in cellular systems as will be described in Chapter 8. It assumes that the distance between the base stations and the measurement station is known in order to eliminate the delays imposed on the pilot signals during transmission. The measured offset, drift, and drift rate are then applied to the range measurements during a position fix. For keeping a terminal synchronized with the base stations, it is possible to model the time offset as an additional unknown in the system of equations used for lateration, which requires measurements to at least four base stations. This is the method applied by GPS and is described in Chapter 7.

6.3.2 Received Signal Strength

An alternative method is to derive the range from the *received signal strength* (RSS), that is, to utilize the path loss or attenuation a pilot signal experiences when traveling from the sender to the receiver. As explained in Section 4.2.4, the attenuation is a function not only of the range between sender and receiver, but also of the wavelength and the path-loss gradient. As also explained in that section, the path loss is determined by more or less complex mathematical models that have been tailored according to the special circumstances of the environment under consideration, for example, indoor versus outdoor and degree of obstacles. The simplest of these models is obviously the Friis free space equation, which, however, allows for considering environmental parameters only in the form of the path-loss gradient. Therefore, a number of dedicated and more accurate models have been proposed for positioning purposes, for example, (Hata and Nagatsu 1980), (Ott 1977), and (Song 1994).

However, RSS-based measurements are in general very inaccurate compared to time measurements. Caffery (2000) reports that variants in the RSS can be as great as 30–40 dB over distances in the order of a half wavelength, and Ott (1977) showed that estimates based on a path-loss exponent of 3.5 may result in range errors of 69% too large or 41% too small compared to the true range. Even worse, many systems use an adaptive power control that dynamically adjusts the transmitted power to a number of parameters, for example, the distance between sender and receiver, the observed bit error rates, or, in CDMA, the number of users in a cell. In order to apply path-loss models for range determination, it is therefore necessary to record and control the transmitted signal strength very accurately and to report it to the site where ranges and positions are calculated. Both may be a very complex and difficult process.

In general, the error potential of signal strength measurements is much higher than that of timing measurements, and hence ranges in most cases are derived by the latter method, especially for satellite and cellular positioning methods. Only in indoor environments, where the signal traveling time is hard to measure due to extremely short distances between target and base stations, some methods fall back on the path loss.

6.4 Accuracy and Precision

Owing to a number of error sources, which will be identified in the next section, positioning suffers from accuracy and precision degradations. The difference between accuracy and precision is illustrated in Figure 6.16. The figure shows two measurement processes that have been carried out with different positioning methods. Each measurement process comprises a number of position fixes, resulting in a number of samples that are represented

FUNDAMENTALS OF POSITIONING

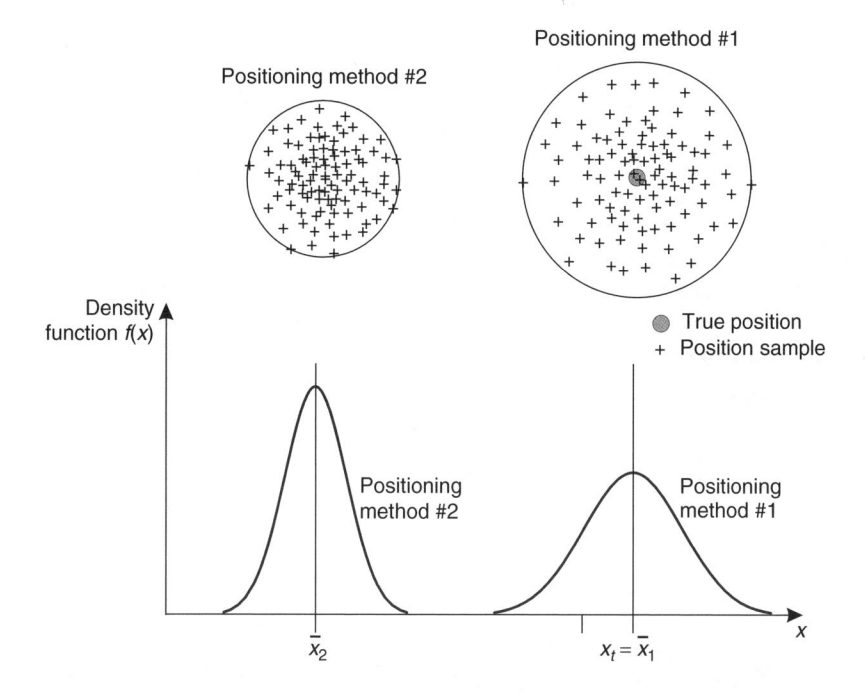

Figure 6.16 Accuracy and precision (modified from (Leick 2004)).

as density functions, one for each method applied. For positioning method #1, the sample mean is at $\bar{x}_1$, which coincides with the true position x_t. The sample mean for method #2 is at $\bar{x}_2$, which is much farther away from x_t. This means that method #1 shows a much higher accuracy than method #2. However, as can also be derived from the figure, the samples of method #2 are much closer, and hence the curve of its density function is taller and more narrow than that of method #1. Thus, method #2 shows a much better precision than method #1, although its accuracy is rather poor.

For accuracy, a number of standards have been defined by various national and international authorities, especially for purposes of surveying and mapping. Most of them are based on the *root mean square* (RMS) or *standard deviation*. The following explanations go back on the standards of the US *Federal Geographic Data Committee* (FGDC) (see (FGDC 1998)).

At first, consider the measure of accuracy in one dimension, that is, on a line. Assuming that x_t denotes the true position of a target on a line, and x_i represents an observed sample for $i = 1, \ldots, n$, the accuracy measure σ_{1D} in one dimension is then given by

$$\sigma_{1D} = \sqrt{\frac{\sum_{i=1}^{n}(x_i - x_t)^2}{n}} \qquad (6.27)$$

Precision can be expressed in a similar way if the true position x_t is replaced by the sample mean $\bar{x}$. Often, accuracy is specified at a certain confidence level, which means that a certain percentage of the measured samples have an error equal or less to the accuracy level reported. For example, the one-σ standard deviation specified in Equation 6.27 corresponds to a confidence level of 68.27%, which means that 68.27% of all samples are centered around x_t in the interval $[x_t - \sigma_{1D}; x_t + \sigma_{1D}]$. The remaining samples lie outside this interval. The so-called two-σ standard deviation results in a 95.45% confidence interval. In most cases, accuracy standards rely on a 95% confidence level, which corresponds to a 1.96-σ standard deviation. The relationship between the standard deviation and some confidence levels is depicted in Figure 6.17.

Table 6.3 shows different accuracy measures for one dimension, which are often used to express the vertical accuracy of a positioning method, that is, the accuracy with regard to elevation. As an example, the table also shows the vertical accuracies provided by GPS for the different measures.

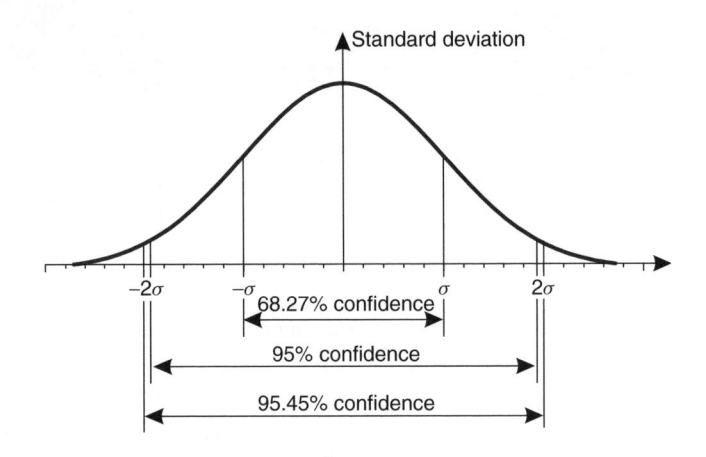

Figure 6.17 Confidence levels for one dimension.

Table 6.3 Accuracy measures

	Confidence level [%]	Relative distance [σ]	GPS accuracy [m]
One dimension			Vertical
One-σ standard deviation	68.27	$1\sigma_{1D}$	±13.8 m
95% probability/confidence	95	$1.96\sigma_{1D}$	±27 m
Two-σ standard deviation	95.45	$2\sigma_{1D}$	±27.7 m
Three-σ standard deviation	99.73	$3\sigma_{1D}$	±42 m
Two dimensions			Horizontal
One-σ standard error circle	39	$1\sigma_{2D}$	±6 m
One deviation RMS	63	$1.414\sigma_{2D}$	±9 m
95% 2-D positional confidence circle	95	$2.447\sigma_{2D}$	±15 m
Two deviation RMS	98	$2.83\sigma_{2D}$	±17.8 m
Three deviation RMS	99.9	$4.24\sigma_{2D}$	±27 m

FUNDAMENTALS OF POSITIONING

Accuracy standards are also specified for two dimensions, which is denoted as horizontal accuracy. In this case, the RMS is composed from the standard deviations σ_N and σ_E of two one-dimensional axes, which span the two-dimensional plane. These axes are commonly referred to as *northing* and *easting*. Here, it is of particular importance whether $\sigma_N = \sigma_E$ or $\sigma_N \neq \sigma_E$. In the former case, the error distribution is the same in both dimensions, and the position samples define a confidence circle around the true position (see Figure 6.18 (a)). In the latter case, the two one-dimensional deviations differ more or less heavily, and hence the samples center in a confidence ellipse around the true position (see Figure 6.18 (b)).

(a) Confidence circles   (b) Confidence ellipses

1DRMS One Deviation Root Mean Square
2DRMS Two Deviation Root Mean Square

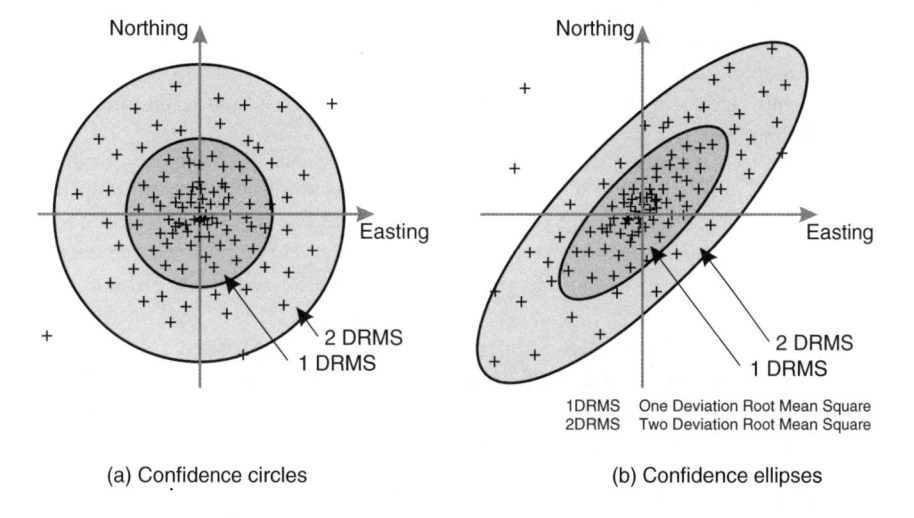

Figure 6.18 Confidence circles and ellipses.

For the circular case, the standard deviation can be composed in the following way:

$$\sigma_{D2} = \sqrt{2\sigma_N^2} = \sqrt{2\sigma_E^2} \qquad (6.28)$$
$$= 1.4142\sigma_N = 1.4142\sigma_E$$

For the elliptical case, the flattening of the ellipse can usually be approximated by

$$\sigma_{D2} = 0.5(\sigma_N + \sigma_E) \qquad (6.29)$$

assuming that the relationship $\sigma_{min}/\sigma_{max}$ is between 0.6 and 1.0 (where σ_{min} is the smaller value of σ_N and σ_E and σ_{max} is the larger one) and the errors of northing and easting are independent (see (FGDC 1998)). The deviation specified in this way corresponds to a 39.35% confidence ellipse. Other confidence levels for horizontal accuracy and GPS examples are listed in Table 6.3.

6.5 Error Sources

Poor accuracy is a result of errors of the measured observables, which have their origin in a number of error sources. Generally, the following error sources can be identified:

- **Clocks.** As discussed in Section 6.3.1, inaccurate and instable clocks of positioning equipment directly lead to errors in the measured ranges and range differences, respectively. Although mechanisms of clock synchronization and mathematical models for eliminating offsets and drifts are used, there always remains a certain error potential. Clocks are an error source for all positioning methods where the ranges and range differences are determined by time measurements.

- **Ionospheric and tropospheric refraction.** Refraction in the Earth's atmosphere is responsible for small deviations of the speed of pilot signals from the speed of light. As stated in Section 4.2.3, problems arise from the fact that these deviations are difficult to predict, since the degree of ionization is a time-varying parameter. Again, it is possible to limit these errors to a certain degree by according mathematical approaches. Ionospheric and tropospheric refraction is of particular importance in satellite systems.

- **Non line of sight.** If line of sight exists between terminal and base station, it can be assumed that the first arriving pulse of a pilot signal has traveled on the line-of-sight path, and subsequent pulses arrive on reflected paths. However, in some cases a line of sight does not exist, and the pulses arrive on reflected paths, which may lead to enormous time delays and subsequent range errors. This is a particular problem for time-based positioning in cellular systems, and is responsible for only moderate accuracy when compared to GPS (which always requires line of sight). Non line of sight also incurs very large errors on AoA position fixes, because a reflected signal may arrive from practically any direction at the base station.

- **Multipath propagation.** As a result of multipath propagation, the different copies of a signal may overlay at the receiver and distort the signal's amplitude and phase (see also Section 4.2.5). This is a particular problem if ranges are determined by the RSS or phase measurements. Also, fingerprints of a certain scene collected for pattern matching may be useless if the configuration of the scene (for example, the arrangement of obstacles) changes.

- **Medium access.** Positioning in integrated systems often suffers from air interfaces that have been designed for communication and therefore suffers from deficiencies when using it for positioning. For example, in FDMA-based system like GSM, a terminal can be received by one base station at a time only, making lateration with at least three range measurements a very tricky matter. CDMA-based systems like UMTS, on the other hand, suffer from a phenomenon known as *hearability*, which leads to similar problems. These particularities will be investigated in Chapter 8. However, note that resolving these problems typically goes along with an additional error potential, which does not occur in stand-alone systems that have been dedicated for positioning.

- **Base station coordinates.** Another error potential results from inaccurate coordinates of base stations. In indoor and cellular systems, this can be simply avoided by accurate surveying. However, in satellite positioning, the general problem is that satellites are exposed to perturbing forces in their orbital motion, which are a result of the nonspherical shape of the Earth, external gravitation, solar winds, and so on. Thus,

FUNDAMENTALS OF POSITIONING

the satellite's true position always slightly deviates from its calculated position as used for lateration and therefore causes degradations of accuracy.

- **Bad geometry.** As already indicated in Section 6.2.6, a bad geometry of base stations involved in a position fix may result in significant range or angle errors. Generally, a bad geometry is given if the base stations are arranged along a line, for example, at highways, or if they are arranged very close to each other. As a consequence, the different radio paths arrive at the terminal at very shallow angles, which increases the error potential. A good geometry is given if these angles are almost right.

Generally, these error sources contribute in different degrees and magnitudes in the different positioning methods and infrastructures. This will be discussed in the following chapters.

Table 6.4 Overview of satellite, cellular, and indoor positioning

Name	Basic method	Mode ta	Mode tb	Mode nb	Type of signal	Measurement	Type of network
Satellite positioning							
GPS	Circ. lat.		×		Radio	Time	
D-GPS	Circ. lat.		×		Radio	Time	
Galileo	Circ. lat.		×		Radio	Time	
Cellular positioning							
Cell-Id	Prox. sens.			×	Radio	Cell-Id (+RTT)	GSM
E-OTD	Hyp. lat.	×	×		Radio	Time	GSM
U-TDoA	Hyp. lat.			×	Radio	Time	GSM
Cell-Id	Prox. sens. (+angul.)			×	Radio	Cell-Id (+RTT + AoA)	UMTS
OTDoA	Hyp. lat.	×	×		Radio	Time	UMTS
E-FLT	Hyp. lat.			×	Radio	Time	cdmaOne/2000
A-FLT	Hyp. lat.	×	×		Radio	Time	cdmaOne/2000
A-GPS	Circ. lat.	×	×		Radio	Time	all
Indoor positioning							
RADAR	Fingerprint.			×	Radio	RSS	WLAN
Ekahau	Fingerprint.	×			Radio	RSS	WLAN
Indoor GPS	Circ. lat.		×		Radio	Time	
RFID	Prox. sens.	×	×	×	Radio	ID	
ActiveBadge	Prox. sens.			×	Infrared	ID	
WIPS	Prox. sens.	×			Infrared	ID	WLAN
ActiveBat	Circ. lat.			×	Ultrasound	Time	418 MHz radio
Cricket	Prox. sens.		×		Ultrasound	ID + time	

6.6 Conclusion

This chapter has introduced the basic positioning methods used for LBSs and identified several criteria for classifying them. Furthermore, it has discussed the pros and cons of each method and identified the major error sources being responsible for accuracy degradations. The following chapters show the realization of these positioning methods in satellite, cellular, and indoor environments. Table 6.4 gives an overview of the different methods implemented in these systems and compares their mode of operation (ta = terminal assisted, tb = terminal based, nb = network based), the type of signals used, the measured or observed parameters, and the type of network the method is used in.

7

Satellite Positioning

The civil use of satellite systems was dominated for a long time by applications like TV broadcasting, telephone backbones, and weather forecasting, whereas satellite positioning had not become popular until the completion of the military-driven GPS in the mid-1990s and its opening for an improved civil use in May 2000. Since then, satellite positioning has been gaining more and more momentum, especially in the areas of car navigation, surveying and mapping, and LBSs.

Starting with a short overview of the historical background, this chapter highlights the technical fundamentals of GPS. To understand the principal working of this system, it necessary to be clear about the orbital constellation of satellite systems in general, and hence the laws of Newton and Kepler and their application for satellite motion in space are briefly introduced in the next section. It is followed by a thorough introduction to GPS, including GPS infrastructure and devices, services and signals, and stages of future expansions planned for GPS. The chapter concludes with a brief introduction to Differential GPS (D-GPS) and an overview of Galileo.

7.1 Historical Background

The historical origin of satellite systems dates back to 1957, during the Cold War, when the Soviet Union launched the first satellite *Sputnik*. Sputnik was designed to broadcast radio signals onto the Earth in order to gather information about the density of the Earth's upper atmosphere. Although its functional range was hence very rudimentary, the launch of Sputnik was an event that shocked the United States and caused the adoption of its own satellite program. This program produced the first US satellite *Explorer I*, which was put into space in 1958 and which was intended to measure cosmic rays. In the ensuing years, both world powers increased their efforts to advance satellite technology and launched a multitude of satellites focusing on different kinds of applications like forwarding telephone calls and television signals, espionage, as well as positioning and navigation.

The first satellite system for positioning, *Transit*, consisted of seven satellites and went into operation in 1964. It was operated by the US Navy. A Transit handheld receiver could

---

Location-based Services: Fundamentals and Operation Axel Küpper
© 2005 John Wiley & Sons, Ltd

derive its position by observing the Doppler shift from signals emitted from the Transit satellites, a process that lasted approximately 15 min for a single position fix – obviously too long for today's applications. Other drawbacks were the temporary unavailability of satellites and the corruption of position estimation when the user traveled at higher velocities, both making Transit impractical for a reliable real-time deployment. Nevertheless, experiences gathered with Transit represented a valuable input for the design of the later GPS.

The activities on GPS were initiated by the US Department of Defense (DoD) in the early 1970s under the term *Navigation Satellite Timing and Ranging System* (NAVSTAR). Primary focus of these activities was the development of a global, all-weather, continuously available, highly accurate positioning and navigation system for military purposes. After considerable research work and various tests, the basic parameters of GPS were fixed. The first of a nominal constellation of 24 GPS satellites was launched in February 1978. In the following years, the GPS program consistently suffered from political and technical issues; one such issue was the accident of the Challenger space shuttle in 1986. Finally, GPS was formally declared ready for full operation by the US Air Force Space Command in 1995.

However, in 1983 the United States, motivated by the bombardment of a Korean airline flight over the territory of the Soviet Union, pronounced to open up GPS for civil use. One of the first civil applications was surveying, followed by aviation and marine. However, the actual potential of GPS for the mass market was not realized before the 1990s when the first low-cost, portable GPS receivers appeared on the consumer markets, for example, as standalone devices for supporting hikers or integrated into vehicles as part of navigation systems.

At this point, it should be mentioned that a satellite-based positioning system was also developed by the Soviet Union and Russia respectively. The GPS counterpart *GLONASS* also consists of nominally 24 satellites and works according to the same principles as GPS. However, unlike GPS, GLONASS has failed in gaining any relevant importance at the consumer market, which is basically due to the fact that the future of GLONASS is generally considered to be rather uncertain. In recent years, several malfunctioning satellites were not replaced in sufficient time, and hence in the meantime the system has shrunk to a number of eight operating satellites or so – too little to guarantee a reliable and worldwide coverage. The situation is made worse by the fact that GLONASS satellites have shown a shorter life expectancy than GPS satellites. Finally, these problems have lead to a lack of trust and interest on the part of potential investors, which, however, would be expected to open up the system for device manufacturers, application developers, and cellular network operators.

In the late 1990s, the European Union decided to launch activities for its own satellite positioning system, which is known as *Galileo*. One obvious reason for this decision was the demand to gain independence from the de facto monopoly of GPS, as many European countries increasingly built up applications on top of GPS, which supported mandatory issues like civil protection, ambulance, fighting crime, and collecting tolls. Besides its utility in LBSs, GPS is also estimated to be an important economical factor in Europe. Another focus is certainly on military applications, although Galileo is officially declared to be a system designed specifically for civil purposes. By its own satellite system, the European Union hopes to fully gain autonomy in satellite positioning without being exposed to decisions made by the United States with regard to GPS quality, reliability, and availability.

According to the latest announcements, the schedule of Galileo foresees to bring 30 satellites into orbit from 2006 to 2008 and to start the commercial operation of the system around 2009. As opposed to GPS, Galileo comes along with various positioning services, differing from each other in their quality regarding accuracy, integrity, and availability. Some of them will be offered free of charge, while others will require the payment of a fee. Another feature of the system is the intended interoperability and compatibility with GPS. This is to be achieved given that Galileo partially operates on the same frequency bands as GPS and by the use of special receivers that are able to observe, analyze, and combine the signals of both types of satellites.

7.2 Orbital Motion of Satellite Systems

The underlying technology of satellite systems is of extraordinary complexity that can hardly be covered in this book. However, there are some general aspects of the orbital motion of satellites that are essential for an understanding of the working of satellite positioning. The two systems covered in this chapter, GPS and Galileo, apply Time of Arrival (ToA), which is based on circular lateration in combination with timing measurements. Recall from Chapter 6 that this requires knowledge about the exact base station coordinates, that is, those of the satellites in this case. While terrestrial base stations are located at fixed points on Earth and hence their coordinates are easy to obtain, satellites continuously circulate the Earth at high velocities. Therefore, knowing where the satellites are at the time of measuring their pilot signals is a crucial point. It is thus necessary to consider the laws that the orbital motion of satellites is based on.

7.2.1 Satellite Orbits

A satellite's motion is described by its orbit, which is a regular, repeating path around the Earth followed by the satellite in certain periods of time. The orbit is arranged on a fixed plane that always passes through the center of the Earth. On the basis of this definition, orbits can be classified according to the following criteria (see also Figure 7.1):

- **Orbit plane.** The orbital plane reflects the position of the orbit with regard to the equatorial plane. It is specified by means of the inclination angle, which denotes the angle between the equatorial plane and the orbital plane. An equatorial orbit always has an inclination angle of 0°, which means that the satellite circulates directly above the equator. Its complement is the polar orbit, where the satellites overfly North and South Poles, and which has an inclination angle of 90°. Anything between equatorial and polar orbit is called *inclined orbit* and needs to be further specified by its inclination angle.

- **Orbit shape.** The shape of the orbit varies from circular to elliptical. If the satellite always keeps the same distance to the surface of the Earth, its orbit has a circular shape with the center of the circle coinciding with the center of the Earth. Alternatively, orbits may have an elliptical shape, where the Earth's center is located at one of the two foci describing the ellipse. In this case, the distance to Earth depends on the satellite's exact position on the elliptical orbit. The point of the ellipse that is farthest from Earth is called its *apogee*, the nearest point to Earth is called *perigee*.

- **Orbit altitude.** Taking into consideration the altitude of satellites, we have *low earth orbit* (LEO), *medium earth orbit* (MEO), and *geostationary earth orbit* (GEO)

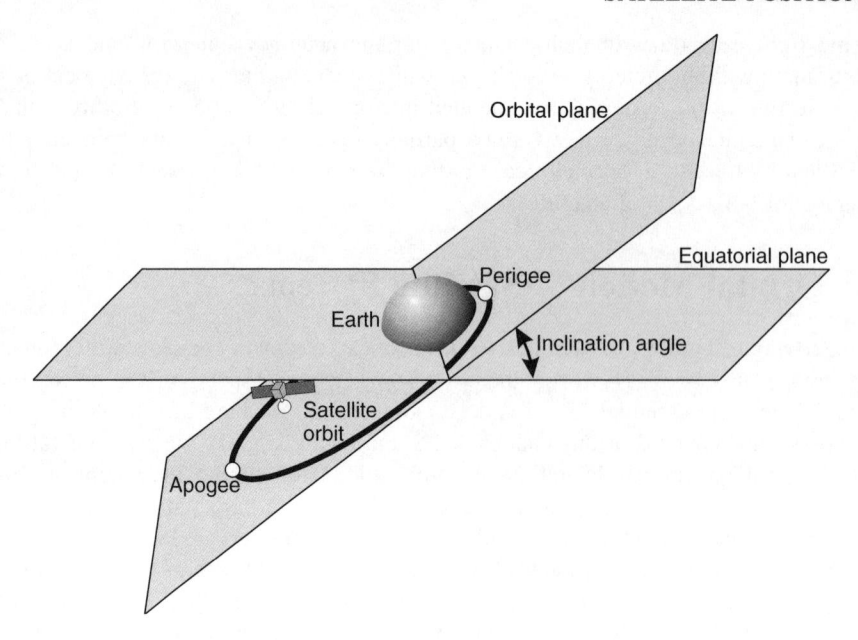

Figure 7.1 Satellite orbits.

satellites. As will be explained in the following text, there exists a fixed coherence between orbit altitude and period. LEO satellites typically have an altitude of under 2,000 km and circulate the Earth in a period in the range of 95 to 120 min. MEO satellites are located at an altitude of 5,000 to 12,000 km and require a period of about 6 h or so to completely circulate the Earth. The altitude of GEO satellites is exactly fixed at 35,786 km. At this altitude, they appear fixed when observed from the surface of the Earth, that is, their period exactly corresponds to the period of Earth rotation of 24 h.

When dealing with the motion of satellites it is important to understand which forces keep them in orbit. It was the English scientist Sir Isaac Newton who first described these forces in his famous laws of motion and gravity in the seventeenth century. Newton's first law of motion states that due to an object's inertia it travels in a straight line at a constant speed unless there is a force acting on it. This means that a satellite circulating the Earth must experience some force that keeps it in orbit, otherwise it would leave the orbit and travel in a straight line away from the Earth. The force that deflects an object from such a straight line is described by Newton's second law of motion and is called *centripetal force*. This force is directed to the center of a circle on which the object travels at a certain point of time, and its magnitude must equal the object's mass multiplied with its centripetal acceleration to keep it on that circle:

$$F_{\text{centripetal}} = m \frac{v^2}{r} = ma \qquad (7.1)$$

In this equation, m and v denote the object's mass and velocity, r is the radius of the circle, and a represents the centripetal acceleration. A special kind of centripetal force

SATELLITE POSITIONING

that applies to satellite motions is the gravitational force of the Earth, which is based on Newton's law of gravity. In general, this law determines the force that acts along the line joining the centers of the mass of two objects that are separated by a certain distance. As an example, consider the gravitational force that appears to an object located at the surface of the Earth:

$$F_{\text{gravity}} = m \frac{G M_{\text{Earth}}}{R_{\text{Earth}}^2} = m g_{\text{Earth}} \quad (7.2)$$

G is the so-called *universal gravitational constant* and its value is $G = 6.67259 \cdot 10^{-11} [\frac{\text{Nm}^2}{\text{kg}^2}]$. The mass of the Earth is given by $M_{\text{Earth}} = 5.98 \cdot 10^{24} [\text{kg}]$ and its radius is $R_{\text{Earth}} = 6.38 \cdot 10^6 [\text{m}]$ on average. The acceleration of gravity g at standard sea level is $g_{\text{Earth}} = 9.8 [\frac{\text{m}}{\text{s}^2}]$.

In order to keep a satellite in orbit one has to establish a balance between its velocity and altitude (see also Figure 7.2). In case of circular orbits, this can be easily done by equating the given gravitational force of the Earth and the centripetal force needed to keep the satellite in orbit:

$$F_{\text{gravity}} = m \frac{G M_{\text{Earth}}}{R_c^2} = m \cdot g_c = m \frac{v_c^2}{R_c} = F_{\text{centripetal}} \quad (7.3)$$

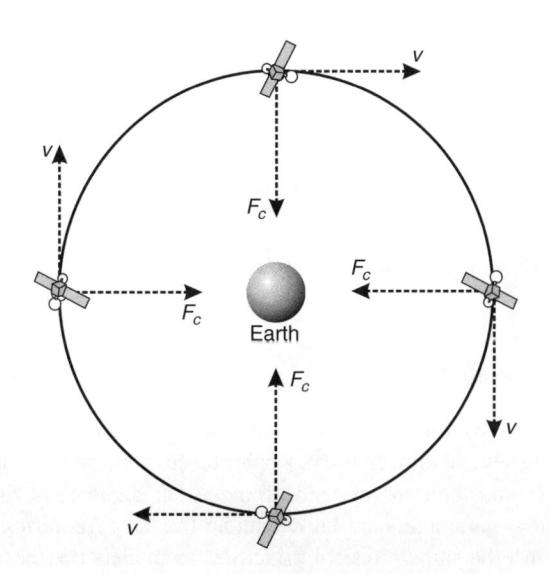

Figure 7.2 Inertia and centripetal force.

This equation can be solved for the radius R_c of the orbit, which yields:

$$R_c = \frac{G M_{\text{Earth}}}{v_c^2} \quad (7.4)$$

Analogously, this equation can be solved for v_c if the radius R_c is given. Thus, it can be concluded that the distance of a satellite from the surface of the Earth depends on the orbit period and vice versa. This is why, for example, GEO satellites operate in an orbit whose height is exactly 35,863 km. At this altitude, the orbit period equals 24 h, which corresponds to a velocity of 3,072.94 m/s, and hence the satellite appears to be stationary if observed from the surface of the Earth. This is useful for many applications like weather or communication satellites.

As stated earlier, Equation 7.3 holds only for circular orbits. To establish an elliptical orbit, the satellite velocity must be modified in a way that it slightly deviates from the velocity where the satellite's inertia and the centripetal force are in balance. As a result of this, a circular orbit changes into an elliptical orbit, which has the following characteristics: first, the distance to Earth is naturally not the same at all points of the orbit. The point of the ellipse that is farthest from Earth is the apogee, and the nearest point to Earth is the perigee. Second, as a result of changing gravity forces along an elliptical orbit, satellites do not travel with constant velocity. A satellite reaches its maximum velocity at the perigee and its minimum velocity at the apogee.

It must be stressed that satellites are always exposed to perturbing forces that stem from the nonspherical shape of the Earth, gravitational attraction of other planets, ocean tides, solar winds, and other phenomena. Therefore, orbits commonly considered to be of circular shape are actually only nearly circular. They are subject to the same characteristics and rules that hold for elliptical orbits. Only a minority of satellite systems are intentionally arranged at elliptical orbits and in order to distinguish them from satellites with nearly circular orbits they are hence often referred to as *highly elliptical orbit* (HEO) satellites.

7.2.2 Keplerian Elements

As stated earlier, for satellite systems in general and navigation satellites in particular it is essential to predict the position of a satellite at a given point in time. For this prediction, it is necessary to express

- the shape and size of the orbit,
- the orientation of the orbital plane, and
- the position of the satellite in the orbit.

The mathematical basis for obtaining a satellite's position in space was founded by Johannes Kepler in the seventeenth century. Kepler discovered three basic laws that describe the motion of planets and their moons and arranged six parameters for determining their position in space. These parameters are known under the term *Keplerian Elements*. Because satellite motion follows the same physical principles as planets do, the Keplerian Elements have become the standard way to express the orbital motion of satellites.

Kepler's first law of planetary motion states that planets always revolve round the sun on elliptical orbits with the sun as the focus. The same law also holds for the motion of satellites. In terms of the Keplerian Elements, the elliptical orbit is defined by the length of the *semimajor axis a* and the *eccentricity e* of the ellipse. The semimajor axis describes the size of the ellipse. It corresponds to one half of the distance between apogee and perigee. Kepler's third law of planetary motion demonstrates how to calculate the orbital period from the length of the semimajor axis.

The eccentricity describes the shape of the elliptical orbit (see Figure 7.3). It defines the ratio between the two foci of the ellipse and the length of the major axis and hence can be

SATELLITE POSITIONING

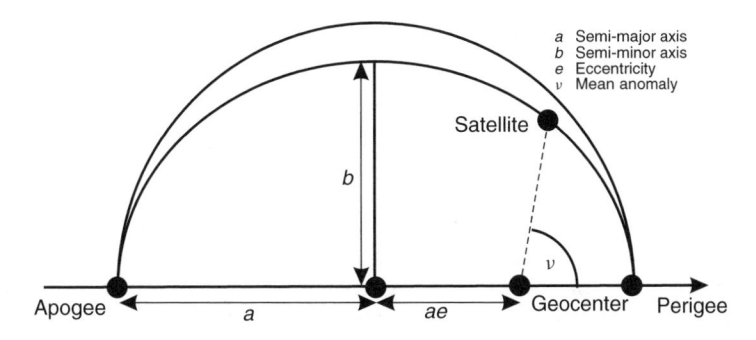

Figure 7.3 Shape and size of orbit.

used to compute the distance between the center of the ellipse and the center of the Earth. For elliptical orbits, the value range of eccentricity is always $0 \leq e < 1$, where $e = 0$ describes an exactly circular orbit, and an eccentricity approaching 1 defines a highly elliptical orbit. Note that orbits with $e = 1$ are of parabolic and those with $e \geq 1$ are of hyperbolic shape, which means that the satellite has overcome the so-called *escape velocity*. At this velocity, the satellite is not exposed to Earth's gravity any longer, that is, it does not revolve round the Earth, but strictly moves away from it along a parabolic or hyperbolic path.

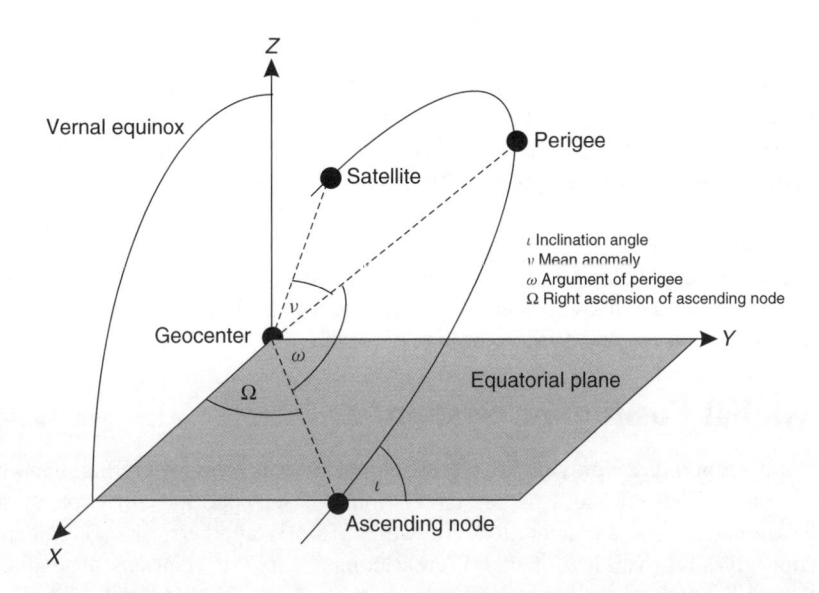

Figure 7.4 Orientation of orbital plane and satellite position.

The orientation of the orbital plane in space is determined by the inclination angle ι, the *right ascension of the ascending node* Ω, and the *argument of the perigee* ω (see Figure 7.4). As stated above, the inclination angle fixes the attitude of the orbital plane with regard to the equatorial plane and is expressed as the angle between both of them.

However, an orbit cannot be unambiguously defined by its inclination angle alone. An important parameter of the Keplerian Elements is therefore the ascending node, which corresponds to the point where the satellite crosses, that is, ascends the equatorial plane when going from the Southern to the Northern Hemisphere.

Unfortunately, the position of the ascending node cannot be expressed by means of the conventional system of latitude and longitude, which is due to the Earth's rotation. Instead, its position is expressed according to the *right ascension system*, which is a fixed astronomical coordinate system used for measuring the position of stars and other celestial objects. The X-axis is the origin of the right ascension system and is fixed by the intersection of the equatorial plane of the Earth (*celestial equator*) and the plane of the Earth's orbit around the sun (*ecliptic*). It is often referred to as the *Vernal Equinox*, which is actually defined as the point on this intersection where the Sun appears to cross the equatorial plane from South to North, which happens each year around March 21. This is why the Vernal Equinox is sometimes defined as the *apparent ascending node* of the Sun. The Y and Z-axes of the right ascension system are then fixed with a right angle to the X-axis.

Having determined a fixed coordinate system in this way, which is independent of the rotation of the Earth, the right ascension of the ascending node Ω is simply defined as the angle between the X-axis and the line that can be drawn from the center of the Earth to the ascending node. The exact position of the ascending node can then be computed from its right ascension and the parameters of the elliptical orbit mentioned earlier. To finally achieve an unambiguous specification of the orbit, the argument of the perigee ω remains to be fixed. It is expressed as the angle between the ascending node, the center of the Earth, and the perigee and measured from the ascending node in the direction of satellite motion.

Finally, the position of a satellite in orbit is specified by the last Keplerian Element, the *mean anomaly* v. Recall from the last section that satellites in elliptical orbits do not travel with constant velocity. The highest velocity appears exactly at the perigee, the lowest at the apogee. The *true anomaly* corresponds to the angle between the perigee and the satellite's position at a particular time, measured in the direction of satellite motion. As suggested by its name, the mean anomaly is based on the average velocity of the satellite and hence reflects its position as if it were in a circular orbit. It is common practice to express a satellite's position by means of its mean anomaly, because it is time independent. The true anomaly can then be derived from it under consideration of the eccentricity of the elliptical orbit.

7.3 Global Positioning System

In its current version, GPS provides two positioning services, one for civil and another for military purposes. These services are based on terminal-based code and carrier phase ranging in combination with circular lateration, for which ToA is another term (see Chapter 6). This section gives an overview of the different components GPS consists of, explains the modulation and structure of pilot signals and assistance data emitted by the satellites, gives an introduction into the properties of the GPS services, and describes the different variants of positioning with GPS.

7.3.1 GPS Segments

The GPS system consists of three segments: the space segment, the control segment, and the user segment. The space segment consists of nominally 24 satellites that circulate the

SATELLITE POSITIONING

Earth every 12 sidereal hours on six orbits. The next section will describe this segment in detail.

The control segment is responsible for monitoring and controlling the satellites of the space segment. It consists of five ground stations around the world, which are located at Colorado Springs, Hawaii, Ascension Island, Diego Garcia, and Kwajalein (see Figure 7.5). These locations have been selected in a way that each satellite can be observed and controlled 92% of the time. All ground stations are *Monitor Stations* that are equipped with GPS receivers for monitoring and tracking the satellites. They control the health of each satellite, their orbits as well as the accuracy of their internal clocks. The data measured by these stations is then passed to the central *Master Control Station*, which is located at Colorado Springs. This station processes the received control data and calculates clock corrections and satellite ephemerides. Furthermore, it centrally coordinates all operations concerning the space segment, ranging from corrections of satellites orbits, clock keeping, and signal encryption. For each satellite a dedicated message with control data is assembled and returned to one of three *Uplink Stations*, which then transmits the control message to the respective satellite. These Uplink Stations are colocated at the Monitor Stations at Ascension Island, Diego Garcia, and Kwajalein.

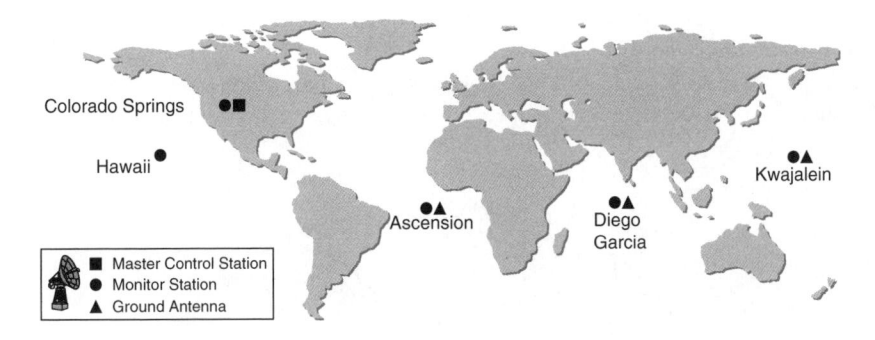

Figure 7.5 GPS control stations.

The user segment is the part of the GPS system the user is most aware of. In the literature, it is often simply equated with a GPS receiver, which is, however, only a part of the user segment. Its configuration actually depends on the application it is used for. Typical areas of deployment are land, sea, and air navigation, surveying and mapping, military applications, and, most important in the context of this book, LBSs. Each application comes along with very different requirements on the user segment, including the range of its functions, accuracy of position data, the TTFF, battery consumption, and so on. For example, navigation applications often require a comparatively modest accuracy, usually in the range of tens of meters, but have strong demands on the TTFF. On the other hand, applications of surveying and mapping require much more accurate position data, but have less priority on the speed with which this data is obtained. In addition, the equipment used in this area usually consists of dedicated, high-cost devices that are all but small-sized and manageable.

For LBSs, the situation is much more diversified. First, the range of applications is much broader, including mobile marketing, community services, health care, and navigation. This manifoldness imposes very different requirements on a positioning service. Second, GPS makes LBSs accessible for the mass market consisting of tens of millions of subscribers. A key prerequisite for a successful mass-market introduction is the availability of GPS receivers that can be integrated into or at least attached to mobile devices and that have only a low power consumption. However, these prerequisites may lead to compromises with regard to accuracy and speed of position acquisition.

7.3.2 Satellite Constellation

A constellation of satellites describes the number and positions of satellites in space. To completely cover the whole surface of the Earth with positioning services, it would be sufficient to operate GPS with a minimum number of 21 satellites. However, the operational GPS constellation consists of 24 satellites, and in general even more of them are launched in order to be able to immediately replace older or malfunctioning ones.

The satellites circulate the earth on six orbits with nominally four satellites per orbit (see Figure 7.6). The orbits are equally spaced 60° apart from each other, and each orbit has an inclination angle of 55°. GPS is a typical MEO system with an orbit altitude of approximately 20,200 km with regard to mean sea level. This altitude has been selected such that each satellite repeats the same ground track exactly each *sidereal day*. Note that a sidereal day exactly corresponds to the time it takes for the Earth to turn 360°. It lasts 23 h, 56 min, and 4.09 s and is thus slightly shorter than a solar day, which is defined as the time interval between two consecutive crossings of the sun at a local meridian.

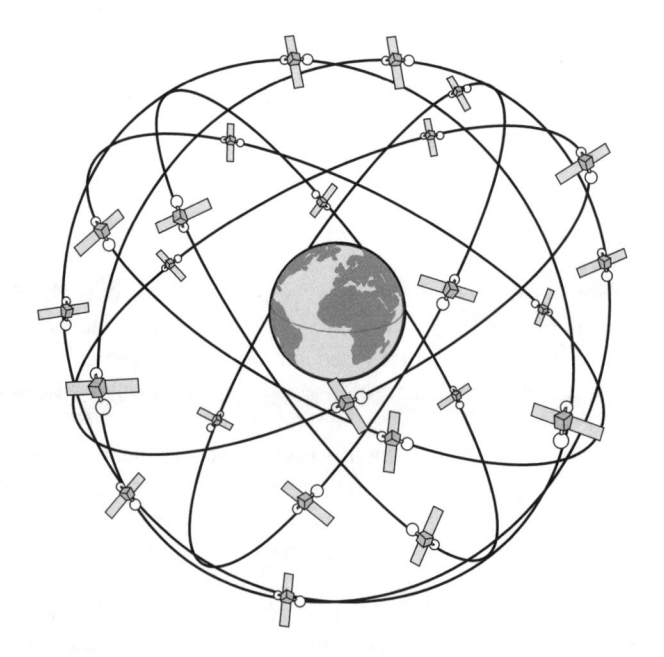

Figure 7.6 GPS constellation.

Given this constellation, it is guaranteed that a GPS receiver is covered by at least four satellites at any point on the Earth's surface, the required minimum number for estimating its position in 3D. However, in most cases, there are between five and eight satellites visible from any point on the Earth, which significantly improves the accuracy of lateration.

Satellites are officially termed *Space Vehicles* (SV) in GPS and are numbered serially by a *Space Vehicle Number* (SVN). The 50th GPS satellite was launched in March 2004. Since the very beginning of GPS, different generations of them have been deployed, referred to as blocks. These blocks are:

- **Block I.** The prototype phase of GPS began in February 1978 when the first Block I satellite was launched. Opposed to later generations, the orbits of Block I satellites had an inclination angle of 63°. They were initially designed for a life cycle of five years, although some of them were operated much longer. Block I comprised a total of 11 satellites (SVNs 1 through 11). In the meantime, all of them were retired, the last one was decommissioned in 1996.

- **Block II.** Block II satellites (SVNs 13 through 21) were the first full-scale operational satellites, and the first of them went into operation in February 1989. They have been designed for a life cycle of 7.5 years and can maintain operation for 14 days without contact from the control segment.

- **Block IIA.** A slightly modified version of Block II satellites was launched between November 1990 and November 1997. These satellites (SVNs 23 through 38) enable 180 days of operation without contact from the control segment. However, this autonomy must be paid with a perceivable reduction of accuracy of position estimation.

- **Block IIR.** At the time of writing this book, Block IIR is the latest version of satellites that are launched. In contrast to their predecessors, they are capable of satellite to satellite communication and are equipped with an autonomous navigation mode for operating up to 180 days without contact from the control segment. They have been designed for an increased life cycle of around 10 years.

- **Block IIF** and **Block III.** These are the future generations of GPS satellites. Again, they have an extended design life compared to their predecessors. Further improvements concern faster processors with more memory, and a new civil signal on a third frequency. Block IIF satellites are scheduled to go into operation in 2005, Block III satellites around 2010.

7.3.3 Pilot Signals and Spreading Codes

As mentioned earlier, GPS applies a terminal-based positioning method based on circular lateration for which the terminal must measure the ranges to the satellites and know the satellites' positions at the time the ranges are measured. Both are enabled by the pilot signals emitted by the satellites. First, they carry two different *ranging codes* that are used to determine the signal traveling time and, taking into consideration the velocity of light, the satellite–receiver range. Second, the signals carry a so-called *navigation message*, which contains assistance data like satellite orbits, clock corrections, and system parameters, and thus enables the receiver to compute the position of each satellite at the time of signal emission. To understand the combination of ranging codes and navigation messages, it is useful to have a closer look at the pilots' parameters and their modulation.

The raw pilot signals are given by two unmodulated carrier waves that are centered on two frequencies of the electromagnetic spectrum. The carrier frequencies have been selected in such a way that the signals are as robust as possible against ionospheric propagation effects and weather conditions. They are generated as multiples of the fundamental frequency $f_0 = 10.23$ MHz of the satellites' clocks. The carriers are termed L1 and L2, and their frequencies are given by $f_{L1} = f_0 \times 154 = 1,575.42$ MHz and $f_2 = f_0 \times 120 = 1,227.6$ MHz. Figure 7.7 gives an overview of the present GPS frequency allocation and also shows the new signals that will be available for advanced positioning services after a modernization of GPS in a few years.

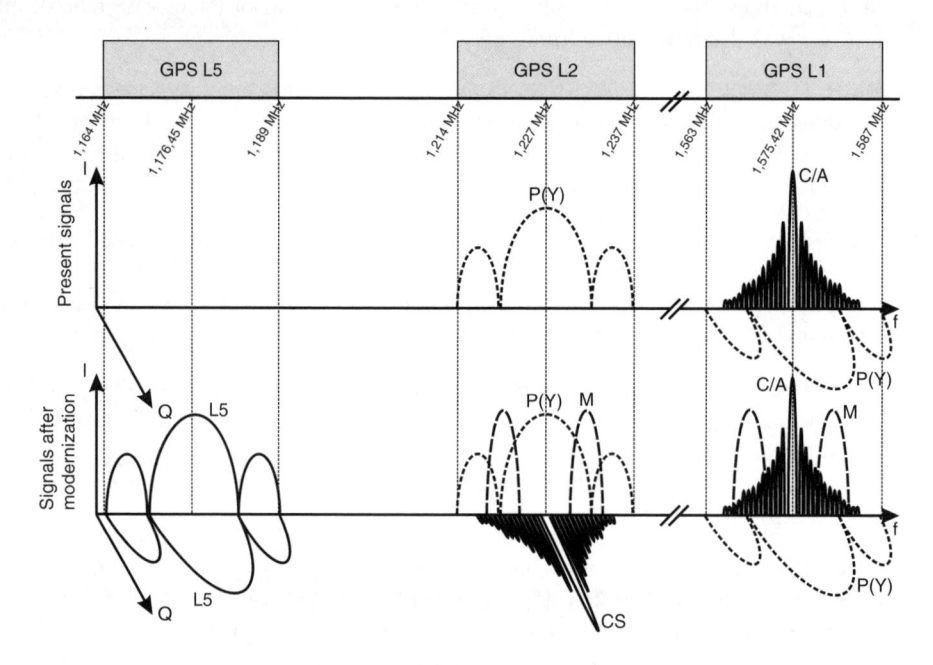

Figure 7.7 GPS frequency spectrum.

These carrier signals are pure sine waves so far, and therefore must first be prepared accordingly for enabling range measurements and transferring messages. In GPS, the range is determined by code phase ranging (see Chapter 6), which is based on the spread spectrum technique (and optionally by carrier phase ranging). The signals are modulated according to a spreading code that spreads the signal over a much wider frequency range than actually required to transmit the bare information that is being sent. Recall from Chapter 4 that as a result of spreading, the signals are less susceptible to distortion and jamming than conventional narrowband signals. This significantly improves the reliability of the GPS service. GPS also utilizes spread spectrum for separating the transmissions of different satellites, that is, GPS uses CDMA as a multiple access scheme.

The official GPS term for spreading code is *Pseudorandom Noise code* (PRN code). There are two kinds of these codes, which are modulated onto L1 and L2 respectively: the *coarse acquisition codes* (C/A-codes) and the *precise code* (P-code). The reason for using two different PRN codes for range measurements can be derived from their names. While the C/A-codes enable only a moderate level of accuracy, the P-code yields a much higher one, which benefits the results of position estimation. The C/A-code is modulated onto the L1 carrier and can be accessed by both civilian and military users. On the other hand, the P-code is transmitted by both the L1 and L2 carriers and can only be accessed by the United States and allied military as well as the US government.

A C/A-code consists of 1,023 chips and is repeated every millisecond, which results in a required bandwidth of 1 MHz. Owing to this chipping rate, it can be concluded that a single chip has a length of 300 m and that the entire C/A-code repeats itself every 300 km between the satellite and the receiver. Besides their usage for code phase ranging, C/A-codes are also the underlying code sequences for CDMA. Each C/A-code is a Gold sequence that has been built according to a certain configuration of linear feedback shift registers. Consequently, all pairs of C/A-codes have a good cross correlation, which predestines them for CDMA. Each satellite is assigned its individual C/A-code and because of the good cross-correlation properties the receivers can distinguish between pilot signals transmitted from the different satellites. Each C/A-code is identified by a *PRN-number*, which is often used to reference a satellite in place of the SVN.

Unlike C/A-codes, the P-code is a very long code consisting of approximately 10^{14} chips. It is simultaneously modulated onto the L1 and L2 carriers with a chip rate of 10.23 MHz, which means that its resolution is 10 times higher than that of C/A-codes. Given the code length and the bandwidth, it can be concluded that the P-code does not repeat for 38 weeks. However, there exists only a single P-code that is shared between all satellites in that each of them is assigned a certain one-week segment of it. The different code segments are then repeated every week by the satellites, always restarting with the beginning of a new week. From all 38 code segments, 32 are intended for use by satellites while the remaining ones are reserved for other purposes, for example, ground transmitters.

Figure 7.8 shows the modulation process. Before modulating the C/A-code onto the L1 carrier, it is first mixed with the navigation message that contains the assistance data for supporting position estimation at the receiver. This is done by a simple modulo-2 operation. To eliminate errors during the transmission as far as possible, the navigation message is modulated with a very high spreading factor of 20,000, leading to a data rate of 50 bps.

The mixed sequence of C/A-code and navigation message together with the P-code is then modulated onto the L1 carrier using binary PSK, which changes the signal's phase by 180° every time the state of the codes changes. C/A-code and P-code can be transferred on the same carrier in that one code is modulated in phase and the other in quadrature, that is, they are shifted by 90° from each other. The L2 carrier is modulated by the P-code only, also by using binary PSK.

For an extension of GPS within the next few years, it is foreseen to establish two additional civil signals, one on the L2 carrier, which is called *civil signal*, and the other on a new carrier denoted as L5 (see Figure 7.7). In this way, it will be possible to perform ranging measurements on two carriers (dual-frequency ranging) and thus to eliminate ionospheric delay. At present, this option is left to military users, which have access to both the L1 and

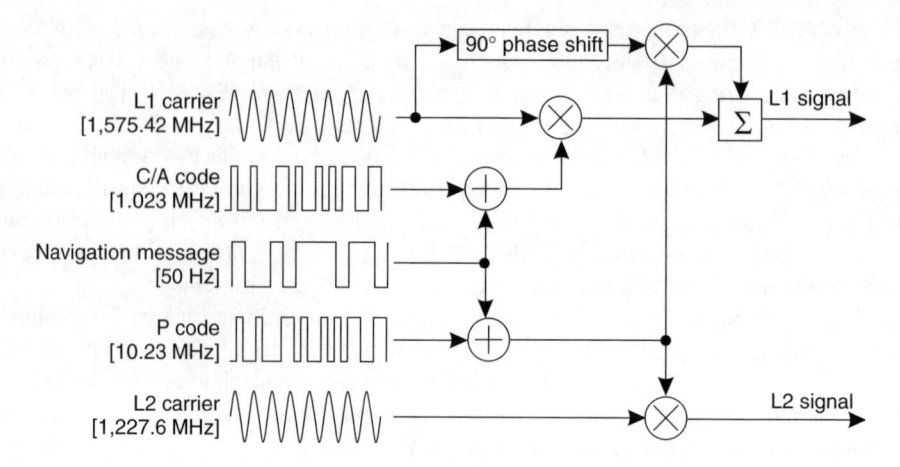

Figure 7.8 Pilot signals and codes.

L2 carriers. Furthermore, it is intended to introduce another military signal, the so-called *M-signal*.

Further details about GPS pilot signals and code sequences can be obtained from (GPS 1995) and (GPS 2000).

7.3.4 Navigation Message

The transmission of the navigation message from the satellites to the receiver is organized in a frame structure that is depicted in Figure 7.9. A frame consists of five subframes, which, in turn, comprises 10 words, each with a length of 30 bits. Thus, the entire frame length is 1,500 bits. As the navigation message is mixed with the C/A-code with a data rate of 50 bps, it needs 30 s to transmit an entire frame, or, in other words, frame transmissions are continuously repeated every 30 s.

The navigation message contains all information necessary for a real-time position estimation at the receiver, especially the position of satellites in orbit, timing information, and data indicating the state of a satellite, the so-called *SV health*. The different information elements are distributed over the five subframes as shown in Figure 7.9.

Each subframe begins with a *telemetry* and a *handover word* (TLM and HOW). The former starts with a preamble of eight bits that indicates the beginning of a new subframe and which is used by the receiver for synchronization purposes. In addition, it carries information about the recent operations that have been performed on the transmitting satellite by the control stations, for example, the upload of a new ephemeris. The HOW contains the so-called *time-of-week* (TOW) of the subsequent sequence, which is required for synchronization with the P-code and which is thus relevant only to users capable of observing the P-code.

The first subframe carries the current GPS week number, the health of the transmitting satellite, and clock correction data. The health gives information about the state of the satellite's transmitted navigation data and signals. For example, it indicates whether navigation data is corrupted, the subframes that are affected by corrupted data, and if the satellite is or will be temporarily out. Investigating the health information, the receiver can thus decide

SATELLITE POSITIONING

Subframe #1	TLM	HOW	GPS week number, SV accuracy and health, clock correction data
Subframe #2	TLM	HOW	Ephemeris data
Subframe #3	TLM	HOW	Ephemeris data
Subframe #4	TLM	HOW	Almanac and health data SVs 25–32, ionospheric data, UTC data
Subframe #5	TLM	HOW	Almanac and health data for SVs 1–24, alamanac reference time, week number

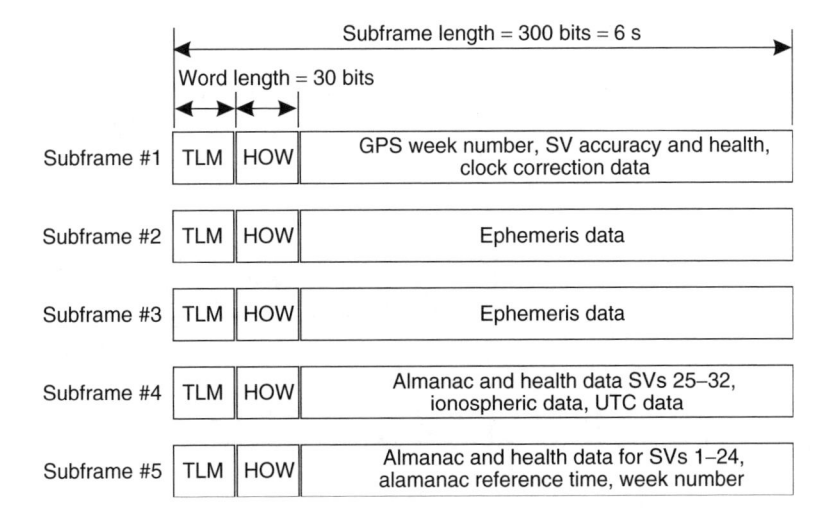

Subframe length = 300 bits = 6 s
Word length = 30 bits

Figure 7.9 GPS navigation message: frame structure.

whether to use navigation and measurement data from this satellite for position estimation. Clock correction data has the purpose of informing the receiver about the amount of the drift of the satellite's clock with regard to GPS time. The correction is specified by means of polynomial coefficients, which are used by the receiver to compute the exact GPS time.

The *ephemeris* of the transmitting satellite is distributed over subframes 2 and 3. It contains all information needed by the receiver to compute the exact satellite position in space when performing measurements. The ephemeris does not reflect the satellite position at the time of transmission. Instead, it has been assembled at a certain reference time and reflects the valid satellite position at exactly that time. The receiver can then estimate the current position taking into consideration the difference between current and reference time. The ephemeris includes the Keplerian parameters, the valid reference time, the satellite's angle velocity, as well as correction parameters that describe the perturbing forces on the satellite. The correction parameters describe deviations of the satellite's latitude, the radius of its orbit, and its angle of inclination over time and must therefore be considered when calculating the current satellite position.

Subframes 4 and 5 contain the *almanac* and health data for all GPS satellites, the almanac reference time, and week number, as well as data for computing the ionospheric delay and relating GPS time to UTC. The almanac is a subset of each satellite's ephemeris and clock data. It contains the Keplerian elements and angle velocity, but does not include the correction parameters described earlier. The almanac helps to speed up the start-up time of the GPS receiver since it enables the obtaining of a rough overview of the current satellite constellation when the receiver is turned on, which replaces the time-consuming identification by means of C/A-codes (which will be described later). Parameters of the ionospheric delay can be used by the GPS receiver to reduce the timely impacts of a signal when passing the ionosphere.

The parameters associated with all these information elements represent a considerable amount of data that does not fit in the fourth and fifth subframes of a single frame. Therefore, their transmission must be distributed over a number of 25 consecutive frames as illustrated in Figure 7.10. The 25 frames connected in that way are referred to as *masterframe*. Within such a masterframe, the subframes 1, 2, and 3 are repeated in each frame, while subframes 4 and 5 of a certain frame carry only a portion of the entire almanac and the other data mentioned earlier. A masterframe has a duration of 12.5 min, and it accordingly lasts that period of time until the GPS receiver has caught a complete navigation message.

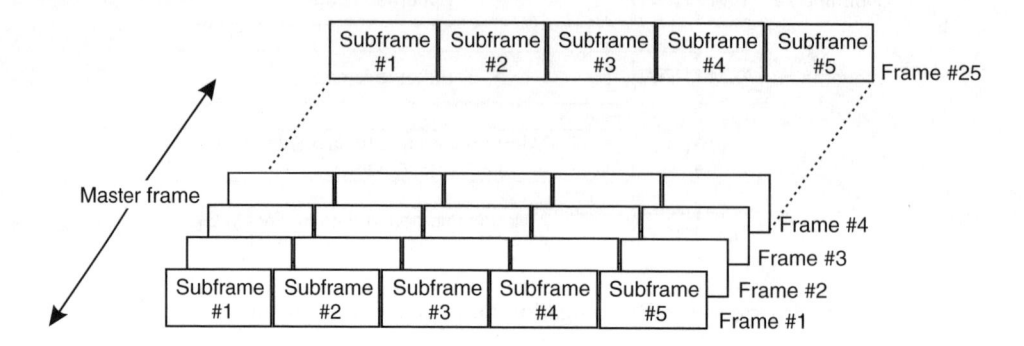

Figure 7.10 GPS navigation message: masterframe structure.

Further details about the structure and content of the GPS navigation message can be received from (GPS 1995) and (GPS 2000).

7.3.5 GPS Services

The are two basic services offered by the GPS system: the *Standard Positioning Service* (SPS) and the *Precise Positioning Service* (PPS). Though both can be requested from all over the world at each time of day without being charged, they differ from each other in the accuracy of delivered position data, the features associated with them, and the groups of users they address.

The SPS is a positioning and timing service focusing on the civilian user. It is based on the C/A-code transmitted at the L1 carrier and the navigation message it transfers. Initially, it was designed to accomplish a predictable position accuracy of 100 m horizontally and 156 m vertically (both with regard to a 95% confidence level). The PPS is a positioning, velocity, and timing service for military applications. It is based on observations of both the C/A and the P-code transmitted on the L1 and L2 carriers. Accordingly, the predictable position accuracy is within at least 22 m horizontally and 27.7 m vertically. The usage of PPS is restricted to authorized users (e.g., United States and allied military and US government) and requires special devices for deciphering and interpreting the P-code.

During the development of GPS it was initially intended to accomplish the different accuracy levels associated with SPS and PPS by various chip lengths of the underlying codes, and hence the C/A-code was designed for rates of 1 Mcps and the P-code for rates of 10.23 Mcps. However, it turned out very soon that SPS users could accomplish accuracy levels very close to that of PPS users. Therefore, the DoD decided to introduce a feature called *Selective Availability* (SA), which made it possible to intentionally degrade the

SATELLITE POSITIONING 171

accuracy of the SPS up to the initially intended level on a global basis. The official reason as given by the DoD was to protect security interests of the United States and allies by denying the full accuracy of the SPS to potential adversaries. SA was implemented in March 1990 on all Block II satellites. The degradation of accuracy was accomplished by falsifying the orbit data of satellites carried in the navigation message as well as by manipulating the satellites' clock frequency. These alterations were done using a well-defined encryption so that it was possible for authorized users of the PPS to remove them.

In May 2000, the President of the United States directed the DoD to discontinue the SA feature on a permanent basis. Instead, it is now intended to deploy mechanisms that, if necessary, will restrict the denial of the civil usage of GPS to a certain region, without affecting the accuracy or availability of the SPS in other parts of the world. As a result of SA deactivation, the SPS users experience a much higher degree of accuracy now, which is specified in the latest GPS SPS Performance Standard (GPS 2001) and which is listed in Table 7.1.

Table 7.1 Positioning and timing accuracy standard

Accuracy standard	95% Vertical error	95% Horizontal error
Global average	≤ 22 m	≤ 13 m
Worst site	≤ 77 m	≤ 36 m

In this table, it is distinguished between the global average and the worst site positioning accuracy, both with regard to a so-called *service volume*. This service volume comprises all points between the surface of the Earth and an altitude of 3,000 km. This means that the SPS user can act on the assumption to obtain at least the worst case accuracy, assuming that he invokes the service from any point of the service volume. However, it must be stressed that the SPS Performance Standard has been specified according to system capabilities the US government has committed to operate GPS with. It does not reflect contributions from outside the system, for example, ionosphere, multipath, or interferences, which have negative impacts on accuracy. On the other hand, the technology of GPS receivers has progressed in recent years, and hence many devices accomplish a level of accuracy that is even much higher than that indicated by the values in Table 7.1.

7.3.6 GPS Positioning

Basically, positioning in GPS comprises three steps: identification of satellites, range measurements, and position calculation. These steps will be sketched in the following text.

Identification of satellites. In the first step, the receiver must identify the satellites to be used for the measurements. The usual number of satellites being in the visible horizon of the receiver is typically in the range of 5 to 10. The procedure used for identifying them depends on the state of the receiver. If it has no information about its last position and no almanac, it must listen to the transmissions on the L1 carrier and compare the received C/A-codes with the entire C/A-code space by autocorrelation. In this way, the associated PRN-numbers are obtained, and the visible satellites are identified. This procedure is called *cold start-up*. On the other hand, if almanac and last position are known, the receiver can perform a *warm start-up*. Having this information, it can roughly estimate the current satellite constellation, which helps to reduce the search space of C/A-codes and thus the time needed for identification. Finally, this time can be further reduced if the receiver holds the valid ephemeris, which allows for calculating the current satellite constellation very accurately, and hence in most cases the identification by means of C/A-codes can be avoided. This procedure is called *hot start-up*. In general, the start-up procedure is complicated in that the receiver must first adjust the Doppler shift the L1 and L2 carriers undergo, which is in addition a time-consuming process that increases the entire acquisition time.

From all satellites identified in this way, the receiver selects a subset of at least four satellites that are actually considered during the measurements. This selection depends on the geometry between each satellite and receiver, which can be quantified by means of the *Dilution of Precision* (DoP). This will be explained in the next section. The duration needed for the different kinds of start-ups heavily depends on the capabilities of the used receiver. Typical durations of low-cost receivers are 40–60 s for cold start-up, 30–40 s for warm start-up, and 5–15 s for hot start-up.

Range measurements. There are different methods for measuring ranges in GPS. The common procedure of conventional GPS receivers for SPS is to apply code phase ranging using the spread pilot signal on the L1 carrier. On the other hand, equipment designed for PPS uses code phase ranging on both carriers (*dual-frequency ranging*) and can thus adjust much better to ionospheric refraction, which significantly increases accuracy of the position fixes. Finally, it is also possible to increase accuracy by performing carrier phase ranging in addition, which, however, is very time consuming and therefore not applicable for real-time applications. Figure 7.11 depicts the GPS measurement procedure.

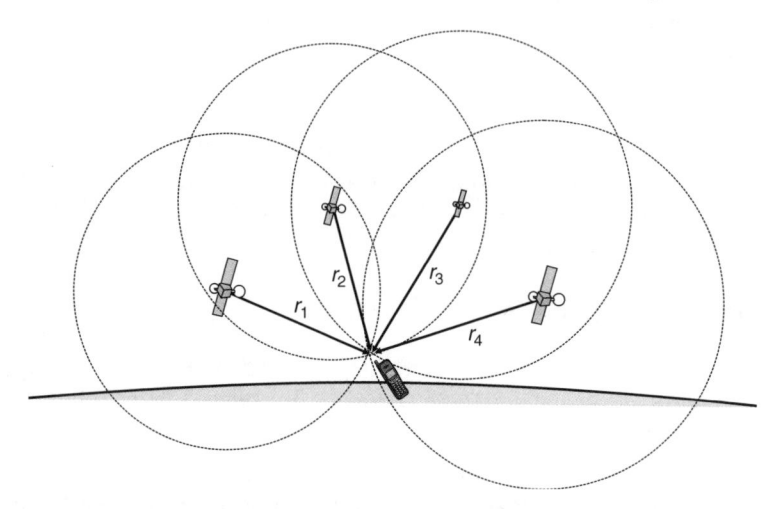

Figure 7.11 Positioning in GPS.

As can be derived from this figure, range measurements must be made to at least four satellites, three for obtaining the position in 3D and the other for time synchronization between the satellites and the receiver. As GPS applies circular lateration, it requires very exact time synchronization (see Section 6.2.2.1). GPS receivers are equipped with low-cost quartz-crystal clocks, which suffer from a considerable drift against the very precise atomic clocks of satellites. The idea for compensating the resulting time offset is to model it as an additional unknown in the system of equations given in Equation 6.2. As a consequence, four equations and thus four range measurements are needed in order to achieve a solution. However, this method is based on the assumption that the time offset of a receiver has the same amount for all satellites, which requires that all satellites operate perfectly synchronized.

As indicated in Section 6.3.1, code phase ranging suffers from integer ambiguity. Compared to the satellite altitude of approximately 20,200 km, C/A-codes have a very short length of 300 km. As a consequence, all measured distances between receiver and satellite appear to have a range of 0–300 km, since the receiver cannot discriminate between different code cycles. However, the integer components of the range, in terms of multiples of the C/A-code length, can be assumed to be constant, as the altitude of the receiver will not vary by more than 300 km anyway (at least for the majority of applications). In contrast to this, integer ambiguity does not appear for measurements based on observing the P-code because it takes one week to transmit the entire P-code segment assigned to a satellite.

Carrier phase ranging is applied to yield very accurate results (theoretically up to the millimeter level), but must be paid with a clearly increased complexity. The increased accuracy results from the comparatively short wavelength of the carriers, which is 19 cm for L1 and 24 cm for L2 carrier signals. Accuracy can be further refined if the phase of the incoming signal can be detected. Modern receivers accomplish a measurement resolution of 1–2% of the wavelength, which leads to a theoretical accuracy in the millimeter range. Unfortunately, accuracies of these degrees must be paid for by the use of very complex and expensive measurement equipment. The major problem of carrier phase ranging is integer ambiguity. As the wavelength of the pilot signals is very short, it is very difficult to discriminate between different signal cycles, that is, to determine the number of cycles between satellite and receiver. To cope with this problem, some receivers use the C/A-code transmissions to determine a rough estimate of the integer number of cycles, which reduces integer ambiguity up to an uncertainty of 100 cycles or so. Resolving this remaining uncertainty is a very complex and time-consuming process. It requires the observation of the transmissions of a satellite over a period of some minutes without interruption and to derive the exact number of cycles by observing the changes in the signals' Doppler shift. Therefore, carrier phase measurements primarily focus on applications in the area of mapping and surveying and are basically not suited for LBSs.

Finally, it must be stressed that the main assumption for performing ranging in GPS is that there is a line of sight between the receiver and the satellites. This is due to the fact that the signals emitted by the satellites hardly penetrate obstacles. While it is sometimes difficult to derive position fixes even outdoors when staying under trees and surrounded by a number of large buildings, it is nearly impossible when staying inside a building. However, in recent times, receiver technology has progressed, and hence modern receivers with high sensitivity can even receive GPS signals with a sufficient strength inside buildings and interpret it. This method is also known as Indoor GPS.

Position calculation. The ranges measured to the satellites are actually pseudoranges, which differ from the true geometric ranges due to a certain error budget. As will be explained in the next section, the largest portion of this budget is ionospheric refraction, which is responsible for slowing down the signals from the satellite and thus for a rough falsification of measured ranges. In order to compensate for these errors, the satellites transmit the coefficients of a correction formula, which the receiver applies on the measured ranges to achieve more accurate results.

In a next step, the receiver must determine the coordinates of each satellite considered during ranging. For each satellite, it extracts the ephemeris (i.e., Keplerian Elements and associated parameters for compensating perturbing forces) from the navigation message and calculates the satellite's position at the time of signal emission. The coordinates of this position are given in the Cartesian ECEF system and serve as input parameters for performing the least square fit according to Equations 6.6 to 6.16. As mentioned earlier, the time offset is modeled as an additional unknown, and hence the system of equations for satellite $i = 1, \ldots, 4$ appears as follows:

$$p_i = \sqrt{(X_i - x)^2 + (Y_i - y)^2 + (Z_i - z)^2} + c\Delta t \tag{7.5}$$

where p_i is the corrected pseudorange, (X_i, Y_i, Z_i) describes the coordinates of satellite i, (x, y, z) represents the wanted coordinates of the receiver, c is the velocity of light, and Δt is the time offset between the receiver clock and GPS time.

After the least square fit, the receiver's position is given in ECEF. In a final step, the position fix is usually converted to an ellipsoidal coordinate system based on WGS-84. Some receivers also offer position fixes for many other local datums. Furthermore, a rough estimation of the degree of accuracy can be determined, which together with the position fix can be indicated to the user.

7.3.7 GPS Error Budget

From the error sources listed in Section 6.5, basically all apply for GPS (apart from medium access), but with different quantities. There are two classes of errors that have an impact on the accuracy of position fixes:

- *User Equivalent Range Error* (UERE) and

- *Dilution of Precision* (DoP).

UERE represents the accuracy of single range measurements to each satellite. It is subject to a number of error sources, which is known as the *GPS error budget*. The UERE is calculated as the square root of the sum of the squares of various error magnitudes caused by these sources. An overview of the more important error sources and their magnitudes for different GPS services is listed in Table 7.2.

As can be derived from this table, the former SA feature, which was canceled in the meantime, was responsible for the highest portion of accuracy degradation for SPS. Today, accuracy for SPS is mainly degraded by ionospheric refraction, which slows down pilot signals when passing the ionosphere. As explained in Section 4.2.3, the amount of delay

Table 7.2 Example of observed GPS error budget (National Research Council 1995)

Error source	Range error magnitude [m, 1σ]		
	SPS (SA on)	SPS (SA off)	PPS
Selective availability	24.0	0.0	0.0
Ionospheric error	7.0	7.0	0.01
Tropospheric error	0.7	0.7	0.7
Clock and ephemeris error	3.6	3.6	3.6
Receiver noise	1.5	1.5	0.6
Multipath	1.2	1.2	1.8
Total user equivalent range error	25.3	8.1	4.1
Typical horizontal dilution of precision	2.0	2.0	2.0
Total horizontal accuracy (2σ)	101.2	32.5	16.4

is a function of the signal's frequency and the density of free electrons. Unfortunately, the latter is not a constant value but varies as a function of the receiver's latitude, the season, and the time of day the range is measured. According to (Rizos 1999), ionospheric delay can be responsible for range errors from about 50 m for signals at the zenith to as much as 150 m for measurements made at the receiver's horizon. To reduce this impact, each satellite executes a rough estimation of the expected ionospheric delay and reports it to the receiver in the navigation message for correcting the measured ranges. However, owing to the dynamic character of ionization mentioned earlier, there remains a certain amount of error potential as listed in the table.

A more effective way to eliminate ionospheric impacts is to use dual-frequency receivers that observe both the L1 and L2 carriers. Since the ionospheric delay imposed on a signal is a function of frequency, it is possible to nearly entirely remove the delay by applying a linear combination on the measurements of both carriers. However, so far, this option is reserved for PPS users only, as it requires decryption of the P-code. For the next generation of GPS satellites, it is envisaged to establish an additional signal, the L5 carrier, which can be accessed by civil users, in addition to L1, for removing ionospheric delays in a similar manner as PPS users.

Furthermore, a signal is exposed to a tropospheric delay. Unlike ionospheric delay, its degree is not a function of frequency, but of pressure, water vapor pressure, and temperature, and hence it cannot be eliminated by observing signals at different frequencies. Other error sources are clock and ephemeris errors, which cannot entirely be removed, the provision of correction data notwithstanding.

DoP is an indicator for an advantageous or disadvantageous geometry of satellite constellation, which can be another important error source. In general, a greater angle between the satellites observed during measurements leads to higher accuracies, while smaller angles cause poor accuracies of position fixes (see Figure 7.12). The DoP is a measure for this geometry and it can be derived from building the covariance matrix of the design matrix

(a) Good DoP

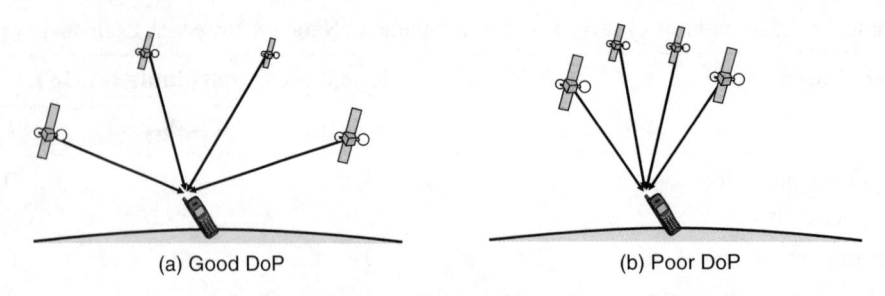

(b) Poor DoP

Figure 7.12 Good and poor DoP.

A from Equation 6.11:

$$\mathbf{Q}_{\text{ECEF}} = (\mathbf{A}^T\mathbf{A})^{-1} = \begin{bmatrix} \sigma_x^2 & \sigma_{xy} & \sigma_{xz} & \sigma_{xt} \\ \sigma_{yx} & \sigma_y^2 & \sigma_{yz} & \sigma_{yt} \\ \sigma_{zx} & \sigma_{zy} & \sigma_z^2 & \sigma_{zt} \\ \sigma_{tx} & \sigma_{ty} & \sigma_{tz} & \sigma_t^2 \end{bmatrix} \quad (7.6)$$

In most cases, it is more convenient to express the standard deviations in terms of latitude, longitude, and altitude instead of referring to the ECEF coordinate system. Therefore, it is required to perform a transformation, which results in the following covariance matrix:

$$\mathbf{Q}_{\text{geo}} = \begin{bmatrix} \sigma_n^2 & \sigma_{ne} & \sigma_{nu} \\ \sigma_{en} & \sigma_e^2 & \sigma_{eu} \\ \sigma_{un} & \sigma_{eu} & \sigma_u^2 \end{bmatrix} \quad (7.7)$$

where σ_n, σ_e, and σ_u refer to deviations in terms of northing, easting, and up. The diagonal elements of Equations 7.6 and 7.7 are then used to calculate different measures of DoP. Table 7.3 gives an overview of them.

Table 7.3 Dilution of precision

DoP measure	Defined by
VDoP	$\sqrt{\sigma_u^2}$
HDoP	$\sqrt{\sigma_n^2 + \sigma_e^2}$
PDoP	$\sqrt{\sigma_n^2 + \sigma_e^2 + \sigma_u^2}$
TDoP	$\sqrt{\sigma_t^2}$
GDoP	$\sqrt{\sigma_n^2 + \sigma_e^2 + \sigma_u^2 + (c\sigma_t^2)}$

The *Geometric Dilution of Precision* (GDoP) is a measure for the general satellite geometry, including both 3-D position and time, and is built from all diagonal elements of the covariance matrix. The remaining measures consider only single standard deviations or combinations of them. The *Position Dilution of Precision* (PDoP) is a measure for accuracy in 3-D position only. It can be further subdivided into the *Horizontal Dilution*

SATELLITE POSITIONING

of Precision (HDoP), which expresses accuracy in 2-D position, and *Vertical Dilution of Precision* (VDoP), which is the accuracy measure for vertical height. Finally, *Time Dilution of Precision* (TDoP) reflects the accuracy of clock synchronization.

The advantage of DoP measures is that they can be calculated by the GPS receiver on the fly in order to select an appropriate constellation of satellites for measurements and to indicate the expected degree of accuracy to the user. Usually, the number of visible satellites is in the range of five to eight. Although, the degree of accuracy is expected to increase with an increasing number of satellites considered for a position fix, it may be degraded if some of them have a bad geometry with respect to the current position of the receiver. Therefore, the receiver determines DoP values for different constellations, and then selects the constellation with the lowest DoP for measurements. In general, PDoP values below four are considered to be very good, while DoP values above five indicate significant errors. It is also possible to have major discrepancies between different DoP measures, for example, a low HDoP indicating a good accuracy with respect to position in 2D, but only high VDoP values resulting in moderate or poor vertical accuracies.

Finally, the relationship between UERE and DoP can be expressed as follows (Leick 2004):

$$\sigma = \sigma_R DoP \quad (7.8)$$

where σ is the overall horizontal or vertical position accuracy and σ_R is the range error caused by UERE at 95% confidence level.

7.4 Differential GPS

Beside carrier phase and dual-frequency ranging, another method for achieving increased accuracy position data is *Differential GPS* (D-GPS). The underlying principle is to observe the pilot signals from the satellites at a well-known position and to determine the difference between the measured range and an approximation of the true range derived from the known position of observation. The difference determined thus can then serve as a correction value for GPS receivers staying close by.

D-GPS requires additional receivers, which are installed at a fixed location on the surface of the Earth and which are called *reference stations*. Basically, a reference station can be seen as a conventional GPS receiver (although it should be equipped with much more sophisticated and accurate receiver technology), but differs from it in that it is stationary and its position has been very accurately determined by surveying. Figure 7.13 demonstrates the interactions between satellites, reference stations, and mobile receivers.

A reference station permanently observes all visible satellites (1) and obtains two ranges $r_{i,rs}$ and $p_{i,rs}$. The former is the range between the accurate position (x_{rs}, y_{rs}, z_{rs}) of the reference station and the position (X_i, Y_i, Z_i) of the ith satellite, which is derived from the ephemeris contained in the navigation message and which might be subject to slight deviations from the satellite's true position owing to perturbing forces. According to Equation 6.2, $r_{i,rs}$ is then given by

$$r_{i,rs} = \sqrt{(X_i - x_{rs})^2 + (Y_i - y_{rs})^2 + (Z_i - z_{rs})^2} \quad (7.9)$$

The other range $p_{i,rs}$ is the measured pseudorange, which suffers from significant errors due to the time offset between the clocks of satellite and receiver, ionospheric and tropospheric refraction, multipath propagation, and other error sources. It therefore differs from

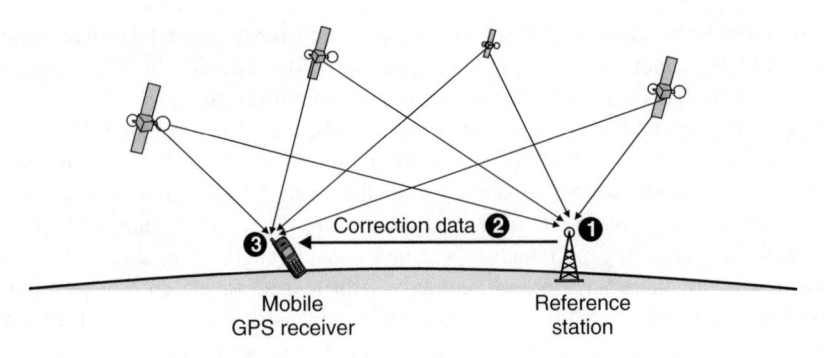

Figure 7.13 Differential GPS.

the true range $r_{i,rs}$ as follows:

$$p_{i,rs} = r_{i,rs} + c \cdot \Delta t_{i,rs} + \varepsilon_{i,rs} \qquad (7.10)$$

where $\Delta t_{i,rs}$ is the time offset and $\varepsilon_{i,rs}$ represents the range difference resulting from the other error sources mentioned earlier. Owing to simplicity, $\varepsilon_{i,rs}$ is not decomposed into the various error fractions here. For a comprehensive overview of this, see Farrell and Givargis (2000). The reference station then calculates the distance

$$\Delta p_i = p_{i,rs} - r_{i,rs} = c \cdot \Delta t_{i,rs} \qquad (7.11)$$

between true range and pseudorange for each visible satellite and composes a set of correction data, which is broadcast to mobile receivers staying close by (2). This does not happen over the GPS carriers but by using a terrestrial wireless transmission technology, for example, radio broadcast or cellular networks. The mobile receiver selects a set of satellites for which correction data is available and performs the conventional range measurements. After that, it corrects each pseudorange $p_{i,mr}$ by the related correction term delivered by the reference station:

$$p_{i,mr} - \Delta p_i = r_{i,mr} + c \cdot \Delta t_{i,mr} + \varepsilon_{i,mr} - (c \cdot \Delta t_{i,rs} + \varepsilon_{i,rs}) \qquad (7.12)$$

The main assumption for the proper working of D-GPS is that errors in the pseudoranges are temporally and spatially correlated if the observations are made in the same local region. Therefore, reference station and mobile receiver should not be separated from each other by distances larger than 60–70 km. Within this distance, ranges observed by the reference station and a mobile receiver usually experience tropospheric, ionospheric, multipath, and noise errors of similar magnitude. Thus, it can be assumed that $\varepsilon_{i,sr}$ and $\varepsilon_{i,mr}$ in Equation 7.12 cancel out each other:

$$p_i^* \approx r_{i,mr} + c \cdot (\Delta t_{i,mr} - \Delta t_{i,rs}) = r_{i,mr} + c \cdot \Delta t_\Sigma \qquad (7.13)$$

D-GPS is not only an efficient means for significantly improving the accuracy of position fixes, but also for reducing the TTFF if the message passed from the reference station to the receiver also carries the current ephemeris and Doppler shift. It is also a promising

SATELLITE POSITIONING

alternative to cellular positioning in that the cellular infrastructure is extended by GPS reference stations and provide mobile terminals equipped with inbuilt GPS receivers with correction and assistance data over signaling channels of the cellular network. This variant is called *Assisted-GPS* (A-GPS) and will be described in the next chapter.

7.5 Galileo

At the time of writing this book, the Galileo satellite positioning system is expected to go into operation around 2009 with the first satellites to be launched in 2006. In contrast to GPS, Galileo is a civil system, that is operated under public control. The Galileo program is managed and financed by the European Commission and the European Space Agency under a mandate from their member states. The basic features of Galileo are worldwide coverage, four positioning services for different user groups, integrity reporting, and increased accuracy compared to GPS (at least with regard to the present GPS). Note that *integrity* refers to the capability of the system to recognize errors in terms of deviations of position fixes from the true position and to report these errors immediately to the users. This is an important feature especially for life-critical services like navigation for aviation.

Like GPS, the Galileo system comprises a space, ground, and user segment. The space segment consists of a minimum number of 27 satellites that circulate the Earth on three orbits, and three additional satellites for replacing malfunctioning ones just in time (see Figure 7.14). The orbits are equally spaced and have an inclination angle of 65°. The semimajor axis of each orbit is 29,994 km, which corresponds to an altitude of 23,616 km.

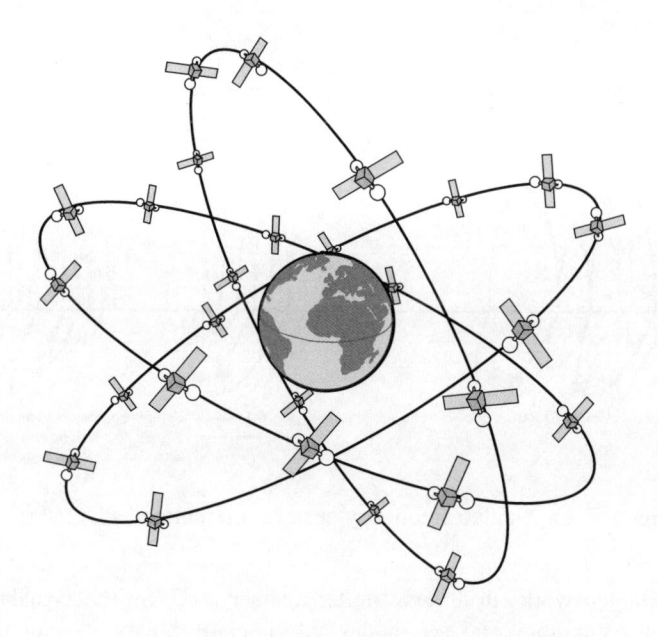

Figure 7.14 Galileo constellation.

The Ground Segment consists of two parts, which are officially denoted as *Ground Control Segment* (GCS) and *Integrity Determination System* (IDS). The former is responsible for the overall control and management of the system as well as for orbit and time determination, while the latter performs integrity monitoring and reporting. At the time of writing this book, the number of GCS and IDS reference stations had not been fixed. According to (Hein and Pany 2002), it is estimated that there will be between 15 and 20 reference stations, five uplink stations, and two control centers. The European part of IDS will include 16–20 monitor stations, three uplink stations for integrity data, as well as two central stations for integrity computations. Like for GPS, the user segment comprises all devices that are capable of observing Galileo signals for deriving position fixes. It is expected that there will be a broad range of devices for different applications like car navigation, surveying, augmentation, aviation, and LBSs for supporting mobile subscribers. Also, it is foreseen that the majority of them will be dual-mode devices, that is, they will be able to access both Galileo and GPS signals.

Galileo realizes 10 pilot signals that are emitted on three carrier frequencies and can be used in various combinations for offering positioning services that are tailored to the special requirements of different user groups. Figure 7.15 gives an overview of the Galileo air interface. The carriers are named E5a-E5b (1,164–1,214 MHz), E6 (1,260–1,300 MHz), and E2-L1-E1 (1,559–1,591 MHz). As can be derived from the figure, some of them overlay the GPS carriers, and hence it is possible to use GPS and Galileo positioning services in conjunction, for example, in order to increase accuracy or availability.

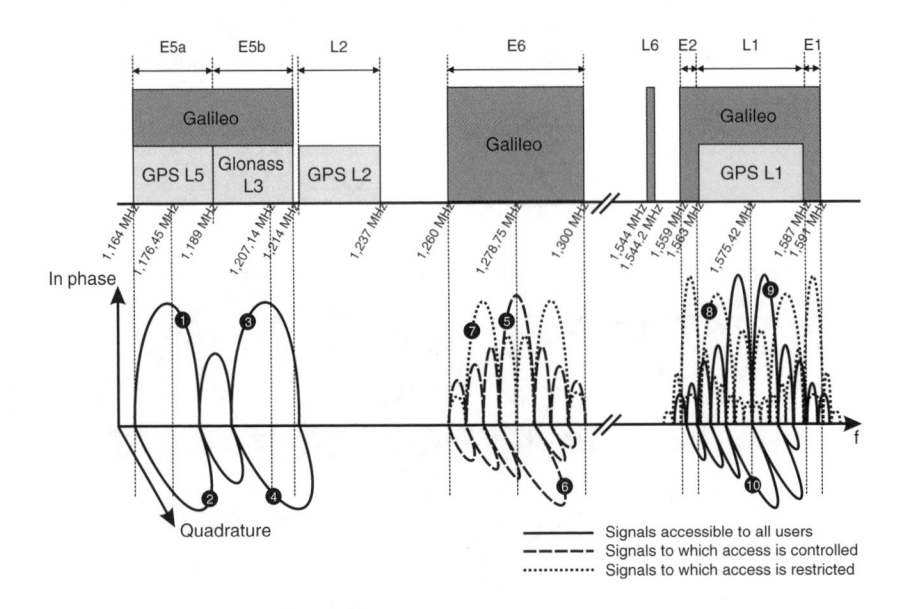

Figure 7.15 Galileo frequency spectrum (Hein and Pany 2002).

Basically, Galileo works in a very similar manner as GPS. The signals are based on spread spectrum technology and are modulated either by binary PSK or another modulation scheme referred to as *Binary Offset Carrier* (BOC) (see (Hein et al. 2002) for an

SATELLITE POSITIONING

introduction). Thus, all satellites operate on the same carriers and access them by CDMA. In this way, it is also possible that GPS and Galileo emit signals in the L1 and L5 bands without significantly interfering each other.

Like for the GPS L1 carrier, each signal that is modulated by binary PSK is subdivided into an in-phase and a quadrature channel. The in-phase channels (1, 3, 5, and 9) carry different kinds of assistance data (navigation messages) that have been spread by a spreading code sequence, while the quadrature channel is a bare pilot that only contains the spreading code but no assistance data (2, 4, 6, and 10). The remaining signals (7 and 8) are modulated by BOC. In addition, some of the signals are encrypted to control or restrict access.

The variety of available pilot signals allows for performing code and carrier phase ranging by observing signals from two (and sometimes even three) carriers. In this way, it is possible to eliminate ionospheric delay to the greatest possible extent and to achieve accuracies that are significantly higher than that of today's GPS signals.

Galileo offers four positioning services that differ in the degree of accuracy, the type of users being able to access them, availability, and integrity. Table 7.4 gives an overview. *Availability* denotes the probability that a positioning service and the associated integrity monitoring are available and provide the service within the specified quality parameters. *Integrity* refers to the capability of the system to immediately report system failures, which appear in the form of deviations of a position fix from the true position, to the user. It is made available in the form of an integrity message composed by the IDS control stations and broadcast by the satellites. As can be derived from the table, integrity is mainly described by

Table 7.4 Quality parameters of Galileo services

	Open service	Commercial service	Public reg. service	Safety of life service	
Level				Critical	Noncritical
Accuracy					
– horizontal(h)	h = 4 m	<1 m	h = 6.5 m	h = 4 m	h = 220 m
– vertical(v)	v = 8 m	(dual frequency)	v = 12 m	v = 8 m	
	(dual frequency)		(dual frequency)	(dual frequency)	
	h = 15 m				
	v = 35 m				
	(single frequency)				
Availability	99.8%	99.8%	99–99.9%	99.8%	
Integrity		optional			
– Alarm limit	–	–	h = 20 m	h = 12 m	h = 556 m
			v = 35 m	v = 20 m	
– Time-to-alarm	–	–	10 s	6 s	6 s
– Integrity risk	–	–	3.5×10^{-7}/ 150 s	3.5×10^{-7}/ 150 s	10^{-7}/ h

three parameters. The *alarm limit* is the maximum allowable error of a position fix before the user is notified about the deviation. *Time-to-alarm* specifies the maximum allowable period between the time an error occurs and the time the associated notification reaches the user. The *integrity risk* defines the maximum allowable probability that the user is not informed about a deviation of position fixes within the specific time-to-alarm.

The *Open Service* (OS) is very similar to the SPS of GPS. It is intended for mass-market applications and provides signals for performing timing and positioning free of charge. The user can access it via small, low-cost receivers that enable either single- or dual-frequency ranging. Single-frequency receivers use the E2-L1-E1 band, and dual-frequency receivers additionally observe the E5a-E5b band. It is also foreseen that most of these receivers can observe both Galileo and GPS signals.

The *Commercial Service* (CS) focuses on applications that require much higher accuracy than offered by the OS, and its usage requires the payment of a fee. Besides this, the CS allows for the dissemination of data with a rate of 500 bps. Services making use of this feature are commonly denoted as value-added services and they can be used, for example, for passing service guarantees (integrity), timing information, data for ionospheric delay models, and local differential correction signals for significantly improving accuracy of position fixes. CS is realized by adding two signals in the E6 band to the free OS signals. Unlike OS signals, however, access is restricted in that these signals are encrypted by protection keys, which will be controlled by the future Galileo operating company, which will in turn be responsible for passing them to authorized users. The CS signals may also be combined by local augmented signals, for example, in order to enable indoor navigation. The expected accuracy is in the degree of 1 m or less and even in the range of centimeters if combined with locally augmented signals.

The *Public Regulated Service* (PRS) is specifically intended for the realization of mandatory applications in the areas of civil protection, emergency response, law enforcement, fighting crime, as well as controlling and supervising external borders. What these applications have in common is that their underlying positioning signals may be subject to an intentional mitigation or even disruption by (economic) terrorists, malcontents, subversives, or hostile agencies. To cope with these concerns, the PRS envisages interference mitigation technologies and limited access to authorized users through key management similar to the CS. The corresponding signals are broadcast on the E6 and L1 frequency bands and are independent of the signals used for the OS and CS in order not to be affected from an intentional degradation of these services. Furthermore, PRS signals occupy comparatively wide bands to be resistant to involuntary interference or malicious jamming.

The *Safety-of-life Service* (SoL) addresses all applications with high demands on safety, above all maritime, aviation, and rail transport. The main difference to other Galileo services is its high integrity level, which guarantees that malfunctions of the system are indicated to its users, which is of high importance for the application areas sketched earlier. The SoL distinguishes two levels of service provisioning: the critical level covers time critical operations, for example, as needed for aviation, while the noncritical level addresses applications being less time critical, for example, sea navigation. As can be derived from the table, the difference between both the levels is reflected by the strictness of integrity parameters. The robustness of SoL is achieved in that its signals are broadcast on three carriers, that is, E5a, E5b, and L1, and in addition, the ranging codes are placed in the in-phase and quadrature

Table 7.5 Mapping of Galileo services onto signals

	1 + 2	3 + 4	5	6 + 7	8	9 + 10
Open service (single frequency)						×
Open service (dual frequency)	×					×
Open service (improved accuracy)	×	×				×
Safety-of-life service	×	×				×
Commercial service				×		×
Commercial service (multiple carrier)	×	×		×		×
Public regulated service			×		×	

of these carriers, where the in-phase contains the navigation message and the quadrature only the pilot. The integrity data is included in the signals of the L1 and E5b bands.

Table 7.5 gives an overview of the different combinations of signals (according to Figure 7.15) the Galileo positioning services make use of.

Finally, it should be noted that Galileo also supports *Search and Rescue* (SAR) services. The satellites are able to detect distress messages emitted by SAR transmitters in the frequency range 406–406.1 MHz and to broadcast them back to Earth on the L6 carrier (see Figure 7.15). In this way, it enables a significant improvement of the terrestrial-based SAR service; for example, it becomes possible to precisely locate emergency situations.

More and up-to-date information about Galileo can be received from the Galileo web site (European Communities) and (Hein and Pany 2002).

7.6 Conclusion

The major benefits of satellite positioning are a comparatively high accuracy (compared to cellular approaches) and a global, worldwide coverage. The drawbacks, on the other hand, are a high battery consumption of receivers, high TTFF values, and the missing support indoors and in close proximity to huge buildings that shadow the signals emitted by the satellites. However, advances in receiver technology achieved in recent years and the combination with terrestrial approaches as applied for D-GPS allow to expect that GPS and maybe Galileo will become the first-class positioning method for future generations of LBSs.

For the first generation of LBSs, satellite positioning was not really an alternative to cellular approaches, which was primarily due to the lack of devices integrating both communication facilities (as provided by cellular phones or PDAs) and GPS receiver functionality. Rather, for a long time the consumer market was dominated by stand-alone GPS receivers, which did a good job with respect to local navigation applications, but which provided no means for realizing more sophisticated services for receiving up-to-date information, for example, about weather conditions on the spot, or for exchanging position fixes between different users. However, in the meantime, the vendors of mobile equipment have recognized the potentials of LBSs, and in some countries, for example, Japan, by law, mobile phones will even have to be equipped with integrated GPS receivers for supporting emergency services. As a consequence, it can be expected in the near future that satellite receivers, besides cameras and local radio technologies, will become another important complement of mobile cellular phones. This development can be seen as one of the most important steps in establishing LBSs on a broad consumer market and making them popular.

In addition, satellite positioning can be easily integrated into cellular networks in that GPS reference stations are integrated into the existing cellular infrastructure. In this way, subscribers equipped with GPS-capable phones can experience a significantly increased accuracy of position fixes compared to conventional GPS, while it is simultaneously possible to reduce the acquisition time. This will be explained in the next chapter. Also, initiatives have started to make GPS signals accessible from indoors.

8

Cellular Positioning

In recent years, GPS is gaining more and more momentum for LBSs. This tendency is certainly fostered by advances achieved in GPS receiver technology during the recent years, which have made it possible to build low-cost GPS receivers that need only low power consumption and to increasingly integrate them into more and more consumer products like cellular phones and PDAs.

However, positioning methods were also developed for cellular networks, for which an important driving force was the E-911 mandate in the United States. The introduction of E-911 was scheduled in two phases, one that obliged operators to provide location data obtained on the basis of radio cells not later than April 1998, and the other that prescribed to enable positioning under much stronger accuracy demands not later than December 2005. This second phase resulted in the development of advanced positioning methods and their installation in cellular networks. Examples are E-OTD, U-TDoA, and A-FLT. An important alternative to these pure cellular methods is the integration of GPS so that the cellular network provides terminals with assistance and correction data of the satellites and thus helps to significantly increase position accuracy and decrease the TTFF. This method is known as *A-GPS*.

This chapter deals with these cellular positioning methods with a special focus on GSM and UMTS networks. At first, the air interfaces of both systems are explained, as this is required for understanding the principles behind the different positioning methods. Subsequently, the chapter introduces the positioning network components and explains the control and coordination of the different methods. The chapter concludes with an overview of positioning in AMPS, CdmaOne, and Cdma2000 and with a discussion and comparison of the presented positioning methods taking into consideration different criteria.

8.1 Positioning in GSM Networks

For GERAN (GSM/EDGE Radio Access Network), different positioning methods have been developed that differ in the degree of accuracy of position fixes and the complexity

Location-based Services: Fundamentals and Operation Axel Küpper
© 2005 John Wiley & Sons, Ltd

of control mechanisms required in the access networks and in the terminals. The different methods are explained as follows:

- **Cell-Id in combination with timing advance.** Cell-Id is based on proximity sensing, that is, the position of the target terminal is derived from the coordinates of the serving base station. To achieve position fixes of higher accuracies, Cell-Id may be combined with the so-called timing advance value, which is a rough estimation of the RTT between base station and terminal and from which the range between them can be derived.

- **Enhanced Observed Time Difference (E-OTD).** This method is based on hyperbolic lateration and is applied in the downlink, that is, the terminal observes the pilot signals emitted by a number of base stations and calculates its position from these measurements.

- **Uplink Time Difference of Arrival (U-TDoA).** This method is also based on hyperbolic lateration, but is applied in the uplink, that is, pilot signals emitted from a terminal are observed by the network, from which the terminal's position is calculated.

- **Assisted GPS (A-GPS).** Positioning follows the principle of D-GPS. The terminals must be equipped with a GPS receiver and are supplied by additional assistance data from the network, which allows to reduce the acquisition time and to increase the accuracy of position fixes.

The introduction of Cell-Id and A-GPS into existing GSM networks is comparatively simple, while E-OTD and U-TDoA require essential modifications and extensions. This section focuses on Cell-Id, E-OTD, and U-TDoA. A-GPS is also available for UMTS and will therefore be covered in a separate section.

8.1.1 GSM Air Interface

To understand the working of positioning in GSM, it is at first useful to have a closer look at the organization of the GSM air interface.

8.1.1.1 Multiple Access and Modulation

Figure 8.1 shows a rough overview of the GSM 900 air interface. As can be derived from Table 5.1 in Chapter 5, GSM also operates in other frequency ranges, but the principle design of the air interface is the same in all of them. GSM uses a combination of FDMA and TDMA for separating the air interface into independent channels and controlling the access to these channels. Uplink and downlink are separated by FDD. A frequency channel (carrier) in each direction has a bandwidth of 200 kHz. Each base station in a GSM network is assigned a limited number of carriers, which is referred to as *cell allocation* (see also Chapter 5).

In the downlink, the terminal scans the signals from all surrounding base stations and monitors their signal qualities. In idle mode, the terminal listens to the broadcast transmissions from the base station with the best signal quality, for example, in order to detect

CELLULAR POSITIONING

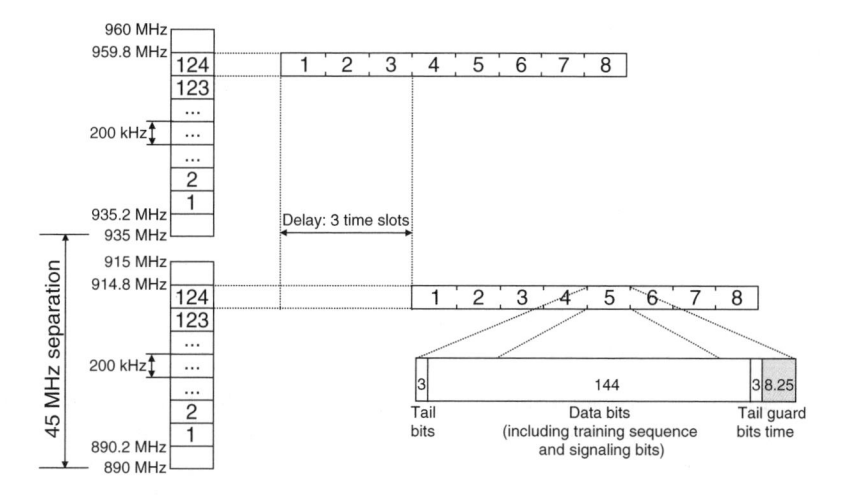

Figure 8.1 Organization of GSM air interface. Adapted from (Eberspächer et al. 2001).

incoming paging requests. In busy mode, it receives the downlink payload data from the base station with the best signal quality. In the opposite direction, it reports measurement parameters describing this signal quality as well as that of surrounding base stations together with the payload data to the network, which is evaluated there for handover decision. Thus, a terminal always observes the signals from all surrounding base stations, which is a useful feature for applying terminal-based positioning using lateration as done for E-OTD. However, in the uplink, base stations receive transmission only from terminals that are in busy mode (or ready state in GPRS). The base stations cannot hear terminals being in idle mode, unless they execute a location update or initialize another data transfer (for example, sending an SMS). Furthermore, if there is a transmission, it can only be received by the serving base station, and not by other surrounding base stations, as these usually operate on another carrier. This makes lateration in the uplink, which is needed for U-TDoA, a complicated matter.

The modulation scheme of GSM is called Gaussian Minimum Shift Keying (GMSK), which is a special version of FSK and belongs to the class of continuous-phase modulation schemes. Its advantage is that it causes only little neighbor channel interferences, that is, the side lobes above and below a certain carrier are rather small when compared to other modulation schemes.

Each carrier is subdivided into eight *time slots* which together constitute a *TDMA frame*. A time slot exactly lasts 15/26 ms (approximately 576.9 μs) and carries 156.25 bits. Thus, a TDMA frame repeats every 4.615 ms and realizes a data transmission rate of 270.83 kbps. A time slot on a certain carrier realizes a *physical channel* that transmits either payload or signaling data. For a number of reasons, TDMA frames are organized in a hierarchical structure consisting of multiframes, superframes, and hyperframes (see Figure 8.2). A multiframe consists of 51 or 26 TDMA frames, and provides a structure for the multiplexing of logical channels onto physical channels (which will be described in the following text). Fifty one or 26 multiframes constitute a superframe, and 2,048 superframes constitute a hyperframe. In this way, a hyperframe lasts 3 h 28 min and 53.760 s. Within a hyperframe,

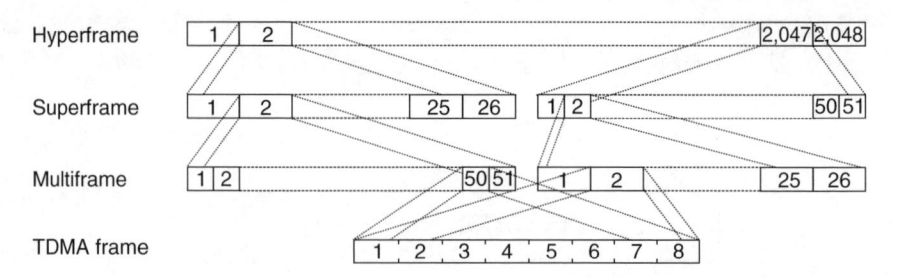

Figure 8.2 Hierarchical GSM frame structure.

all TDMA frames are numbered serially, and the resulting TDMA frame number is used for data encryption as well as for synchronization purposes during positioning.

As can be derived from Figure 8.1, the transmission of frames in the uplink is shifted by three time slots with regard to the transmission of the downlink. Because of this shift and because a terminal is assigned the same time slot in the downlink and uplink, it does not have to send and receive simultaneously and does not therefore require an expensive duplex unit. However, this advantage might be lost in GPRS if the multislot operation is configured in a way that bundled time slots for reception and transmission overlap.

The amount of data transferred in a time slot is referred to as *data burst*. Its main structure can be derived from Figure 8.1. Each start and end of a burst is indicated by three so-called *tail bits* during which transmitter power is ramped up or down and during which no data transmission is possible. Between starting and ending tail bits, there remain 142 bits for data transmission. However, it must be stressed that not all of them are available for exchanging user or signaling data. A considerable amount is reserved for achieving synchronization and frequency correction between base station and terminal. Strictly speaking, GSM defines five different kinds of bursts, which are called *normal, frequency correction, synchronization, dummy,* and *access burst*. All of them differ in their purpose, their internal structure, and the amount of user and control data they carry. For detailed information on that, see (Eberspächer et al. 2001).

It is important to note that there is no timing synchronization between base stations, which, with respect to positioning, is the major drawback of GSM. Each base station emits the downlink structure of TDMA frames and time slots according to its internal clock, which is usually realized by an OCXO (see Chapter 6) and thus suffers from significant drifts. For realizing positioning with lateration, it is hence necessary to install additional components in the network dealing with synchronization issues. This is the task of the so-called *Location Measurement Units* (LMUs), as will be explained later.

8.1.1.2 Guard Times and Timing Advance

Neighboring time slots must be separated by a guard time, which lasts exactly 8.25 bit periods. Remember from Section 4.3.3 that guard times are always necessary in TDMA systems to avoid interferences between adjacent time slots. This can be best explained by means of the scenario depicted in Figure 8.3. Two terminals A and B have been assigned time slots 1 and 2 on the same frequency channel. Terminal A stays in a large distance d_1 to the base station, while terminal B is close by with distance d_2. The base station sends TDMA frames

CELLULAR POSITIONING

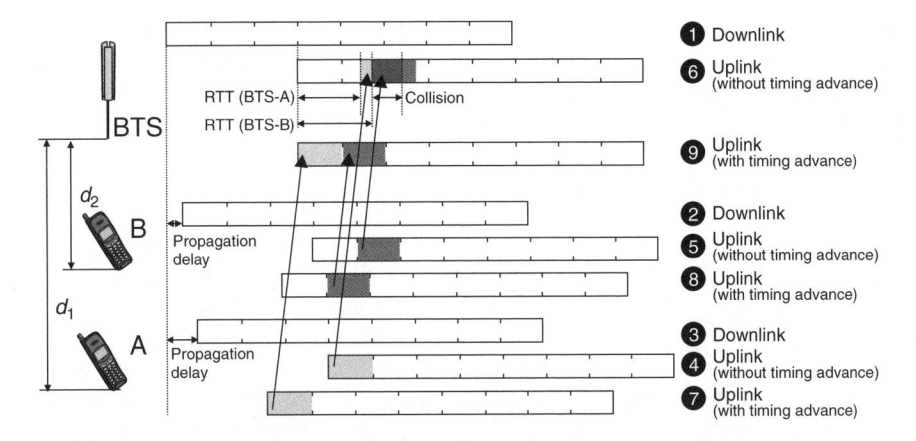

Figure 8.3 Propagation delay and timing advance.

on the downlink (1) and due to the propagation delay the downlink frame arrives much later at terminal A than at B (2, 3). Both terminals then start their transmissions with a delay of three time slots with respect to the arriving downlink frame (4, 5), and are consequently not properly synchronized. In addition, the propagation delay doubles when time slot 1 arrives at the base station (6), which corresponds to the so-called *round trip time* (RTT). As a consequence, from the point of view of terminal B, the transmission of A in time slot 1 is delayed and not finished when B starts to transmit in time slot 2, that is, both transmissions partly overlap. In order to avoid such collisions, a guard time must be arranged between time slots during which transmission is not allowed, and this guard time must be dimensioned to the maximum RTT that may occur. However, GSM has been designed for cell radii of up to 35 km, which corresponds to an RTT of 234 μs or 63 bit periods. Compared to the total length of a time slot, a guard time of this length would lower the transmission capacity by approximately 40%, which is too much and therefore not acceptable.

To cope with this problem, GSM applies a technique known as *timing advance* (or *adaptive frame synchronization*). The basic principle behind this mechanism is to advance the transmission of a terminal by an amount that corresponds to its distance to the serving base station (see Steps (7), (8), and (9) in Figure 8.3). For this purpose, it is necessary that the base station permanently measures the RTT in the course of a connection. This can easily be managed in that it determines the time period between the start of a time slot in the downlink and the arrival of the corresponding time slot from the terminal in the uplink. It then calculates the timing advance value and reports it to the terminal on a signaling channel. The timing advance value has a resolution of 64 steps, and each step corresponds to one bit period or 3.6 μs by which the terminal has to advance its transmission. For example, step 0 means that no timing advance takes place, that is, the terminal transmits with a time shift of exactly three time slots with regard to the downlink, which corresponds to 468.75 bit durations. Step 63 means that the terminal has to advance its transmission by 63 bit periods, that is, the delay with regard to the downlink is only 405.75 bit durations. In this way, it is possible to significantly reduce the guard time between slots to 8.25 bit periods.

In this way, the timing advance value also enables to calculate a rough estimation of the distance between base station and terminal, which is useful for refining position data obtained by Cell-Id positioning. However, the timing advance value is only available for terminals in busy mode, as it is always calculated by the serving base station. Therefore, it is not possible to adopt this value for replacing timing measurements to three or more base stations as required for lateration.

8.1.1.3 Physical and Logical Channels

Each time slot on a certain carrier realizes a *physical channel* that transmits either user or signaling data. The transmission of both types of data is organized by means of *logical channels* that are mapped onto physical channels according to a well-defined scheme. For circuit-switched services, speech and data are transmitted in a *Traffic Channel* (TCH), which is able to realize data rates of up to 13 kbps. A subscriber is assigned a TCH on the downlink and uplink carrier for the duration of the connection. For the packet-switched GPRS, data is carried in a *Packet Data Traffic Channel* (PDTCH), which is dynamically assigned to subscribers as long as the transmission lasts. In order to achieve higher data rates, multiple PDTCHs can be bundled, which is referred to as *multislot operation*. The logical TCHs and PDTCHs are mapped onto physical channels so that there is one time slot per TDMA frame for each TCH and PDTCH respectively.

The transmission of signaling data is much more complicated. GSM defines ten signaling channels and GPRS requires additional seven channels. They are used for paging subscribers, location updates, exchange of measurement data for handover decision, and frequency correction, to name only a few. In general, it is categorized as broadcast and point-to-point signaling channels. The former are only deployed on the downlink carrier. Examples are the *Paging Channel* (PCH) for paging a subscriber in the cells of his location or routing area and the *Broadcast Control Channel* (BCCH). The latter informs all terminals in a cell about the configuration of the cell's air interface. It contains information such as cell allocations of the current and all neighboring cells, LAI and CI, and structural organization of other signaling channels. Point-to-point channels, on the other hand, are defined for uplink and downlink directions. They are used for preparing a radio transmission or controlling an ongoing one.

The mapping of logical signaling channels onto physical ones is a complex matter. Basically, it can be stated that signaling channels require much lower data rates than traffic channels, and hence it would be a waste of resources to reserve an entire time slot per frame for each of them. Therefore, multiple signaling channels share a time slot according to well-defined patterns that are based on the multiframe structure mentioned earlier. There are various options to do so, and the configuration where to find the signaling channels in the constellation of the current cell's carriers and time slots is contained in the BCCH. Thus, all terminals in a cell must receive the BCCH and analyze the data it contains. The BCCH is always transmitted in the first time slot on the first downlink carrier of the cell allocation. Finally, it should be stated that both physical and logical channels are organized and controlled by the respective BSC the base station is connected to.

8.1.2 GSM Positioning Components

There are two additional components for positioning in GERAN, which are called *Serving Mobile Location Center* (SMLC) and *Location Measurement Unit* (LMU). Figure 8.4 shows the integration of them into GERAN and also highlights the official terminology

CELLULAR POSITIONING

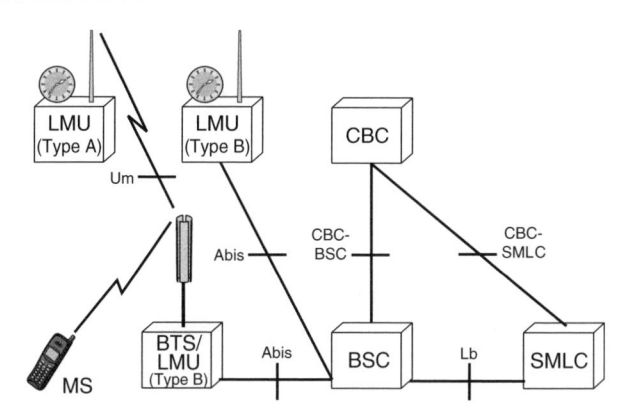

Figure 8.4 GSM positioning architecture.

of interfaces as used in the 3GPP specifications. Note that positioning is controlled by additional components in the core network, which will be covered in Chapter 11.

An LMU observes the transmission of data bursts either from different base stations or from terminals in its surrounding area in order to perform timing measurements. The observation of downlink transmissions of base stations is used for detecting time offsets between the time slots of different base stations and is used for achieving an a posteriori synchronization as needed for E-OTD. Observations of uplink transmissions of terminals are used for U-TDoA. Like for base stations, the coordinates of LMUs must be well known in order to relate timing measurements with the ranges to the respective base station or terminals.

The SMLC controls the entire positioning process, including allocation of resources, evaluation of timing measurements, and calculation of position fixes. Depending on the positioning method used, it controls one or several LMUs and advises them how to measure transmissions in the uplink or downlink. The measurement results returned from an LMU are used to compile assistance data (E-OTD) or to calculate the terminal's position (U-TDoA). In addition, it is required that SMLCs belonging to different access networks interact with each other in order to coordinate positioning of subscribers moving between these access networks.

An access network usually comprises several LMUs depending on the positioning method that is to be supported and a single SMLC. There are two types of LMUs, which are referred to as *Type A LMU* and *Type B LMU*. The former is connected to the access network over the air interface, that is, there is no wired connection and the LMU basically appears to the network as a conventional terminal and only differs from it in that it is stationary and its position has been surveyed. On the other hand, a Type B LMU is connected over a wired link to the access network and may be either a stand-alone component or integrated into a base station. Furthermore, timing measurements are done either against the signals of a selected base station or against an absolute time source. In the latter case, the preferred option is to use GPS time, for which the LMU has to be equipped with a GPS receiver. An SMLC may be a stand-alone component or it may be an integral part of the BSC.

An additional component needed for supporting positioning is the *Cell Broadcast Center* (CBC). It is used for broadcasting positioning assistance data to the terminals for E-OTD

Table 8.1 Signaling protocols for positioning in GSM

Components	Signaling protocol	3GPP spec.
SMLC ↔ BSC	Base Station Subsystem Application Part (BSSAP)	3GPP TS 48.008
	Base Station Subsystem Application Part LCS Extension (BSSAP-LE)	3GPP TS 49.031
SMLC ↔ MS	Radio Resource Link Protocol (RRLP)	3GPP TS 44.031
SMLC ↔ LMU	LMU LCS Protocol (LLP)	3GPP TS 44.071
SMLC ↔ SMLC	Serving Mobile Location Center Peer Protocol	3GPP TS 48.031
SMLC → CBC		3GPP TS 44.035
CBC → BSC		3GPP TS 23.041

and A-GPS. The assistance data is compiled by the SMLC and delivered to the responsible CBC, which then forwards it to the respective BSCs for transmission. Note that a dedicated set of assistance data has to be compiled for each cell separately.

Within an access network, positioning must be coordinated by signaling between SMLC, BSC, LMUs, CBC, and the target terminals. Table 8.1 gives an overview of the protocols the exchange of signaling messages is based on and lists the associated 3GPP specifications. Note that the listed protocols are arranged on top of a complex signaling protocol stack. An overview of this stack is given in (3GPP TS 43.059).

The following sections describe the various positioning methods in detail and describe the deployment of the protocols listed in the table.

8.1.3 Cell-Id Combined with Timing Advance

Cell-Id is a method based on proximity sensing and optionally the consideration of the timing advance value that is continuously measured by a base station during a connection or packet transfer. The method is being standardized since Release 98 for the GSM CS domain, while a version for GPRS is supported since Release 5. Both versions are network-based approaches. Interestingly, a pure Cell-Id method without timing advance is not specified by 3GPP, but many operators offer it and deploy proprietary solutions from different vendors for it.

As can be derived from Figure 8.5, the timing advance value allows to identify a ring of potential positions of the target terminal with the serving base station in its center. In case of sectorized antennas, the potential position can be further narrowed down to one-third or one-fourth of a ring, depending on the angle of sectors.

The control flow for Cell-Id positioning with timing advance in the CS domain and the target terminal being in dedicated mode is demonstrated in Figure 8.6(a). It is realized between SMLC and BSC using the BSSAP-LE protocol (Base Station Subsystem Application Part LCS Extension) (3GPP TS 49.031). In dedicated mode, the terminal transmits a data burst in a certain time slot of each TDMA frame (1). Upon arrival of a position request, the SMLC sends a TA request to the BSC that currently serves the terminal (2). Note that this requires that the SMLC has to first identify the base station the

CELLULAR POSITIONING

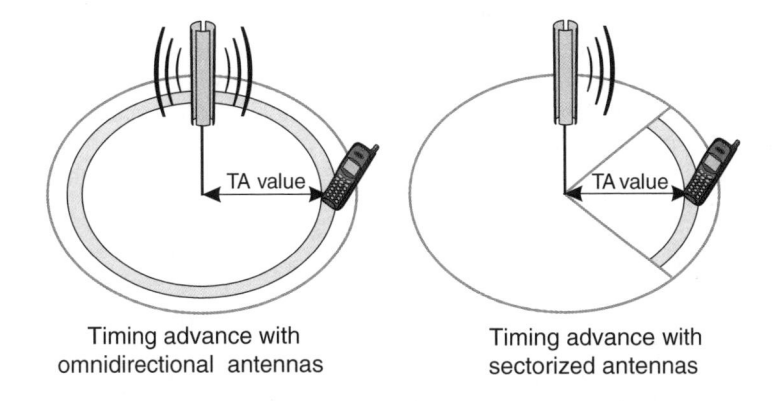

Figure 8.5 Timing advance with omnidirectional and sectorized antennas.

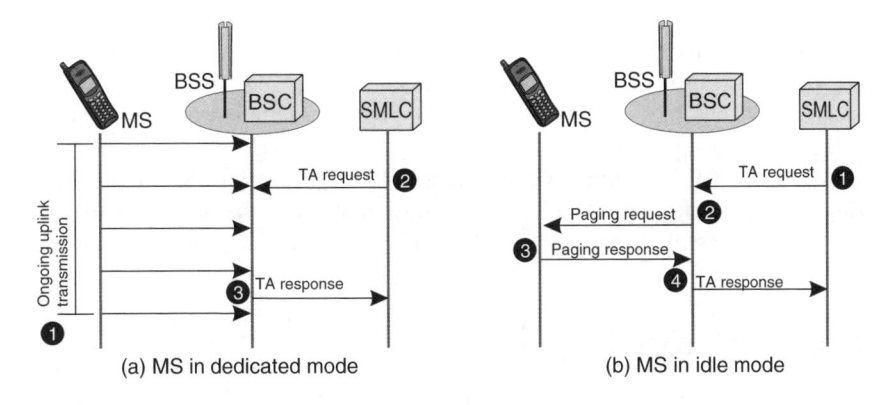

Figure 8.6 Control of TA in the circuit-switched domain.

terminal is currently attached to, a procedure that will be described in Chapter 11. As a result, the BSC then returns the CI of the serving base station and the current timing advance value (3). In addition, the TA response may also include the latest radio measurements of serving and neighboring base stations, which are periodically transferred from the terminal to the BSC for supporting the handover decision. Among other things, these measurements comprise the signal strength and signal quality of serving and neighboring base stations as experienced by the terminal, which can be utilized for refining the position estimation. Note that in GSM a terminal is only in connection with one base station at a time, and hence timing advance values for neighboring base stations are not available.

Figure 8.6(b) shows the control flow for locating terminals in idle mode, that is, when no connection or data transfer is in progress and the timing advance value is unknown. After receiving the TA request (1), the BSC has to pretend an incoming call for the terminal by initializing a paging request at all base stations of the terminal's registered location area. The terminal, which always listens for paging requests at the base station with the best signal quality, then responds to this request by transferring a data burst in a predefined time slot of a signaling channel (3). Measuring the time of arrival of this burst at the base station then delivers the desired timing advance value. The ongoing setup of the false connection is interrupted in due time, so that the subscriber does not recognize the paging request. In the last step, the parameters mentioned earlier are transferred to the requesting SMLC (4).

The procedure for the PS domain is very similar and can be obtained from (3GPP TS 43.059).

8.1.4 E-OTD

For E-OTD, the target terminal applies circular or hyperbolic lateration based on pulse ranging in the downlink. Remember from Chapter 6 that this requires the observation of the time of arrival or the time difference of arrival of pilot signals emitted from a number of base stations in the surrounding area of the terminal. In GSM, no dedicated pilot signals are used for positioning. Instead, the structure of time slots and TDMA frames serve as a basis for timing measurements. Strictly speaking, the terminal observes the first arriving pulse of a predefined time slot, which is referred to as data burst in the following text, and timestamps it. Furthermore, for circular lateration, it is required to have the terminal synchronized with the base stations, while for hyperbolic lateration the base stations must be synchronized among each other. Unfortunately, as mentioned earlier, both of these are not fulfilled in GSM. Therefore, GSM requires LMUs for measuring time offsets and achieving an a posteriori synchronization. Before explaining the details of executing E-OTD, it is at first useful to have a closer look at the basics of synchronization needed for E-OTD.

8.1.4.1 Basics of E-OTD

The basic principle behind E-OTD is depicted in Figure 8.7. The time slot structure generated by different base stations is observed at the terminal as well as at an LMU. Measurements at the terminal are needed in order to derive the range difference of the base stations at the terminal's position and the ranges between terminal and base stations respectively, while measurements at the LMU are needed in order to determine time offsets resulting from missing synchronization. E-OTD is available in two configurations: hyperbolic and circular lateration. They differ in that the former observes the time difference of the arrival of data bursts at the terminal and an LMU, while the latter merely records their time of arrival. However, both of them are based on the same infrastructure components and signaling messages.

Hyperbolic E-OTD. Hyperbolic E-OTD works as follows: given a pair of two base stations (BTS_1, BTS_2) with r_1 being the geometrical range between the terminal and BTS_1 and r_2 being the geometrical range between the terminal and BTS_2, the terminal's position can be narrowed down to all coordinates for which the difference $r_2 - r_1$ has the same amount. By definition, these points lie on a hyperbola (see Figure 8.8; see also Chapter 6). If the hyperbola of another pair of base stations is built in the same way, for example, of the pair (BTS_2, BTS_3), then both hyperbolas will intersect at a certain point, which represents the desired position.

CELLULAR POSITIONING

BS Base Transceiver Station
LMU Location Measurement Unit
MS Mobile Station
OTD ObservedTime Difference
RIT Radio InterfaceTiming

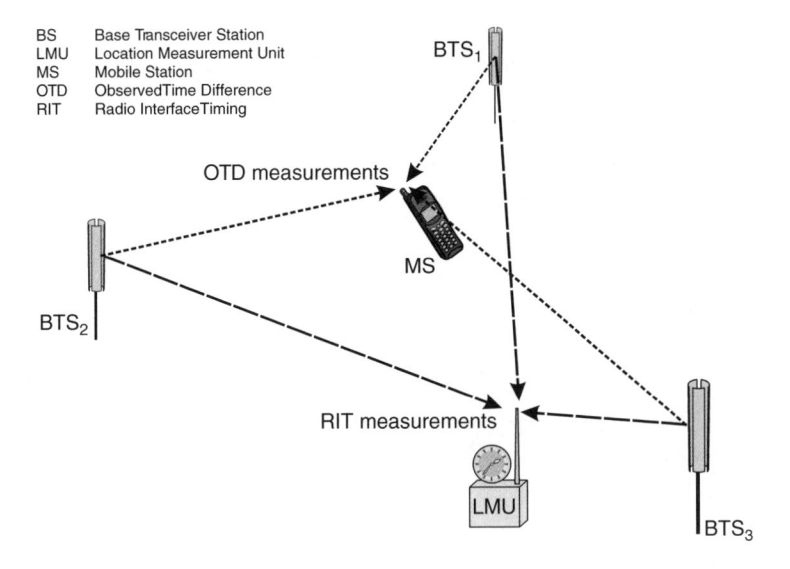

Figure 8.7 RIT and OTD measurements.

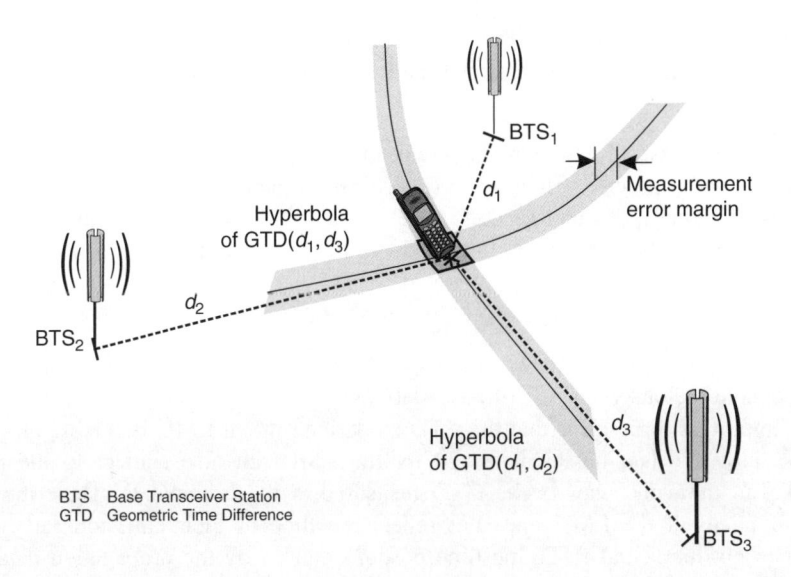

Figure 8.8 Hyperbolic D-TDoA.

The difference $r_2 - r_1$ is proportional to the time period between the arrival of data bursts in the downlink of different base stations, assuming that they have been emitted there at exactly the same time. This time period is called *Geometric Time Difference* (GTD), and it is given by $GTD = (r_2 - r_1)/c$, where c denotes the speed of light. However, owing to the absence of time synchronization, it is necessary not only to measure the time period between the arrival of data bursts at the terminal, but also to determine the time offset between their emissions at the base stations. This gives rise to two other timing quantities that are called *Observed Time Difference* (OTD) and *Real Time Difference* (RTD).

Figure 8.9 demonstrates the relationships between GTD, OTD, and RTD. The OTD refers to the time period that is observed at the terminal between the arrival of data bursts from two different BTSs. If t_{a1} denotes the time of arrival of a data burst from BTS_1 and t_{a2} denotes the time of arrival of a subsequent data burst from BTS_2, then $OTD = t_{a2} - t_{a1}$. Note that t_{a1} and t_{a2} are measured against the terminal's internal clock.

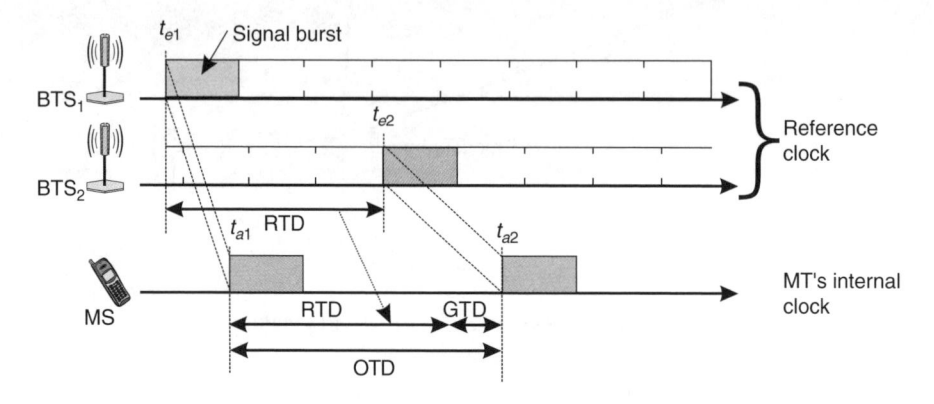

Figure 8.9 Relationship between GTD, OTD, and RTD.

The RTD expresses the time offset between the emission of signal bursts at two distinct base stations and it is observed at an LMU against its internal clock. If t_{e1} is the emission time of the signal at BTS_1 and t_{e2} that of BTS_2, then $RTD = t_{e2} - t_{e1}$. Having determined OTD and RTD values in this way, the GTD can then be determined as follows:

$$GTD = OTD - RTD \tag{8.1}$$

The GTD determined in this way corresponds to the variable d_{ij} in Equation 6.18, where i and j denote the observed pair of base stations.

It is important to be clear that the RTD as derived by an LMU is merely an *observed RTD*, because the data bursts measured by the LMU are also subject to the propagation delay in the same way as the OTD measured at the terminal is. Thus, the time of arrival of bursts at the LMU does not reflect the time of their emissions at the BTSs. In order to get the actual RTD, the time portions caused by the propagation delays must be removed afterward. This assumes knowledge of the exact coordinates of LMU and all base stations involved in the measurements. From the geometrical range between them, the propagation delay can be estimated and removed afterwards from the observed RTD. This process is required by the coordinating SMLC. Figure 8.9 shows the actual RTD that has been corrected by the propagation delay. It must be stated that, besides the measurements of OTDs, this correction procedure may represent an additional error source in E-OTD if positions of LMUs and base stations have not been accurately surveyed or the data bursts do not arrive via line of sight at the LMU, but are subject to multipath propagation.

CELLULAR POSITIONING

Circular E-OTD. Observations made at the terminal and the LMU can also be used for applying a circular lateration. The ranges r_1, r_2, and r_3 here represent circles that intersect at the terminal's position. In order to obtain them, it would be necessary to know the traveling period of signals from the base stations to the terminal. However, the traveling period cannot be determined, because the time of emission of a data burst is not recorded at the base station, and even if so, it would not be possible to relate it with the time of arrival at the terminal as clocks of terminals are not synchronized with those of the base stations.

Therefore, another approach is required. Let $t_{i,MS}$ denote the time of arrival of a burst from base station BTS_i measured against the terminal's internal clock and $t_{i,LMU}$ the time of arrival of a burst from BTS_i measured against the internal clock of the LMU. Furthermore, let r_i represent the searched geometrical range between BTS_i and terminal, and $r_{i,LMU}$ the well-known geometrical range between BTS_i and LMU. Then, the following relationship holds:

$$r_i - r_{i,LMU} = c(t_{i,MS} - t_{i,LMU} + \varepsilon) \qquad (8.2)$$

where c represents the speed of light and ε the time offset (clock error) between the clocks of the terminal and the LMU. Solving this equation for r_i and inserting it into Equation 6.2 leads to a system of three equations with three unknowns, that is, the terminal's position (x, y) and the clock error ε, which can be similarly solved as for GPS (see Chapter 7). Note, however, that this approach only works if the time offset ε is determined for the same clock pairs. This means that all base stations are measured by the same LMU or against the same reference time of different LMUs, which is possible if they are equipped with synchronized GPS clocks.

For both types of E-OTD, timing measurements have to be done at the LMU and the terminal. Measurements made by an LMU are called *Radio Interface Timing* (RIT) measurements, while those performed by the terminal are denoted as *OTD measurements*. Because they are executed at two different sites, it is necessary to carefully coordinate them and to align the results achieved during the measurements. This task is adopted by an SMLC, which communicates with the LMU and the terminal in order to instruct them how to exactly perform measurements and to exchange RTD and OTD values. Figure 8.10 provides a rough overview of this coordination process. RIT measurements are initialized by an `RIT Request` message, which contains the pairs of base stations to monitor and the instructions to perform the measurements. The results are returned in a `RIT Response` message, which can be sent on demand, periodically, or on the occurrence of predefined events. Analogously, the SMLC communicates with the terminal via `Measure Position Request` and `Measure Position Response` messages for controlling OTD measurements. The following sections describe both types of measurements as well as their underlying control flows in detail.

8.1.4.2 RIT Measurements

RIT measurements are always related to a *reference cell*, which is chosen under consideration of hearability and line-of-sight conditions between the cell's base station and the LMU. The arrival of data bursts from base stations of the remaining *neighbor cells* are then measured against the arrival of data bursts from the reference cell. If *RC* denotes

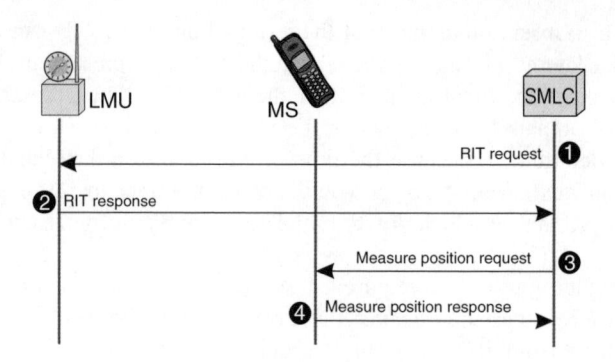

Figure 8.10 Coordination of measurements.

the reference cell and $NC_1, \ldots, NC_m$ represent the m neighbor cells, then RTD values are recorded for the pairs $(RC, NC_1), \ldots, (RC, NC_m)$. Reference and neighbor cells are usually selected by the SMLC, but can optionally be chosen autonomously by an LMU taking into consideration the signal conditions on the spot.

For RIT measurements, the BCCH carrier is observed, which is always arranged on the first frequency of a cell allocation. An LMU monitors the beginning of time slots at the BCCH carriers of reference and neighbor cells. The BCCH carrier has been chosen for measurements since it is the only carrier that guarantees a continuous transmission of data bursts in all time slots, which is an essential prerequisite for performing measurements at all. Even if there is no traffic to be transmitted, a base station always transfers dummy bursts in the idle time slots of the BCCH carrier, which, apart from positioning, is required for frequency synchronization between base station and terminal. In general, normal, dummy, and synchronization bursts can serve as a reference for RIT measurements (see Figure 8.11).

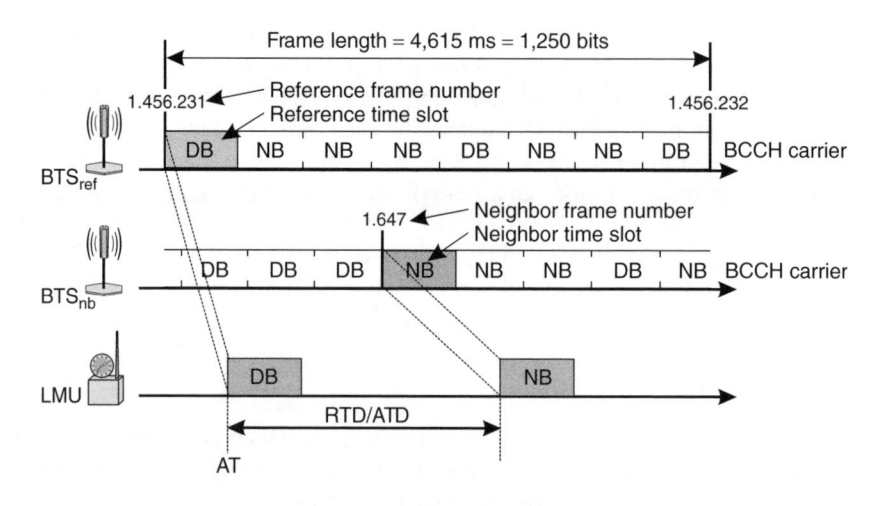

Figure 8.11 Determination of RTD.

CELLULAR POSITIONING

In addition, the BCCH carrier transfers important information elements that characterize the organization of the radio link. For example, from the BCCH carrier of the reference cell, the LMU can obtain cell identifiers and BCCH carriers of neighbor cells, which are transmitted in the so-called *System Information List*. The information of this list can be utilized to choose neighbor cells for RIT measurements if they have not been prescribed by the SMLC.

Generally, the RTD is calculated by observing the timing shift of the arrivals of time slots from reference and neighbor cells. The incoming transmission stream is scanned for the *starting moment* of a time slot, where the starting moment is defined as the first detected pulse of a signal that belongs to a time slot. A time slot selected for measurements in this way is called *reference time slot* (if received from the reference cell) or *neighbor time slot* (if received from a neighbor cell). Analogously, the TDMA frames carrying these time slots are called *reference frame number* and *neighbor frame number*, where the number of a frame is related to its position within a hyperframe.

The RTD is reported as the timing offset between reference and neighbor time slots. It is defined as $t_{nb} - t_{ref}$, where t_{nb} is the starting moment of the neighbor time slot and t_{ref} that of the reference time slot. The RTD may cover the duration of several time slots, that is, the neighbor time slot is not necessarily the slot that is closest in time to the reference time slot (see Figure 8.11). It is reported to the SMLC with a value range of 0–923,200 and a resolution of 0.005 µs, which approximately corresponds to the length of a TDMA frame of 4.615 ms.

The LMU not only computes RTD values, but also provides measurement data for detecting drifts between the base station clocks and to predict future RTD values. Such a prediction is useful for optimizing the reporting between LMU and SMLC and to determine an expiry of RTD values. The recent specifications envisage two options for this:

- **Reporting of frame numbers.** For each RTD value, an LMU records the associated frame numbers of the data bursts from reference and neighbor cells. This frame number corresponds to the position of a TDMA frame within a hyperframe. The number of the measured time slot within a frame is also recorded.

- **Reporting of absolute time.** This option assumes that the LMU is equipped with a GPS receiver, whose clock allows a very accurate and precise chronometry by receiving timing information from a GPS satellite. The arrival of a burst is recorded against this GPS clock. The timestamp of arriving bursts is officially denoted as *absolute time* (AT). If the RTD is also calculated against a GPS clock, then the term *absolute time difference* (ATD) is used instead of RTD.

Measurements at the LMU are typically organized in *monitor periods*, which are frequently repeated and whose length is prescribed by the SMLC. Typical lengths are between 10 and 60 s. During a monitor period, an LMU continuously measures AT and ATD/RTD values and computes their changes. They are caused by the drifts of the internal base station clocks and are reported to the SMLC. An important measure is the *reference AT change*, which is defined as the first time derivative of the AT value of the reference cell during the monitor period. It expresses the portion by which the timing relation between the GPS clock and the clock of the reference cell has been changed. The reference AT change can be described by

$$C_n = C_{n,n-1} + C_{n-1} \qquad (8.3)$$

where C_n and C_{n-1} express the change that have been accumulated during a monitor period after n and $n-1$ measurements and $C_{n,n-1}$ represents the clock drift that has occurred between the nth and $n-1^{th}$ measurements. This clock drift is given by

$$C_{n,n-1} = AT_n - AT_{n,\text{ estimated}} \qquad (8.4)$$

where $AT_{n,\text{ estimated}}$ is the estimated arrival time of the burst and AT_n is the time when the burst actually arrives at the LMU. The change of AT values C_m that has been obtained after m measurements of a monitor period is reported to the SMLC.

The *ATD/RTD change* has a similar meaning. It is the change of an ATD/RTD value that has been observed during a monitor period and represents the accumulated clock drift of reference and neighbor cells. If the AT change is known in addition, and thus the clock drift of the reference cell, the RTD change can be used to separate the clock drifts of neighbor cells from the accumulated drift. Reference AT change and RTD change are used by an SMLC for controlling clock stability of base stations and adjusting outdated RTD values, and also for optimizing the number and frequency of RIT measurement reports to the SMLC. In general, an SMLC can instruct an LMU when to deliver measurement reports. The following instructions are possible:

- **Reporting period.** The reporting period refers to the frequency of periodical measurement reports. It should be defined with regard to clock stability of measured base stations and can vary between 10 s and 2 h.

- **Change limit.** The change limit defines a threshold for the change of AT and ATD/RTD values. If any AT or RTD/ATD value has changed more than this threshold compared to the last reported values, the LMU sends a measurement report with the latest results to the SMLC (see Figure 8.12(a)).

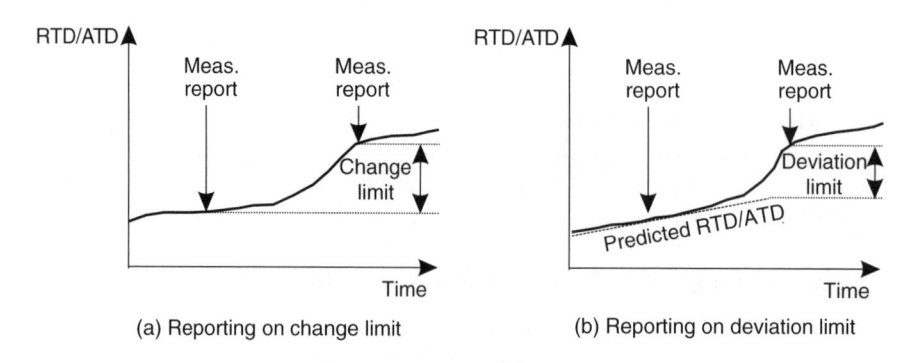

Figure 8.12 Reporting on change and deviation limit.

- **Deviation limit.** The deviation limit is a threshold for triggering measurement reports depending on the deviation of AT and RTD/ATD values. If the deviation of any AT or RTD/ATD value exceeds this threshold, the LMU reports the latest results to the SMLC (see Figure 8.12(b)).

Table 8.2 Parameters of RIT measurement request in GSM (extract)

Parameters	Description
Measurement type	Indicates whether the SMLC wants the LMU to perform ATD or RTD measurements.
Reporting period	Instructs the LMU to send periodic reports with the frequency indicated by this parameter.
Change limit	Instructs the LMU to send extraordinary reports if AT or ATD/RTD values (with regard to the last reported values) has changed more than the limit specified by this parameter.
Deviation limit	Instructs the LMU to send extraordinary reports if the deviation of AT or ATD/RTD values exceeds the value specified by this parameter.
Environment characterization	Provides information about the expected multipath and line-of-sight conditions on the spot.
Reference and neighbor BTSs	Reference and neighbor cells to be monitored.

Table 8.3 Parameters of RIT measurement response in GSM (extract)

Parameter	Description
Reference BTS	Reference cell that has been used for measuring.
Reference frame number	Frame number of the last measured burst from the reference cell.
Reference time slot	Time slot of the last measured burst from the reference cell.
Response type	Indicates whether or not AT and ATD values of the reference BTS are reported.
Reference AT	GPS time of the last reference time slot.
Reference AT change	AT first time derivative of the AT value.
Reference RX level	Received signal strength of the reference cell.
Neighbor BTSs	Neighbor cells that have been used for measurements.
Neighbor frame number	Frame number of the last measured burst from the neighbor cell.
Neighbor time slot	Time slot of the last measured burst from the neighbor cell.
ATD/RTD value	Timing offset of reference and neighbor time slot.
ATD/RTD change	First time derivative of ATD/RTD since last measurement.

All variants can be combined with each other, that is, regular reports as defined by the reporting period may be complemented by extraordinary reports if exceptional changes of AT and RTD/ATD values have occurred.

Tables 8.2 and 8.3 summarize the parameters of RIT measurement requests and responses that are exchanged between SMLC and LMU. For a complete and detailed list, see (3GPP TS 44.071).

8.1.4.3 OTD Measurements

The SMLC collects the measurement reports received from all LMUs it is responsible for, and compiles the so-called *assistance data*, which is passed to the target terminal in order to configure it for OTD measurements and, if requested, for deriving position fixes. Essentially, this data consists of a list of pairs of base stations for which to determine OTD values, RTD values of these pairs, and, optionally, the coordinates of all base stations. Two options exist for delivering assistance data to the target terminal:

- **Dedicated point-to-point signaling.** A dedicated signaling channel is established between the SMLC and the terminal. Using this channel, the SMLC delivers assistance data to the terminal and explicitly requests OTD measurements and, optionally, position estimation. As illustrated in Figure 8.10, assistance data is passed to the terminal via a Measure Position Request message. The measurement results are then returned to the SMLC in a Measure Position Response message.

- **Broadcast signaling.** Assistance data is broadcast by a base station and is thus available to all terminals in the corresponding cell. Thus, there is no explicit request from an SMLC for performing measurements. Basically, the terminal can decide autonomously whether to perform positioning and how often to perform it. The results may be processed internally, for example, by a client application running at the terminal, or they may be passed to an external entity that has requested for positioning and which is located outside the cellular network, for example, a third party service provider. The SMLC has to compile the assistance data for each base station individually and regularly update this data with the latest RTD values as derived from the measurement reports received from its associated LMUs.

In the point-to-point approach, positioning is initialized upon arrival of a location request at the SMLC that contains the identifier of the target terminal and its rough position, and hence it might be possible to page the terminal. The latter is needed to identify reference and neighbor base station in the surrounding area of the terminal, which are supposed to be used for OTD measurements. With knowledge of the rough position, the SMLC is able to compile the relevant assistance data and to pass it in a Measure Position Request message to the target terminal. Among other things, this message instructs the terminal to operate in one of the following two modes:

- **Terminal-assisted mode.** The terminal only measures OTD values and returns the results to the SMLC in a Measure Position Response message, where the terminal's position is estimated by lateration taking into consideration the OTD values (as observed by the terminal), RTD values (as derived from the measurements of the LMU), and the coordinates of the involved base stations. This is the easiest mode, as the bulk of the work, that is, the position estimation, is done at the SMLC and not at the terminal.

- **Terminal-based mode.** In this mode, the terminal not only measures OTD values, but also performs the position estimation. This requires that assistance data contains the coordinates of reference and neighbor base stations as well as the associated RTD values. After this self-estimation of the terminal's position, the derived coordinates can optionally be returned to the SMLC in a `Measure Position Response` message.

Which kind of signaling and which mode to prefer may depend on the application that has requested positioning, and also on the policy of the network operator that offers it and the capabilities of the target terminal. In general, broadcast signaling makes sense in conjunction with the terminal-based mode if the derived position data is processed by a client application that is executed at the terminal. In this case, the terminal can carry out positioning autonomously without causing additional signaling traffic at the air interface. The SMLC tailors a set of position data for each base station individually and instructs this base station to broadcast it, for example, by using the SMS cell broadcast service in GSM networks (see (3GPP TS 23.041)). Thus, the waste of radio resources caused by positioning is actually independent of the number of positioning requests, and hence this approach scales very well for any number of terminals.

On the other hand, point-to-point signaling in combination with the terminal-assisted mode should be preferred if position data needs to be processed outside the terminal, for example, at a service provider. Unlike broadcast signaling, this configuration wastes additional radio resources for each position request. In addition, the SMLC must coordinate and control each request separately, which makes it necessary to carefully dimension its computational resources depending on the expected request rate. Alternatively, point-to-point signaling can also be applied in conjunction with the terminal-based mode. In this case, position estimation has to be done by comparatively complex software at the terminal, which may overburden especially low-cost or outdated terminals, that is, those with very limited computational resources.

An advantage of point-to-point signaling from the operator's point of view is that it has control over the entire positioning process and can charge for it. This is an important side effect if it wants to gain revenues by selling location information to third parties. However, in order to control the access to assistance data when using broadcast signaling, the GSM specifications envisage to cipher relevant parts of the broadcast message. In this case, the terminal must possess a key for deciphering, and the subscriber may be charged for obtaining this key.

Having explained the different options with regard to signaling assistance data and the terminal's operation modes, it is important to get an overview of the elements that the assistance data consists of:

- **Mode of operation.** The SMLC instructs the terminal whether it should operate in terminal-assisted or terminal-based mode. This instruction is only included if assistance data is delivered by point-to-point signaling. If it is however transferred by broadcast, positioning is always terminal based.

- **Reference and neighbor base stations.** The SMLC may prescribe the base stations for which the OTD is to be measured. Like for RIT measurements at the LMU, the base stations are arranged pairwise, each pair consisting of the reference and one neighbor base station.

- **Measurement support data.** To speed up the measurement process, the SMLC can precalculate an *expected OTD* and report it to the terminal. Furthermore, measurement support data contains the RTD values for each pair of reference and neighbor base stations if terminal-based positioning has been requested.

- **Base station coordinates.** Also for supporting the terminal-based mode, assistance data contains the coordinates of reference and neighbor base stations. With knowledge of these coordinates, the RTD values, and the measured OTD values, the terminal can estimate its position by itself.

- **Response time and accuracy.** These parameters refer to the quality of measurement and estimation processes. The SMLC may specify a maximum response time, after which the terminal should have finished its measurements in order to return in time the results to the SMLC. In the terminal-based mode, the SMLC may request the terminal to estimate its position with a minimum level of accuracy. These quality specifications make sense only in conjunction with point-to-point signaling.

The parameters response time and accuracy need to be further explained. Basically, a terminal performs not only one, but several measurements (similar to an LMU during a monitor period). The results of the different measurements are then used to compute OTD values and to determine their quality in terms of their standard deviation. Thus, the higher the number of measurements, the more reliable is the calculated deviation of OTD values, which benefits the accuracy of the subsequent position estimation. However, as each measurement lasts a certain time period, a long monitor period would dramatically increase the response time of the position request. Thus, response time and accuracy behave in an antagonistic way. Whether high response times are acceptable depends on the requirements of the respective applications. Some applications may be very time critical, while for others the availability of very accurate position data might be preferred.

OTD measurements at the terminal are very similar to RIT measurements carried out at an LMU. The parameters associated with OTD measurement requests and responses are listed in Tables 8.4 and 8.5.

OTD measurements can be performed when the terminal is in idle or dedicated mode. In both modes, the terminal listens to the BCCH carriers of reference and neighbor cells and records the timing offset between them. However, owing to lack of precise and accurate clocks, the terminal receives the clock cycles from the reference cell, and hence the OTD is not expressed in milliseconds (like the ATD/RTD), but in bit durations. It is defined as $(t_{nb} - t_{mod})$ modulo burst length and is recorded with a resolution of 1/256 bit and a value range of 0–39,999, which results in a maximum offset of 156.25 bits, that is, the exact number of bits carried in a time slot.

The SMLC supports OTD measurements in that ATD/RTD values as received from an LMU are converted into an appropriate form and an expected OTD is estimated. Both help to speed up measurements at the terminal since they make it easy to scan BCCH carriers of neighbor cells and detect the beginning of 51-multiframes and TDMA frames. If operating in the terminal-based approach, ATD/RTD values are also needed for position estimation.

The conversion of ATD/RTD values at the SMLC includes the removal of time portions resulting from the propagation delays between base stations and LMU and the conversion from the milliseconds timescale to a three-step timescale of multiframes, bits, and portions of a bit. The *multiframe offset* describes the difference between the start of 51-multiframes

Table 8.4 Parameters of OTD measurement request in GSM (extract)

Parameters	Description
Method type	Indicates whether terminal-assisted or terminal-based positioning is requested.
Response time	Specifies the desired response time.
Accuracy	Specifies the desired accuracy in case of terminal-based positioning.
Environment characterization	Indicates multipath and line-of-sight conditions on the spot.
Reference BTS	Specifies the reference cell to be monitored.
Reference BTS position	Coordinates of the reference BTS specified in latitude/longitude.
Neighbor BTSs	Neighbor cells to be monitored.
Multiframe offset	Difference in frames between the start of 51-multiframes of reference and neighbor cells.
Rough RTD	Beginning of reference and neighbor frames in terms of bits.
Fine RTD	Timing offset in 1/256 bit steps of reference and neighbor cell with regard to rough RTD.
Expected OTD	Rough estimation of the OTD based on the coordinates of the reference BTS, timing advance, and other data.
Relative north, east, and altitude	Relative position of neighbor BTS with respect to reference BTS.

of reference and neighbor cells. It can be derived from the reference and neighbor frame numbers as delivered by an LMU. The *rough RTD* is related to the difference in the beginning of reference and neighbor frames in terms of bits and consequently has a value range of 0–1,249 bits. To further refine the resolution, the *fine RTD* expresses the timing offset in terms of 1/256 bit with respect to the rough RTD.

The expected OTD is a rough estimation of the OTD values as the SMLC expects that it is observed at the terminal at its current position. It is only available if assistance data is passed via point-to-point signaling to the terminal. The estimation of the expected OTD is based on the coordinates of the reference base station, which is always the serving base station used for passing the Measure Position Request message to the terminal, the timing advance value measured at the serving base station, and other information.

If terminal-based positioning is applied, the SMLC also passes the coordinates of relevant base stations to the terminal. The position of the reference base station is expressed in latitude and longitude according to WGS-84. The positions of neighbor base stations are then specified by their distance to the reference base station in north and west directions. Latitude and longitude are also used for expressing the estimated position of the terminal.

Table 8.5 Parameters of OTD measurement response in GSM (extract)

Parameters	Description
Reference LAC and Cell-Id	Reference cell that has been used for measurements.
Reference time slot/Frame number	Time slot and frame number of the last measured burst from the reference cell.
Number of measurements	Indicates the number of measurements performed.
Number of measured neighbors	Number of different neighbor cells.
Neighbor LAC and Cell-Id	Neighbor cells that have been used for measurements.
Neighbor multiframe offset	Frame difference between the start of a 51-multiframe of the reference cell and each neighbor cell.
Neighbor time slots	Time slots of the last measured bursts of each neighbor cell.
OTD	Derived OTD for each pair of reference and neighbor cells.
Position estimation	Calculated terminal in latitude/longitude (in terminal-based mode only).

8.1.4.4 Overview of E-OTD Control Flow

After having explained RIT and OTD measurements in detail, Figure 8.13 provides an overview of the entire E-OTD control flow between SMLC, LMU, and the terminal.

Steps (1)–(4) cover the coordination of RIT measurements between an SMLC and an LMU, whereas the actual positioning of a target terminal is reflected by steps (A)–(G) and (a)–(e) respectively. Note that both processes are decoupled from each other. RIT measurements are usually performed periodically in the network (regardless of whether positioning has been requested at all), while positioning is only performed on demand, for example, when requested by an external client (see also Chapter 11).

In the first step, the SMLC sends an RIT Request message to an LMU (1). After receiving this message, the LMU starts measurements (2) and, in most cases, periodically reports the measurement results in an RIT Response message to the SMLC (3). Additionally, it may send extraordinary reports if ATD/RTD values have exceeded a predefined change limit or if their deviation has exceeded the deviation limit. If the SMLC wants an LMU to abort measurements, for example, in order to reconfigure it for measurements, it sends an RIT Stop message to the LMU, whereupon all measurements and reporting activities are canceled there (4).

The SMLC collects the measurement reports received from all LMUs it is responsible for and compiles assistance data. In the point-to-point approach, positioning is initialized upon arrival of a Location Request message at the SMLC (A), which contains the identifier of the target terminal and its rough position. The latter is needed to identify reference and neighbor base stations that are located in the surrounding area of the terminal and are supposed to be used for OTD measurements. With knowledge of the rough position,

CELLULAR POSITIONING

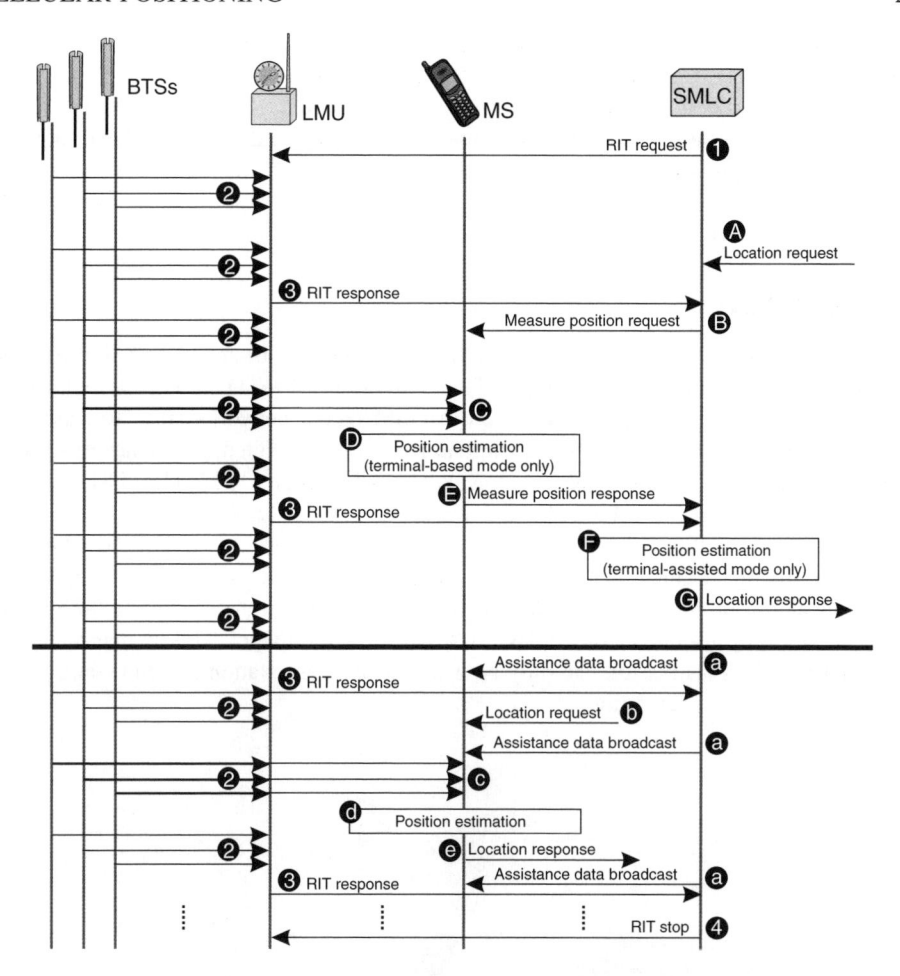

Figure 8.13 Overview of the E-OTD positioning process.

the SMLC is able to compile the relevant assistance data and to pass it in a Measure Position Request message to the target terminal. This message instructs the terminal whether to operate in terminal-assisted or terminal-based mode. In the former case, the terminal only measures OTD values (C) and returns the results to the SMLC in a Measure Position Response message, where the terminal's position is estimated using lateration (F). In case of the terminal-based mode, the terminal not only measures OTD values (C), but also performs the position estimation (D). After that, the derived coordinates are returned to the SMLC. In the last step, the SMLC passes the estimated position to the requesting entity (G).

In the broadcast approach, the SMLC advices the base stations to periodically broadcast assistance data to the terminals (a). The SMLC has to compile the assistance data for each base station individually and regularly update this data with the latest ATD/RTD values as derived from the measurement reports received from its associated LMUs. Positioning at the terminal is activated upon arrival of a Location Request message (b). In contrast to the point-to-point approach, this request may be directly passed to the terminal from an external LCS client, for example, via GPRS, or it may originate from an internal client application. After receiving and evaluating the assistance data from the currently serving base station, the terminal starts the measurement of OTD values (c) and subsequently estimates its position (d). Unlike point-to-point signaling, the broadcast approach works only in conjunction with the terminal-based mode. In the last step, the position is returned to the requesting entity (e).

8.1.5 U-TDoA

Like E-OTD, *Uplink Time Difference of Arrival* (U-TDoA) utilizes hyperbolic lateration based on observing the arrival of time slots, but unlike E-OTD, it is a network-based method. U-TDoA is specified for operation in GSM/GPRS networks since Release 6. At the time of writing this book, the specifications were not completed, and hence this section only sketches the basic idea behind U-TDoA, which is based on (3GPP TR 45.811). Further details can be received from the Release 6 version of (3GPP TS 43.059).

For U-TDoA, distance measurements are carried out in the uplink as shown in Figure 8.14. In this figure, the terminal is in busy mode, that is, being in an ongoing connection with a base station. Besides this base station, the transmission of the terminal is also observed by a number of LMUs for timing measurements. This is due to the fact that the terminal's transmission can only be heard at one base station at a time, and at least three are necessary for lateration.

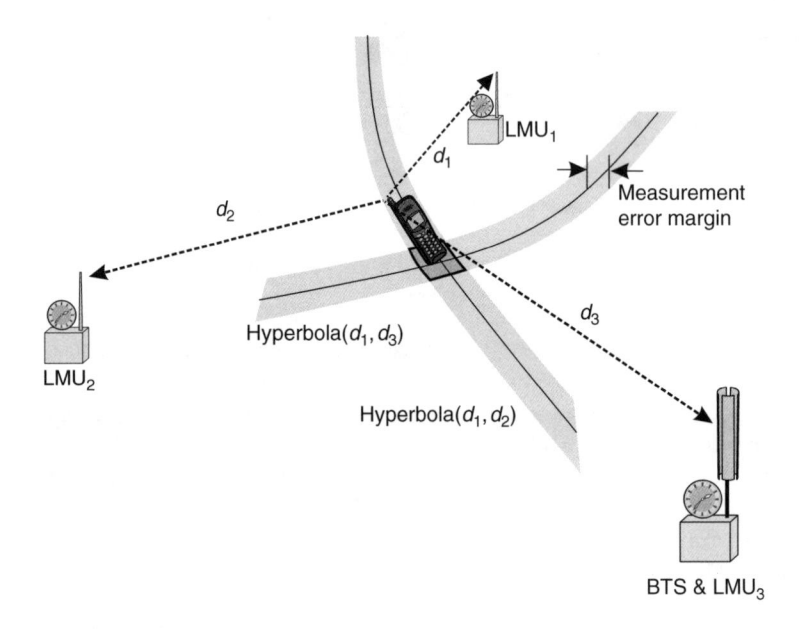

Figure 8.14 U-TDoA Configuration.

An important prerequisite of U-TDoA is therefore to have a sufficient number of LMUs in close proximity to the terminal. Another prerequisite is that the terminal is in busy mode

CELLULAR POSITIONING

during the measurements, as the LMUs can only perform measurements when the terminal is actually transmitting. In order to perform positioning of an idle terminal, nonetheless, it must be stimulated by the network to transmit data, which may go along with additional overhead for paging and authentication. Once a data transmission is in progress, it must last for a certain duration to achieve a minimum degree of accuracy. The recommended range of measurement duration is 20–80 frames. This procedure depends on whether positioning is done in the circuit or packet-switched mode.

Figure 8.15 gives an overview of the procedure of the circuit-switched version when the terminal is in an ongoing uplink transmission (1). In a first step, the SMLC must discover the serving base station and the physical uplink channel, that is, the carrier and the time slot the terminal uses for uplink transmission. As channel allocation for each access network is coordinated by the BSC, the SMLC can query this BSC for the terminal's serving base station and its physical channel (2).

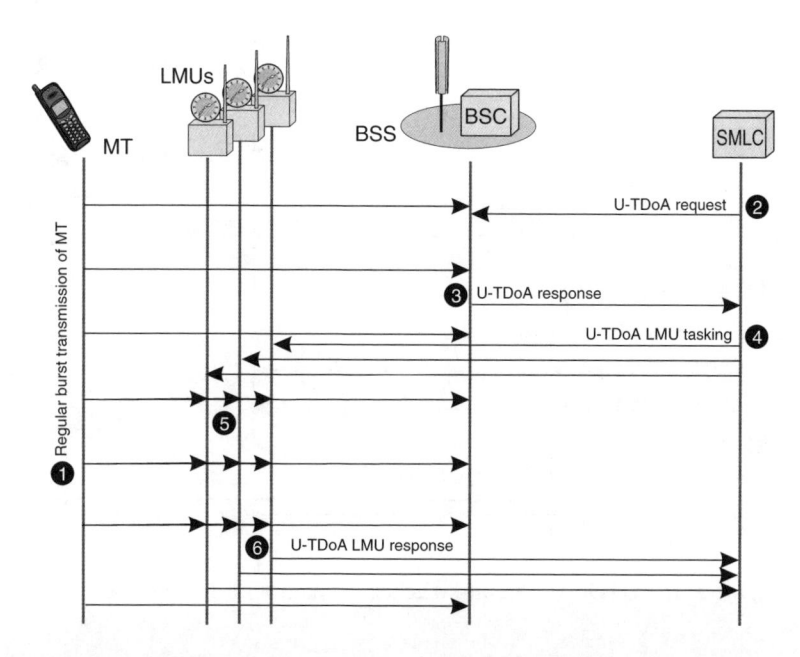

Figure 8.15 Overview of the U-TDoA positioning process.

After receiving this information (3), a set of LMUs in close proximity to the terminal must be identified. The rough position of the terminal is known from the coordinates of the serving base station, maybe in conjunction with the timing advance parameter. Both can be used to select a number of LMUs that are able to hear the data bursts of the terminal with sufficient strength. At least three LMUs at distinct sites are required. After selecting appropriate LMUs, they have to be configured for measurements (4). This configuration comprises the measurement method (GPS time or GSM time of serving base station), the measurement windows, consisting of the starting time of measurements and their duration, as well as the physical channel to be monitored. During the measurement window, the

LMUs then listen to the incoming bursts from the terminal and record their times of arrival (5). In the last step, the results are returned to the SMLC (6), which derives the time differences of arrival and estimates the position of the terminal.

The control flow for locating idle terminals in circuit-switched mode is shown in Figure 8.16. First, the SMLC requests the BSC to initialize a connection setup with the target terminal (1). The BSC then pages the terminal and runs the usual authentication and ciphering procedures (2). As a result of this process, the serving base station is known and a physical channel is allocated for the uplink transmission. Both are returned to the SMLC (3), which then configures the LMUs in the same way as described earlier (4).

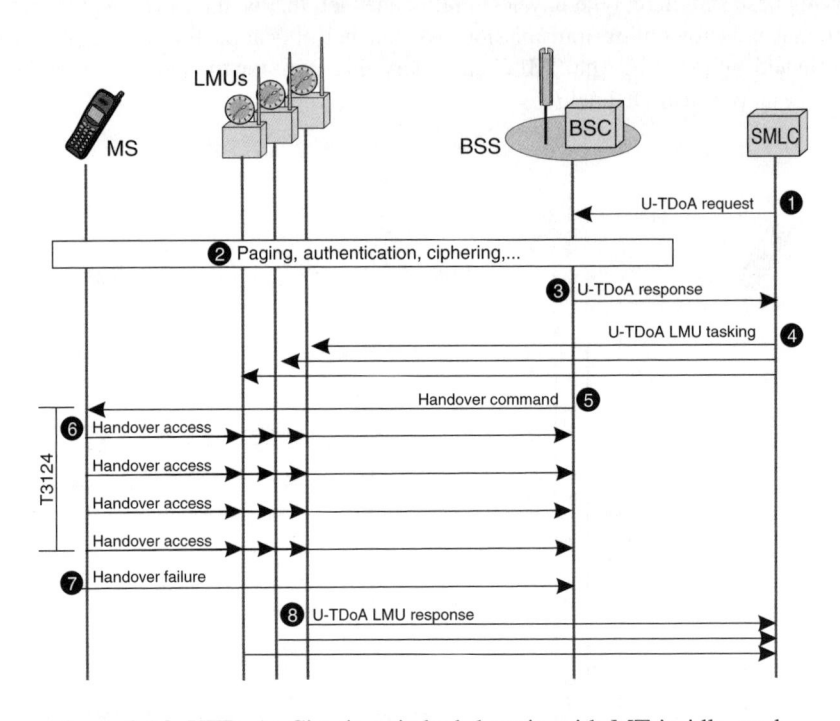

Figure 8.16 UTDoA: Circuit-switched domain with MT in idle mode.

However, up to this time the terminal has no data to transmit. Therefore, to stimulate a data transmission, the BSC simulates the necessity of a handover and sends a handover command to the terminal (5). This handover command contains the physical channel to which the terminal is supposed to switch over and which has been reported to the SMLC in step 3. In the next step, the terminal tries to complete handover by accessing this new channel via access bursts (6). The time of arrival of these bursts at the different LMUs are then recorded there. The terminal repeats the handover access for a certain period of time, whose maximum length is prescribed by the so-called *T3124 timer*. In case of a real handover, the BSC would acknowledge a successful handover within this time period. However, as the handover is only faked here, this acknowledgement from the BSC is suppressed and the handover access is repeated until the T3124 timer expires. The repeated transmissions are used by the LMUs to justify measurements and to increase accuracy.

After T3124 expires, the terminal informs the BSC about the failure of the handover (7). Finally, measurement results are transferred from the LMUs to the SMLC (8).

For terminals in packet-switched mode the procedure is very similar, that is, during a packet transfer, the LMU observes the arrival of time slots carrying the packets. If there is no packet transfer in progress in the uplink, the network must cause the terminal to send packets, similar to the false handover command in the circuit-switched mode. For this purpose, a message is sent to the terminal, which is called *packet polling procedure* and which is usually used for coordinating the medium access in GPRS (see also (3GPP TS 44.060)). Upon arrival of this message, the terminal answers with a packet control acknowledgement, which is then observed by the LMUs. In order to achieve a sufficient accuracy, this procedure must be repeated 4 to 20 times.

8.2 Positioning in UMTS Networks

The following three positioning methods have been specified for UTRAN networks:

- **Cell-based methods.** Proximity sensing is available with various options, which are subsumed under the term *cell-based methods*. In its simplest form, the target's position corresponds to the coordinates of a nearby base station, that is, a Node B. If required, position data can be refined by taking into account the distance between terminal and base station, which can be derived from the RTT or the so-called *Rx timing deviation*, the AoA of signals from the terminal at the base station, or both of them.

- **Observed time difference of arrival with idle period downlink (OTDoA-IPDL).** This positioning method is basically the UTRAN counterpart of E-OTD and follows the same principles, that is, terminal-based lateration. However, as will be described later, timing measurements at the terminal suffer from the so-called *hearability problem*, and hence it is necessary to periodically interrupt downlink transmissions during the measurements. The term IPDL refers to this feature.

- **Assisted GPS (A-GPS).** Like for GERAN, A-GPS is also available for UTRAN.

Note that a counterpart of the GERAN U-TDoA is missing in UTRAN.

8.2.1 UMTS Air Interfaces

Before explaining the UTRAN cell-based and OTDoA-IPDL positioning methods, this section introduces the UTRAN air interface in brief.

8.2.1.1 Multiple Access and Modulation

Different configurations of the air interface are envisaged for UMTS. The feature common to all of them is that they are based on WCDMA, and each carrier has an effective bandwidth of 3.84 MHz. With guard bands, the total bandwidth required is 5 MHz. Unlike GSM, neighboring base stations need not be separated by frequency, though they may operate on the same 5 MHz carrier and in this case are separated by different sets of chipping sequences. Thus, a complete network can be operated with a single 5 MHz carrier (or with a single pair of carriers if downlink and uplink are separated by FDD, see the following text). Within a cell, different subscribers and different channels of the same subscriber are also separated by different chipping sequences. Alternatively, neighboring cells can also use different carriers, which is especially the case if they belong to the networks of different operators.

UTRAN consists of two modes, which are denoted as UTRAN-FDD and UTRAN-TDD and which differ in the way multiplexing and duplexing are arranged. Figure 8.17 provides an overview of both of them.

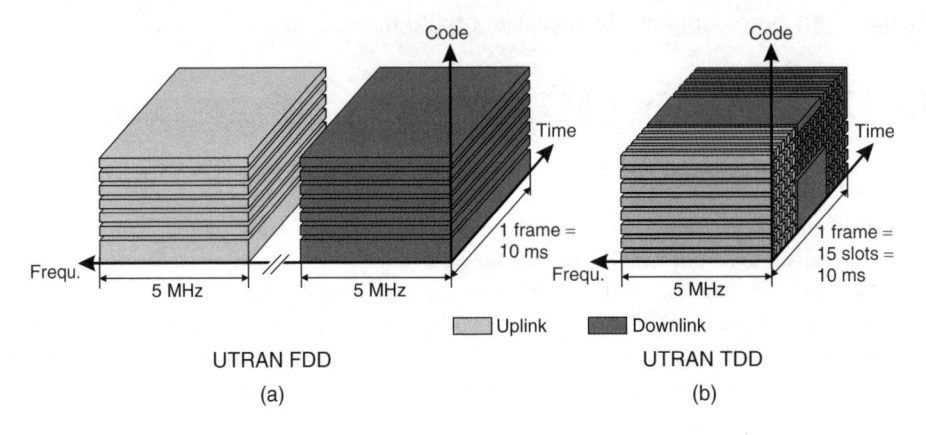

Figure 8.17 Organization of UMTS air interface.

In UTRAN-FDD (see Figure 8.17(a)) downlink and uplink are separated by frequency, that is, in each direction a 5 MHz carrier is required. In UTRAN-TDD (see Figure 8.17(b)) downlink and uplink use the same carrier, but are separated in the time domain. For both air interfaces, the time domain is divided into frames of length 10 ms and each frame in turn into 15 time slots. In UTRAN-FDD, each channel occupies all slots of a frame. The division of frames into slots only serves for synchronization purposes, performance, and handover control here. In UTRAN-TDD, on the other hand, each slot is reserved either for downlink or uplink direction. The number of slots per direction can be dynamically configured depending on the data rates required in each direction. In addition, the slots of a frame can be assigned to different subscribers. Thus, multiple access in UTRAN-TDD is actually a combination of CDMA and TDMA.

Unlike GSM, the UTRAN air interface is not structured according to a complex hierarchy of super, hyper, and multiframes. Instead, each frame is only serially numbered by a so-called *System Frame Number* (SFN), which is used to identify the framing and timing of a base station. The value range for the SFN is from 0 to 4,095.

The modulation scheme used in UTRAN is QPSK, that is, each signal change or each symbol represents two bits. Chips are modulated with a rate of 3.84 Mcps on each carrier. Thus, a frame carries 38,400 chips and a slot contains 2,560 of them. In order to support variable data rates, data can be encoded with spreading factors of $k = 2^i$ with $i = 2, \ldots, 8$ in UTRAN-FDD and $i = 0, \ldots, 4$ in UTRAN-TDD. The spreading factor must be chosen in such a way that its multiplication with the data rate always yields 3.84 Mcps. For example, a spreading factor of $k = 8$ results in a symbol rate of 960 ksps. Increased flexibility can be achieved in that each subscriber is assigned multiple codes (*multicode subscribers*) and, in UTRAN-TDD, in addition, multiple slots (*multislot subscribers*). Thus, a slot in UTRAN-TDD can simultaneously be accessed by several subscribers assuming that they use different chipping sequences. The maximum data rates supported are 384 kbps in UTRAN-FDD and 2 Mbps in UTRAN-TDD.

The frequency spectrum for UMTS has been dimensioned for competition of four to six operators in the same coverage area. For UTRAN-FDD, two blocks of 60 MHz have been allocated, one block between 1,920 and 1,980 MHz for uplink and the other one between 2,110 and 2,170 MHz for downlink direction. For UTRAN-TDD, a total amount of 35 MHz has been reserved, which is located between 1,900 and 1,920 MHz as well as 2,010 and 2,025 MHz.

In order to achieve a separation between different channels, subscribers, and cells, UTRAN applies a two-step encoding of data by using two classes of chipping sequences, which are referred to as *channelization* and *scrambling codes*. This process is depicted in Figure 8.18. Channelization codes are built according to a so-called *Orthogonal Variable Spreading Factor* (OVSF) code tree, which delivers codes with the spreading factors mentioned earlier and full orthogonality assuming that they are correlated against each other unshifted. In the first step, the data stream to be transmitted is encoded with such a channelization code. As can be derived from Table 8.6, channelization codes are used to distinguish between different channels originating from the same terminal in the uplink and between different subscribers of a cell in the downlink.

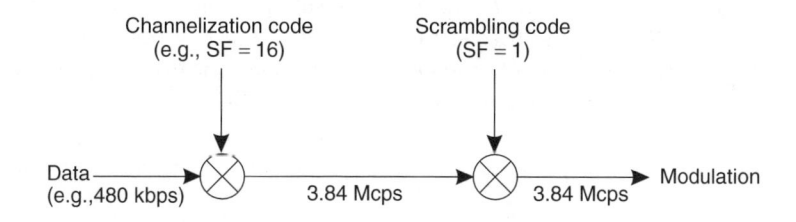

Figure 8.18 Encoding of data in UTRAN.

Table 8.6 UTRAN code types and their usage

	Uplink	**Downlink**
Scrambling codes	Subscriber separation	Cell separation
Channelization codes	Separation of channels from the same UE	Subscriber separation

In a second step, the resulting chip stream is additionally encoded with a scrambling code. It is built by using *Linear Feedback Shift Registers* (LFSRs) and is referred to as *Gold sequence*. During this second encoding, no additional spreading occurs, that is, the spreading factor is $k = 1$. Scrambling codes are not fully orthogonal, but have low cross-correlation and good autocorrelation properties. They are used to separate transmissions from different terminals in the uplink and from different base stations in the downlink. In the downlink, each base station is assigned a set of 16 scrambling codes and each scrambling code can be combined with up to 256 channelization codes. Neighboring base stations must use different sets of scrambling codes. In the uplink, there are up to 2^{24} scrambling codes available.

Finally, it is important to note that, similar to GSM, neighboring base stations in UTRAN-FDD are not synchronized and they emit the structure of frames and slots according to their internal clock. In contrast to this, base stations in UTRAN-TDD are very well synchronized, which significantly eases time measurements as needed for OTDoA-IPDL.

8.2.1.2 Active Sets, Near–far Effect, and Hearability

Because neighboring base stations in a UMTS network may operate on the same carriers, it is possible that a terminal maintains *signal branches* to several base stations simultaneously. This means that it can receive the transmissions of several, close by base stations and vice versa, and several base stations can listen to the transmissions of the terminal, both at the same time. This feature can be utilized to significantly improve the quality of data transmission and to support the handover process.

The set of base stations to which a terminal maintains signal branches is referred to as its *active set*. The composition of an active set depends on the signal qualities, in particular the signal strengths, of close by base stations. A base station is added to the active set if the signal quality exceeds a well-defined threshold and is removed if it falls below a threshold. Typically, an active set contains between one and three base stations and may comprise a maximum number of six. As can be derived from Figure 8.19, the quality of data transmission can be increased in that data streams resulting from multiple signal branches are combined at the RNC. If the base stations of an active set are connected to different RNCs, it is distinguished between a *Serving RNC* (SRNC) and one or several *Drifting RNCs* (DRNCs). The DRNCs forward the data streams to the SRNC where they are correlated, and a combined, proper stream is then forwarded to the core network.

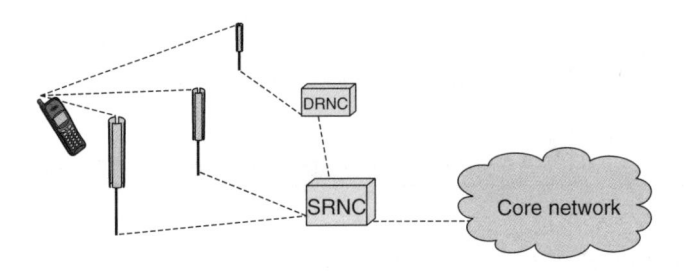

Figure 8.19 Signal branches between terminal and base station.

Maintaining signal branches to several base stations simultaneously also benefits the handover process. In particular, it is easily possible to establish a new connection to another base station before releasing the connection to the old one. This special form of handover is called a *soft handover* and has the advantage that it does not cause a loss of quality and thus cannot be recognized by the subscriber. UMTS envisages several other types of

CELLULAR POSITIONING

handover, among them hard, interfrequency, intrafrequency, and intersystem handovers. For a comprehensive introduction, see Kaaranen et al. (2001).

However, the fact that several subscribers within a cell and the base stations of neighboring cells are simultaneously transmitting on the same carrier is also responsible for two problems, which are known as *near–far effect* and *hearability problem*. Both of them occur owing to the differences in the path loss of signals, the former in the uplink and the latter in the downlink.

The near–far effect appears if the signals of a terminal that is close to the serving base station drowns out the signals of terminals located far away from the same base station. As a result, the base station may not properly reconstruct or even recognize the signals from terminals being farther away, which is illustrated in Figure 8.20(a). In the worst case, signals from terminals in large distances may appear as background noise or interferences when arriving at the base station. To cope with this problem, each terminal must adjust its transmitted power in a way that the signals from all terminals in a cell are of similar strength when arriving at the base station, which is illustrated in Figure 8.20(b). Thus, terminals located far away must transmit with much more power than those located close by. The adjustment of power is a complicated process for which different methods exist. One approach is to measure the signal strength in the downlink and to derive the uplink power from that, which is referred to as *open-loop power control*. Another approach is *closed-loop power control*, where the serving base station instructs the terminal to either increase or decrease transmission power based on the signal-to-interference ratio measured at the base station.

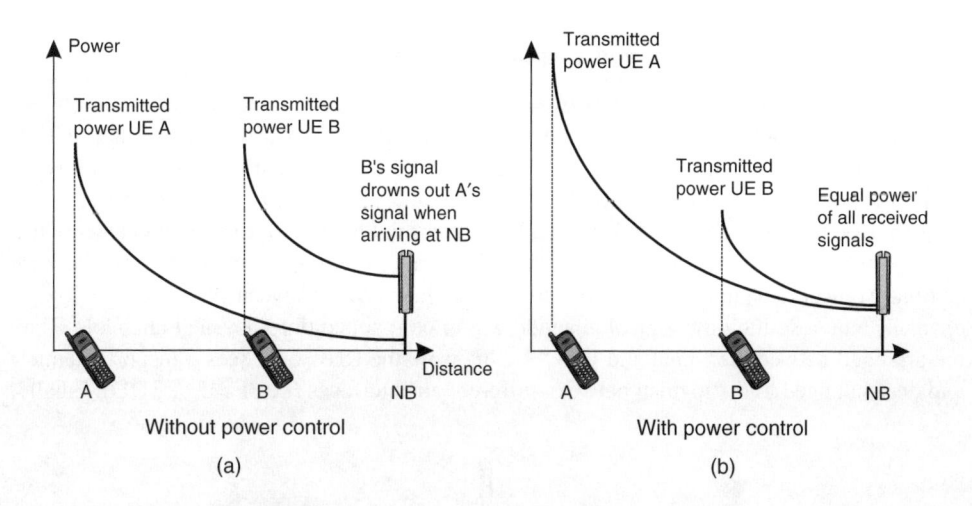

Figure 8.20 Near–far effect and power control.

The hearability problem is similar to the near–far effect, but appears in the downlink. If a terminal stays in close distance to a base station, the signals from this base station may drown out signals from base stations located farther away. Consequently, the terminal cannot properly receive their signals or even does not *hear* them as depicted in Figure 8.21. While this problem is of minor concern for data transmission, it makes terminal-based lateration

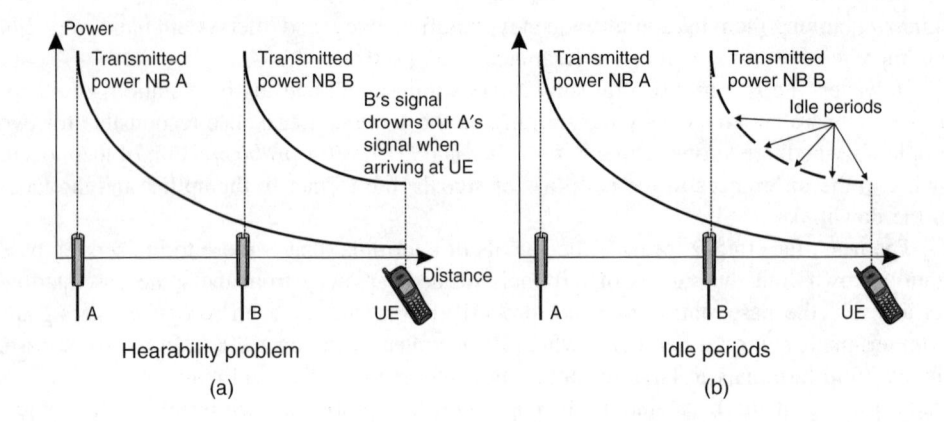

Figure 8.21 Hearability problem and idle periods.

a complicated matter. To solve this problem, power control at the base station, similar to that of terminals for compensating the near–far effect, is not possible because signal strength of a base station has to be adjusted in that it guarantees optimal coverage for all terminals in a cell and not only a single one. Therefore, an option for OTDoA positioning is to regularly interrupt a base station's transmission for short periods of time during which timing measurements for lateration can be made. These periods are denoted as idle periods and are depicted in Figure 8.21(b).

8.2.1.3 Physical, Transport, and Logical Channels

In UMTS, there is a distinction between *physical*, *transport*, and *logical channels* for transferring payload and control data over the air interface. This is in contrast to GSM, which defines only physical and logical channels. The arrangement of these channel types in the RNS is shown in Figure 8.22. Physical channels are basically represented by certain structures of frames and slots, which contain different fields for carrying payload and control data, pilot bits, and data for power control. Physical channels are organized by the base station. Transport channels represent an intermediate layer between physical and logical channels and describe how logical channels are to be mapped onto physical channels. They are arranged between terminal and RNC. In this way, the RNC only sees transport channels and does not need to distinguish between different air interfaces (i.e., FDD or TDD). Finally,

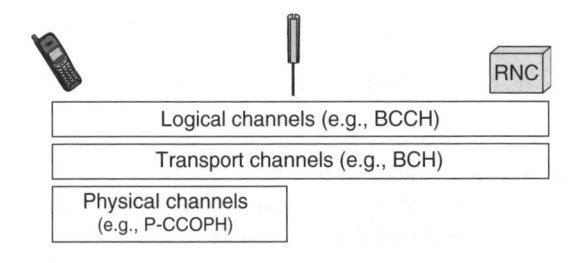

Figure 8.22 Arrangement of physical, transport, and logical channels.

logical channels describe the types of information to be transmitted. Similar to GSM, it can be distinguished between traffic channels for circuit and packet-switched data and control channels for broadcast and point-to-point signaling. Some of them are defined either for uplink or downlink and others for both directions.

Similar to GSM, there is a *Broadcast Control Channel* (BCCH) in downlink direction, which is a logical channel and which informs all terminals about the organization of the air interface in a cell and neighboring cells. It is mapped onto a logical channel called *Broadcast Channel* (BCH), and this in turn, is carried by a physical channel called *Primary Common Control Physical Channel* (P-CCPCH). The P-CCPCH is available in a way that all terminals are able to demodulate its content, that is, it uses fixed and well-known channelization and scrambling codes. In this way, this channel corresponds to the BCCH carrier of GSM. BCCH and P-CCPCH respectively are accessed by terminals in the context of OTDoA positioning for receiving information about neighboring cells that are to be considered for lateration.

Another physical channel accessed for positioning is the *Common Pilot Channel* (CPICH), which is arranged in the downlink. It does not carry any transport channels, but is merely an unmodulated code channel carrying a well-known scrambling code. It is permanently monitored by all terminals in a cell for performing signal strength and timing measurements that are needed for power control, handover decision, positioning, and other reasons. In the context of positioning, it is used for observing the so-called *SFN–SFN observed time difference* between neighboring cells, the RTT between a base station and terminal, and the *Rx timing deviation*. These parameters will be explained in the following text.

8.2.2 UMTS Positioning Components

Similar to GSM, UMTS requires dedicated components for controlling positioning. However, unlike GSM most of them do not appear as stand-alone components, but in most cases are integrated into the base stations or RNCs. Base stations operating under UTRAN-FDD are equipped with an *associated LMU* that provides for timing synchronization with neighboring base stations. However, if bad or non line-of-sight conditions between the measuring and neighboring base stations prevent accurate measurements, operators can alternatively establish *stand-alone LMUs* at sites with better radio conditions. These LMUs are usually connected to the access network over the air interface. Neighboring base stations of UTRAN-TDD are synchronized a priori and therefore do not need an LMU. The SMLC is usually integrated into the RNC, but optionally operators may also install *stand-alone SMLCs* (SASs).

Figure 8.23 shows the UTRAN positioning architecture and also highlights the official interface terminology as used in the 3GPP specifications. Like for GSM, positioning needs to be controlled by additional components in the core network, which will be covered in Chapter 11. Note that the broadcast of assistance data as required for OTDoA-IPDL and A-GPS is coordinated by an SMLC inside the RNC. Unlike GSM, it is thus not required to access a CBC for this purpose.

Table 8.7 gives an overview of positioning-related protocols applied between RNC, SAS, stand-alone LMU, and terminal and lists the associated 3GPP specifications. The control flows for UMTS positioning methods can be obtained in all details from the listed documents.

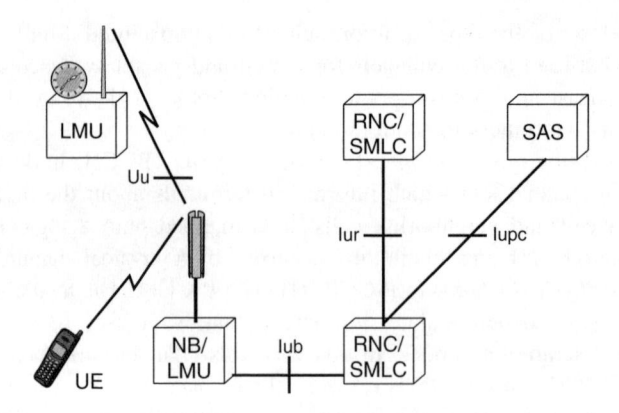

Figure 8.23 UMTS positioning architecture.

Table 8.7 Signaling protocols for positioning in UMTS

Components	Signaling protocol	3GPP spec.
RNC ↔ UE	Radio Resource Control (RRC) protocol	3GPP TS 25.331
RNC ↔ LMU(stand-alone)	Radio Resource Control (RRC) protocol	3GPP TS 25.331
RNC ↔ LMU(associated)	UTRAN Iub interface	3GPP TS 25.430
RNC ↔ SAS	UTRAN Iupc interface	3GPP TS 25.453
RNC ↔ RNC	UTRAN Iur interface	3GPP TS 25.423

8.2.3 Cell-based Methods

Similar to the Cell-Id method in GSM, cell-based methods in UMTS derive a terminal's position from the coordinates of the serving base station. In its simplest form, the position is derived from mapping the cell identifier to the coordinates of the serving base station. If desired, the position can be further refined by taking into account the distance between terminal and base station and the AoA of incoming signals at the base station. The way the distance is determined depends on the UTRAN mode. In UTRAN-FDD, the *Round Trip Time* (RTT) is continuously measured in the UTRAN for all terminals being in a PMM CONNECTED state. According to (3GPP TS 25.215), which summarizes all parameters measured in UTRAN-FDD, the RTT is defined as $RTT = t_{RX} - t_{TX}$, where t_{TX} is the time of transmission of the beginning of a downlink frame to a terminal, and t_{RX} the time of reception of the beginning of the corresponding uplink frame from the terminal. In UTRAN-TDD, the distance is derived from the so-called *received timing deviation*, which, besides other parameters, is specified in (3GPP TS 25.225). It is defined as $t_{RXdev} = t_{TS} - t_{RXpath}$, where t_{RXpath} is the time of arrival of an uplink slot at the base station, and t_{TS} denotes the reference time the slot should have arrived according to the base station's internal timing. The received timing deviation is not only used for positioning, but especially for implementing a timing advance mechanism similar to that in GSM. The determination of the AoA is a feature that is only available for UTRAN-TDD if the respective base station

CELLULAR POSITIONING

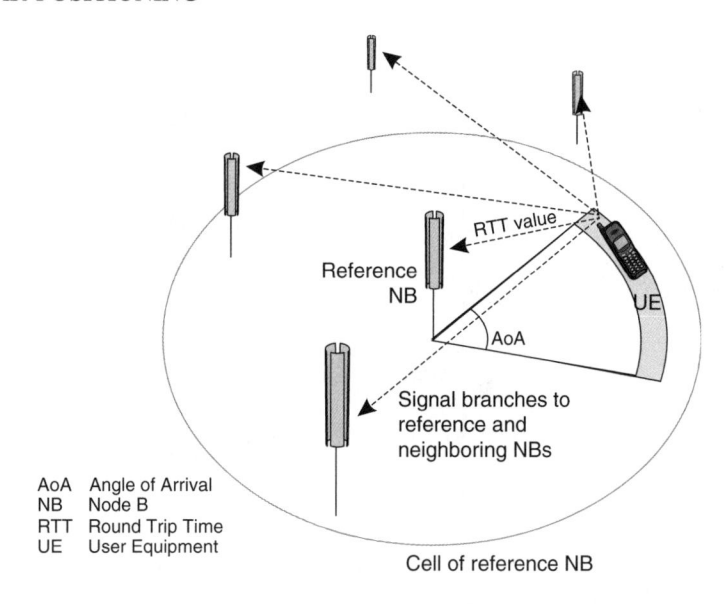

Figure 8.24 Cell-based positioning methods.

is equipped with an antenna array. Figure 8.24 gives an overview of the different options for cell-based methods.

As shown in the figure, a distinctive UMTS feature, which makes the actually simple cell-based methods more difficult than in other networks, is that the terminal may have signal branches to several sectors of a base station or to several base stations simultaneously. Consequently, it is necessary to select from the terminal's active set a reference base station that best corresponds to its current position. This selection process may happen according to one or several of the following criteria:

- Parameters describing the quality of signals received at the base station, for example, signal strength and error rates,
- the base station that was used during connection setup with the terminal,
- the base station the terminal was most recently associated with,
- the latest base station that has been added to the active set,
- the base station from the active set with the shortest distance to the terminal, and
- the base station the terminal has an active connection with at the time the position request arrives at the SRNC.

The exact way how these criteria might be applied for position estimation is not part of the 3GPP specifications and is implemented by each vendor of UMTS location equipment individually.

The control flow of cell-based positioning is shown in Figure 8.25. It depends on the state of the terminal upon arrival of a position request at the responsible SRNC (1) (for an

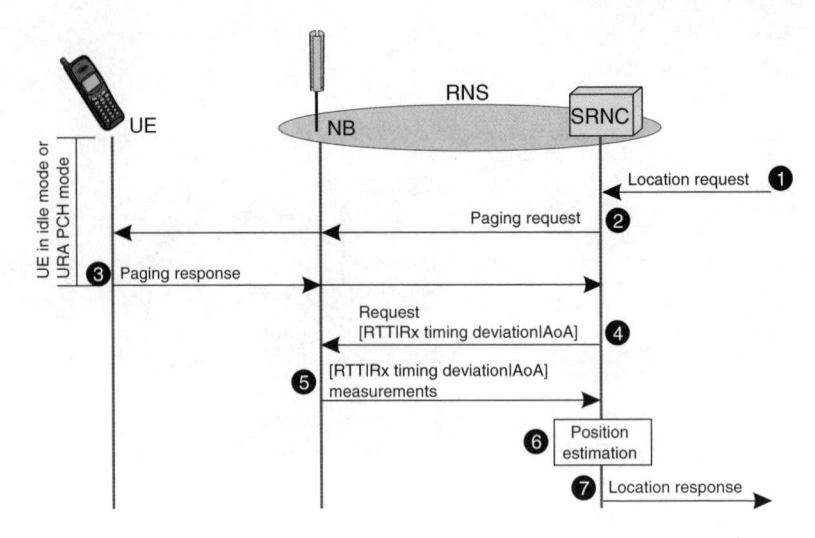

Figure 8.25 Control of cell-based positioning.

overview of the different terminal states, see Section 5.6.3.2). In the PMM DETACHED and IDLE states, that is, when no connection between terminal and network exists, the network only knows the terminal's current location or routing area, and hence the SRNC has to page it first in order to find out the most nearby base station (2). If the terminal is in the PMM CONNECTED state, the necessity of paging depends on the concrete substate. In the CELL CONNECTED state, there is an ongoing data transfer, either circuit or packet switched, and hence the cell identifier can be easily obtained from the serving base station and paging is not required. On the other hand, in the URA CONNECTED state, the access network only knows the terminal's position on the level of URAs and therefore has to page it in its current URA in order to determine a nearby base station.

Once a cell identifier has been derived by the SRNC, either by paging or due to an ongoing data transmission, it can optionally request RTT values (UTRAN-FDD only) or the received timing deviation and the AoA (UTRAN-TDD only) from the serving base station (3). After these values have been delivered (4), the position is estimated and returned to the requesting entity.

8.2.4 OTDoA-IPDL

OTDoA-IPDL is the counterpart of E-OTD of GSM and follows the same principles. It is available with the same options, that is, circular or hyperbolic lateration, dedicated point-to-point or broadcast signaling of assistance data, and terminal-assisted or terminal-based positioning. Consequently, the control flows explained in Section 8.1.4 are nearly the same. The main difference is the way RIT and OTD measurements are performed. Both have been specified taking into consideration the structure of the UMTS air interface. Since the general working of OTDoA is the same as for E-OTD, this section only focuses on the particularities of RIT and OTD measurements in UTRAN.

8.2.5 RIT Measurements in UMTS

RIT measurements performed by an associated or stand-alone LMU are only necessary for base stations operating in the UTRAN-FDD mode, as UTRAN-TDD base stations emit frames and slots according to a reference clock that each base station has to synchronize with. Similar to E-OTD in GSM, an LMU determines RTD and AT values of neighboring base stations. The timing measurements are based on observations of the CPICH. Similar to GSM LMUs, the timing measurements can be performed against the LMU's internal clock or GPS time. The RTD is obtained by computing the so-called *SFN–SFN observed time difference* (see also (3GPP TS 25.215)). It is defined as $t_{CPICH,Rx_{nc}} - t_{CPICH,Rx_{rc}}$, where $t_{CPICH,Rx_{rc}}$ denotes the starting moment of receiving a frame (i.e., the first detected path) on the CPICH from the reference cell, and $t_{CPICH,Rx_{rc}}$ denotes the starting moment of receiving a frame from the neighbor cell that is the closest in time to the frame from the reference cell. The SFN–SFN observed time difference is reported to a precision of one chip, leading to a value range of 0–38,400, which stems from the number of chips carried by a frame.

Alternatively, a base station may only measure the AT of its own frames against a GPS clock. This mode is called *UTRAN GPS Timing of Cell Frames*. In this case, the ATs of all measured neighbors are collected by the SMLC, which then computes the corresponding RTD values.

Tables 8.8 and 8.9 list the parameters of RIT measurement requests and reports in UMTS. Another overview is given in (3GPP TS 25.305). The RRC protocol, which is used for exchanging measurement requests and reports and from which all parameters can be obtained in detail, is specified in 3GPP TS 25.331.

Table 8.8 Parameters of RIT measurement request in UMTS (extract)

Parameters	Description
Measurement type	Indicates whether the NB should measure the SFN–SFN observed time difference or UTRAN GPS timing of cell frames.
Report characteristics	Instructs the NB whether to return measurement reports on demand periodically or according to a change or deviation limit of either SFN–SFN observed time difference or UTRAN GPS timing of cell frames.
Environment characterization	Provides information about the expected multipath and line-of-sight conditions on the spot.
Neighbor cells	Defines the neighbor cells to be monitored.
Measurement accuracy	Defines the required accuracy by means of three classes if UTRAN GPS timing is activated.

8.2.5.1 OTD Measurements in UMTS

Like RIT measurements, OTD measurements at the target terminal are also based on the SFN–SFN observed time difference. However, the definition of this parameter slightly

Table 8.9 Parameters of RIT measurement response in UMTS (extract)

Parameter	Description
SFN	SFN of measured frame.
AT	GPS time of frame emission.
AT quality	Standard deviation of AT in 1/16 chip.
AT drift rate	Indicates the AT drift rate in 1/256 chip per second.
RTD	SFN–SFN observed time difference in chips.
RTD quality	Standard deviation of SFN–SFN observed time difference in 1/16 chip.
RTD drift rate	Indicates the drift rate of the SFN–SFN observed time difference in 1/256 chip per second.

differs depending on whether it is observed in UTRAN-FDD or TDD mode (see (3GPP TS 25.215) and (3GPP TS 25.225)). Similar to GSM, the terminal is supplied with assistance data, whose content depends on whether positioning is performed in the terminal-assisted or terminal-based mode. However, in contrast to GSM, the reporting facilities between terminal and SMLC are much more sophisticated. Measurement responses are not only transmitted on demand. Like for RIT measurements, the SMLC may configure a terminal to send extraordinary responses that are triggered upon exceeding a certain change or deviation limit of the measured values, or, in the terminal-based approach, if the estimated position has changed by more than a predefined value. The latter is very useful for supporting proactive LBSs, which are initialized if the subscriber enters a certain position or if his position has changed. Tables 8.10 and 8.11 provide an overview of the most important parameters associated with OTD measurement requests and responses in UMTS (see also (3GPP TS 25.305) and (3GPP TS 25.331)).

As mentioned earlier, a basic concern of CDMA systems is the hearability problem, and hence the terminal might be unable to detect a sufficient number of neighbor base stations for OTD measurements. In the worst case, OTDoA would only work if the terminal were located on the edge of its serving cell, which is of course not tolerable if seamless and reliable positioning is desired. To overcome this problem, each base station must cease its transmission for short periods of time during which the terminal is able to detect the CPICH from neighbor base stations and synchronize to them. These periods are termed *idle periods* and the method for coordinating the temporary interruption of a base station's transmission is called *Idle Period Downlink* (IPDL). Idle periods are inserted into a downlink transmission stream in a predetermined pseudorandom fashion in order to avoid uniform interruption patterns in the regular data transmission. They can be included in a base station's transmission stream in two ways:

- **Continuous mode.** In this mode, idle periods are continuously sprinkled from time to time into the downlink transmission.
- **Burst mode.** Idle periods occur in bursts, where each burst covers a limited number of frames and contains enough idle periods to allow a terminal to perform a reasonable

Table 8.10 Parameters of OTD measurement request in UMTS (extract)

Parameters	Description
Method type	Indicates whether terminal-assisted or terminal-based positioning is requested.
Response time	Specifies the desired response time.
Accuracy	Specifies the required accuracy of location estimation in case of terminal-based positioning.
Environment characterization	Expected multipath and line-of-sight conditions on the spot.
Reporting period	Instructs the UE to send periodic reports.
Change limit	Triggers extraordinary reports if OTD values have changed more than the limit specified by this parameter.
Position change	Triggers extraordinary reports if the current UE position has changed more than the value of this parameter.
Reference cell	Specifies the reference NB to be monitored.
Reference cell position	Coordinates in latitude/longitude of the reference NB.
IPDL	IPDL configuration of reference NB.
Neighbor cells	Specify the neighbor NBs to be monitored.
Neighbor cell position	Relative position of neighbor NB with respect to reference NB.
SFN offset	Difference in frames relative to the SFNs between the beginning of frames from the reference and a neighbor NB.
SFN–SFN relative time difference	Difference in chips between the beginning frames from reference and neighbor NBs in chips.
SFN drift	Indicates the clock drift of a reference NB in 1/256 chips per second.

number of measurements for position estimation. Consecutive bursts are separated by longer duration of frames where no idle periods occur.

The placement of idle periods into a base station's transmission stream is illustrated in Figure 8.26. In UTRAN-FDD, the idle period has typically a length of five or ten chips. There is at most one idle period per frame and its position within a frame is determined by a pseudorandom generator. As depicted in the figure, an idle period may span the boundaries of consecutive slots. In contrast to that, idle periods in UTRAN-TDD have always the duration of a time slot (not illustrated in Figure 8.26) and are not allowed to span the boundaries of time slots.

Table 8.11 Parameters of OTD measurement response in UMTS (extract)

Parameters	Description
Reference cell	Specifies the measured reference NB.
Reference Rx–Tx time difference	Time difference between the UE uplink frame transmission and the first detected path of a downlink frame from the reference NB.
SFN	SFN of the frame from the reference NB that has been last measured.
Number of measurements	Indicates the number of measurements performed at the UE.
Neighbor cells	Specify the measured neighbor NBs.
Neighbor Rx–Tx time difference	Time difference between the UE uplink frame transmission and the first detected path of a downlink frame from a neighbor NB.
Measurement quality	Denotes the standard deviation of measured SFN–SFN OTD values and of Rx–Tx time difference that has been observed during the measurement period.
SFN–SFN OTD	OTD between the reference and a neighbor cell.
Position estimation	Calculated position in latitude/longitude of the UE if in terminal-based mode.

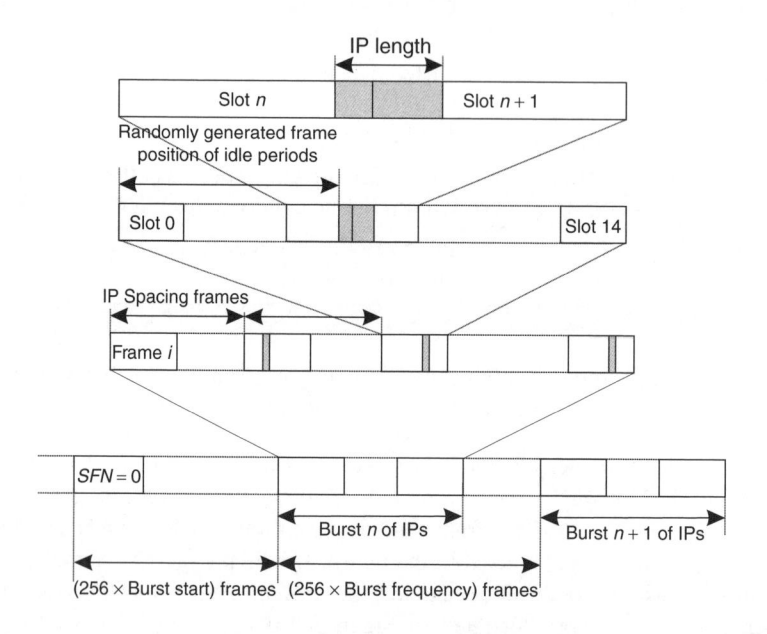

Figure 8.26 Placement of idle periods.

The idle period spacing refers to the number of frames between the start of a frame that contains an idle period and the next frame that contains an idle period. If idle periods are included in burst mode, then additional parameters describe the structure of a burst. The burst start marks the beginning of a burst of idle periods and is specified as a multiple of 256 with regard to the SFN. The length of a burst is described as the number of idle periods it contains, whereas the occurrence of bursts in a transmission stream can be controlled by the burst frequency. The continuous mode is a special case of the burst mode with only one burst spanning the whole SFN cycle of 4,096 radio frames

The control of IPDL is with to the SMLC, which configures base stations for inserting the idle periods. The parameters of an IPDL configuration are also passed to the target terminal as part of assistance data.

8.3 Assisted GPS in GSM and UMTS

Owing to its global coverage and high accuracy, GPS represents a promising alternative to cellular positioning methods introduced earlier. However, with regard to the requirements of LBSs, GPS also has some drawbacks such as high power consumption of the GPS receiver, a long TTFF, and bad indoor coverage. Furthermore, by its very nature, GPS is a stand-alone positioning system and does not provide any communication facilities, and hence its usage today is mostly limited to applications installed locally at the GPS receiver. In order to solve these problems, it is envisaged to integrate GPS into cellular networks so that GPS positioning is supported by additional D-GPS reference stations as integral part of the cellular infrastructure and by additional signaling procedures between network and terminal. The resulting positioning method is commonly known as *Assisted GPS* (A-GPS) and is specified for nearly all cellular systems. This section deals with A-GPS for GSM and UMTS networks only, but its working in other systems like D-AMPS, CdmaOne, or Cdma2000 is very similar.

Compared to conventional GPS positioning, A-GPS provides the following improvements:

- improved accuracy,

- reduction of position acquisition time,

- less power consumption at the GPS receiver, and

- increase of receiver sensitivity.

To be pinpointed by A-GPS, the target terminal must be equipped with a GPS unit for receiving pilot signals and assistance data from the satellites. In order to achieve the improvements mentioned earlier, this GPS unit is supported by additional assistance data and control information from the cellular network. Thus, the GPS unit inside a terminal is actually supported by two sets of assistance data, one compiled by the network and the other delivered by the satellites.

Besides GPS-capable terminals, another prerequisite is the availability of reference stations inside the cellular infrastructure for calculating correction data and compiling the raw material for assistance data. The reference station is connected to an SMLC (either stand-alone or associated with a BSC/RNC), which, like for other positioning methods, coordinates the A-GPS process. It is not necessary to equip each radio cell with its own reference station. Instead, one reference station for areas with a radius of approximately 200 km is sufficient (3GPP TR 25.850). As a possible configuration, there might be a dedicated reference station for an area served by a single BSC and RNC respectively. The infrastructure of D-GPS reference stations inside a cellular network is also known as *wide-area D-GPS network*. Figure 8.27 gives an overview of the A-GPS architecture and the signaling between its components.

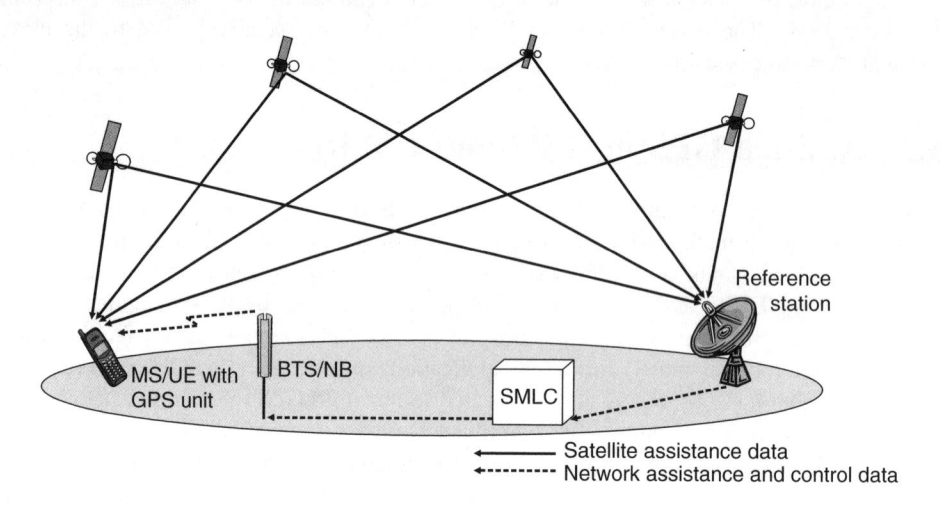

Figure 8.27 A-GPS infrastructure.

As can be derived from the figure, additional signaling is required between reference station and SMLC and between SMLC and target terminal. The former interface is not covered in the current GSM/UMTS specifications and it is therefore obligatory for the vendors of network and GPS equipment to develop and implement proprietary solutions for the integration of reference stations. For the signaling between SMLC and terminal, the specifications follow an integrated approach that uses the same protocols as E-OTD (in GSM) and OTDoA-IPDL (in UMTS) respectively. As a consequence, A-GPS is available in the same configurations and modes as E-OTD and OTDoA, that is, signaling can be done by point-to-point or broadcast transmission and positioning is available in a terminal-based or terminal-assisted approach.

Figure 8.28 shows the A-GPS positioning process with point-to-point signaling in a sequence diagram. The reference station observes the signals from all visible satellites, extracts assistance data from the navigation message, assembles acquisition data (see the following text), and calculates D-GPS correction data (1). These sets of data are subsequently used to support positioning in different ways. For this purpose, they are first sent to the RNC (UMTS) and SMLC (GSM) respectively, which may happen upon request from the RNC/SMLC or periodically. Upon arrival of a location request (3), the RNC/SMLC sends a measure position request to the terminal (3), which partly contains the assistance, acquisition, and correction data obtained previously from the reference station. The amount

CELLULAR POSITIONING

of data carried in this request depends on whether A-GPS is executed in terminal-assisted or terminal-based mode. Table 8.12 gives an overview of the parameters contained in the measure position request in both modes. At the terminal, assistance and acquisition data is used to identify the satellites that are to be taken into account for pseudorange measurements (5). Subsequently, if terminal-based positioning is activated, the terminal estimates its position and returns it to the RNC/SMLC (7). For terminal-assisted positioning, only the measured pseudoranges and related parameters are returned in Step (7), and the position estimation is done in the network (8). Table 8.12 shows the parameters of the measure position response for both modes. Finally, the terminal's position is returned to the entity that has initialized the location request (9).

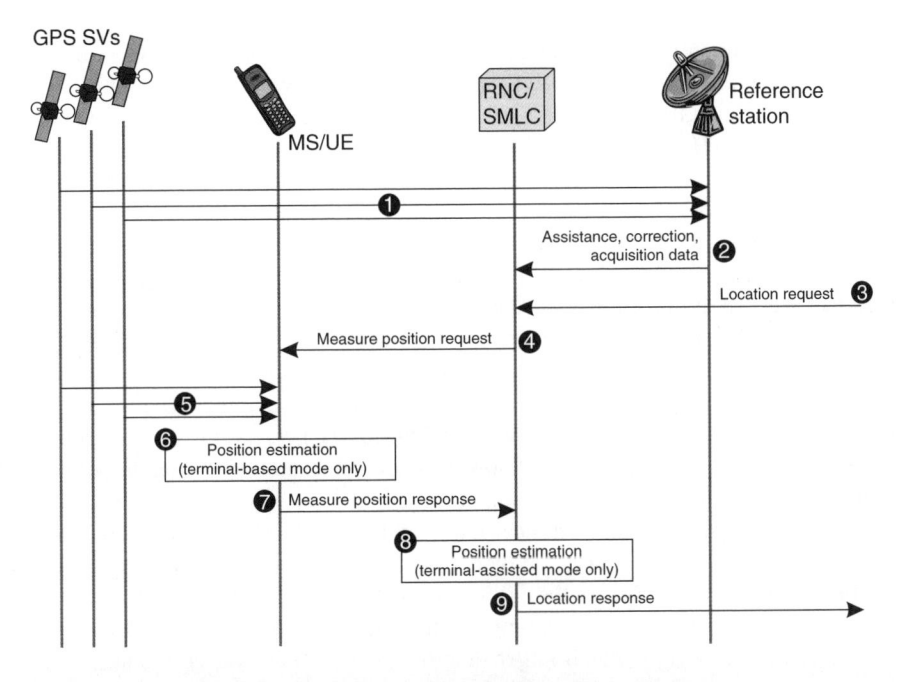

Figure 8.28 Overview of the A-GPS positioning process.

As mentioned earlier, the terminal can alternatively be supported by broadcast signaling, which, like for E-OTD, is managed by a CBC or, like for OTDoA-IPDL, by the RNC. In this case, positioning is always terminal-based and the position is calculated in the terminal. Thus, the broadcast approach enables an autonomous self-localization independent of any control procedures in the SMLC and is the preferred method for supporting local applications inside the terminal, for example, navigation. For charging the subscriber for this service, the network operator may optionally encrypt the broadcast message, similar to the broadcast messages dedicated to E-OTD and OTDoA-IPDL. Decryption is then only possible if the subscriber possesses a key, which he might obtain against payment of a fee.

Finally, it is necessary to have a closer look at the parameters transferred as part of a measure position request and response respectively. As shown in Table 8.12, the

Table 8.12 A-GPS parameters of a position measure request

Parameters	Terminal-assisted	Terminal-based
Reference time	×	×
Visible satellite list	×	×
Acquisition assistance data	×	
Reference location		×
Satellite ephemeris and clock corrections		×
Almanac		×
D-GPS corrections		×

request contains the reference time and the visible satellite list regardless of whether positioning is terminal based or assisted. The reference time specifies the mapping between GERAN/UTRAN and GPS time and is used for time recovery at the terminal when predicting satellite signals. The satellite list identifies the satellites for which the assistance data in subsequent parameters is available. In terminal-assisted mode, the terminal is provided with the so-called *acquisition assistance data*, which consists of the code phase and Doppler shift values of the satellites. Both of them appear as a range of several subparameters (see, for example, (3GPP TS 44.031)) and enable the fast acquisition of GPS signals. It must be stressed that the validity of acquisition data is limited to a few minutes, and hence the GPS unit in the terminal has to perform the measurement immediately after receiving the request.

The set of parameters of terminal-based positioning is much more complex. The reference location is a rough estimation of the position of a terminal. It may be derived from other positioning methods, for example, Cell-Id. The intention of this parameter is to provide the terminal with a priori knowledge of its location in order to improve the performance of the GPS unit. The remaining parameters are ephemeris, clock corrections, and almanac as recorded by the reference station and the calculated D-GPS correction data. Compared to the acquisition assistance data, ephemeris and almanac are valid for a much longer time. They can be used at the GPS unit up to 12 h after they have been received there. However, note also that the amount of data carried in a measure position request for terminal-based positioning is much larger than for the terminal-assisted mode.

The parameters of the measure position response are listed in Table 8.13. The reference time refers to the time of measurements and position calculation respectively. In terminal-based mode, the response carries the terminal's position and velocity as calculated in the terminal. The latter may be derived from the Doppler shift or various position fixes made over a period of time. In terminal-assisted mode, the response contains a set of parameters that allows calculation of a position fix at the SMLC, among them being the measured pseudoranges as well as the Doppler shift, signal-to-interference ratio, and multipath conditions experienced during the measurements. On the basis of these parameters, the SMLC can derive a position fix and also estimate its accuracy. In addition, it may take D-GPS correction data from the reference station into account. In contrast to the measure position request, the bulk of data carried in a response accumulates in the terminal-assisted mode.

To conclude, the terminal-assisted mode should be preferred if the terminal is to be located for a single time, because the acquisition assistance data is only valid for a very short period. On the other hand, the terminal-based mode is suitable if the terminal is

Table 8.13 A-GPS parameters of a position measure response

Parameters	Terminal-assisted	Terminal-based
Reference time	×	×
MS/UE position		×
MS/UE velocity		×
Pseudorange	×	
Doppler shift	×	
Signal-to-interference ratio	×	
Pseudorange RMS error	×	
Multipath indicator	×	
Number of pseudoranges	×	

to be tracked over a longer period of time, especially if location updates are to be sent periodically to the network (or according to a distance or zone-based strategy, see also Chapter 10). The additional signaling overhead, which is caused by the transmission of ephemeris and almanac data in the measure position request, is compensated by comparatively short measure position responses. However, it must be stressed that the terminal-based mode requires a fully functional GPS receiver inside the terminal, which is able to compute position fixes from the measurements, while terminal-assisted A-GPS requires only a slim and less expensive GPS unit consisting of antenna, receiver, and digital processor for making measurements.

8.4 Positioning in other Cellular Systems

Positioning methods applied in other cellular systems such as AMPS, D-AMPS, CdmaOne, and Cdma2000 follow the same principles as those described here for GSM and UMTS. They differ in the components in the access networks coordinating positioning, for example, the counterpart of the SMLC is called *Positioning Determining Entity* (PDE), and the related protocols. As the systems mentioned are predominantly deployed in the United States, the development of positioning methods for them was mainly fostered by the E-911 mandate. All of them support positioning by Cell-Id with different variants, for example, in combination with RTT, which was sufficient for meeting Phase I of the E-911 mandate. For meeting the demands of Phase II, all of them fall back on A-GPS, may be in combination with other methods.

For AMPS and D-AMPS, A-GPS has been standardized by the TIA. Similar to A-GPS in GSM and UMTS, it requires dedicated terminals with GPS units as well as with protocol implementations for exchanging assistance and position data between terminal and network. A problem arises from the fact that AMPS is an analogous system that increasingly becomes obsolete, and D-AMPS is currently being replaced by GSM and 3G systems. Therefore, vendors of network and terminal equipment as well as operators mostly avoid investments in the implementation of A-GPS in AMPS and D-AMPS. The A-GPS signaling protocols for AMPS can be obtained from (TIA/EIA IS-817 2000), and that for D-AMPS from TIA136–740.

CdmaOne and Cdma2000 systems envisage a combination of A-GPS and the so-called *Forward Link Trilateration* (FLT). FLT is basically the counterpart of E-OTD and OTDoA, that is, it applies hyperbolic lateration in the downlink. However, the major benefit in CdmaOne and Cdma2000 is that the base stations in these networks operate synchronously. They emit CDMA code sequences at exactly the same time, and hence a cumbersome a posteriori synchronization by LMUs is not necessary. FLT is based on the observation of the difference in the code phase of pilot signals from different base stations and transforms them into range differences for hyperbolic lateration. It is available as *Enhanced FLT* (E-FLT) and *Advanced FLT* (A-FLT). The former method evaluates the so-called *pilot strength measurement* messages transferred from the terminal to the network, which are needed for handover purposes there. Though it is not a technique developed for positioning (similar to timing advance in GSM), it can be utilized for positioning. The advantage of E-FLT is that it works with all legacy terminals without modifications. However, code phase differences are only reported with a resolution of one chip, and hence accuracy is restricted to 250–350 m and thus does not meet the E-911 demands. A-FLT, on the other hand, has been specifically developed for positioning and reports code phase differences with a resolution of 1/8 chip, leading to higher accuracies in the range of 50–200 m. Similar to E-OTD, OTDoA-IPDL, and A-GPS in GSM/UMTS, A-GPS and A-FLT in CdmaOne and Cdma2000 are available in a terminal-based and a terminal-assisted mode and hence, in contrast to E-FLT, require dedicated terminals. Both modes are realized by point-to-point signaling; an approach for the broadcast of assistance data as in GSM/UMTS is still missing. The signaling protocols for CdmaOne, which is a 2G system, are standardized by TIA (see (TIA/EIA IS-801-1 2001a) and (TIA/EIA IS-801-1 2001b)). The protocols for the 3G Cdma2000 system are specified by 3GPP2 (see (3GPP2 C.S0022-0-1 2001)).

8.5 Conclusion

The first generation of LBSs was primarily based on Cell-Id positioning. The main reason for this was the fact that Cell-Id required only minor modifications on existing network infrastructures. Basically, it is sufficient to maintain a simple database for mapping cell identifiers onto geographic coordinates as well as to install signaling and SMLC functionality as has been explained in this chapter. Another major advantage is that Cell-Id positioning, which is almost always implemented as network-based approach, does not require dedicated functionality in the mobile terminals and thus also works with legacy terminals. However, it soon became apparent that position fixes obtained by this method do not meet the accuracy demands of many LBSs, and especially not those claimed for E-911 Phase II. This was the main motivation behind the specifications of more sophisticated methods like E-OTD, U-TDoA, and A-GPS.

Table 8.14 gives an overview of the accuracy, consistency, and yield performance of the presented positioning methods. This table has been generated taking into consideration several sources, like (CGALIES 2002), (BWCS 2002), (SnapTrack 2003), and (TruePosition 2003). It must be stressed that some of these or similar sources stem from the vendors of equipment for positioning, and hence they may not be objective as anticipated. Furthermore, performance parameters as depicted in the table can be measured in several ways, but, in most cases, the conditions of these measurements are not published. Therefore, accuracy

Table 8.14 Performance characteristics of cellular positioning methods

	Accuracy			Consistency	Yield
	Rural	**Suburban**	**Urban**		
Cell-Id	>10 km	2–10 km	50–1,000 m	Poor	Good
E-OTD & OTDoA	50–150 m	50–250 m	50–300 m	Average	Average
U-TDoA	50–120 m	40–50 m	40–50 m	Average	Average
A-GPS	10–40 m	20–100 m	30–150 m	Good	Good

declarations in the table are listed as value ranges derived from the references mentioned earlier and consistency and yield are rated as poor, average, or good.

Not surprisingly, Cell-Id shows the poorest accuracy and consistency of all, as these parameters significantly depend on the cell size. On the other hand, the attainable yield is very good, because Cell-Id requires only a single base station for positioning and this condition can almost always be fulfilled owing to excellent coverage conditions in most cellular networks today.

The performance of lateration-based methods was under special investigation during the introduction of Phase II of E-911. In general, it can be stated that E-OTD, OTDoA, and U-TDoA show much better results for accuracy and consistency than Cell-Id. However, they provide only a moderate yield, because lateration involves at least three base stations and LMUs respectively to be in range, which is not the case everywhere, especially not in rural areas. GSM operators in the United States decided at an early stage to implement E-OTD for fulfilling the Phase II accuracy demands. However, it turned out that, when first implemented, E-OTD seemingly failed to meet these demands. This realization resulted in many activities for improving E-OTD performance, including the development of new algorithms and technologies for calculating position fixes and rejecting multipath, the increase of measurement periods in the terminals, or changes in antenna configurations (BWCS 2003). However, these efforts apparently turned out to be useless and therefore many operators have decided to switch to U-TDoA in the meantime. According to (BWCS 2003), the reason for accuracy degradation in E-OTD is that timing measurements made at the terminal, as required for E-OTD, suffer from larger clock drifts than those made at an LMU for U-TDoA. The magnitude of these drifts is declared to be in the range of one microsecond during a ten second measurement period, compared to a drift of about only 20 nanoseconds in the same period at an LMU.

A-GPS provides the best performance characteristics of all positioning methods. Compared to stand-alone GPS, accuracy is significantly improved due to the consideration of D-GPS correction data during the calculation of position fixes. The best accuracy is achieved in rural areas, while in urban areas A-GPS suffers from shadowing effects near huge buildings, but much less than conventional GPS. And, the yield of A-GPS is quite acceptable, because acquisition assistance data delivered by the network helps to improve the sensitivity of the GPS receiver inside the terminal.

Apart from performance characteristics, position methods should also be compared with regard to other features, which are listed in Table 8.15. An important parameter, especially for E-911, is the TTFF, which is very low for Cell-Id and in a moderate range of equal magnitude for the remaining methods. Note that the TTFF values listed in the table

Table 8.15 Other features of cellular positioning methods

	TTFF	Terminal	Overhead	Costs
Cell-Id	approx. 1 s	No changes	Very low	Very low
E-OTD & OTDoA	5–10 s	Dedicated software	Medium/high	High
U-TDoA	5–10 s	No changes	Medium	Medium
A-GPS	5–10 s	Dedicated software and hardware	Medium/high	Low to medium

refer to the mere time needed for performing measurements and calculating position fixes. Additional latencies of high variances are imposed by the exchange of signaling messages for controlling the positioning process.

With regard to terminals, it is of major importance whether positioning is terminal or network-based. Network-based methods offer the great advantage that they do not require any dedicated software or hardware installed in the terminals. Thus, it is even possible to serve legacy terminals without any modifications. This does not hold for terminal-based or terminal-assisted methods. For example, E-OTD requires the installation of software upgrades at legacy, or if dedicated hardware is required as for A-GPS, positioning does not work at legacy terminals at all. The consideration of legacy terminals is of major importance in the transition time when operators introduce positioning methods and want to serve as many subscribers as possible from the beginning.

The overhead in Table 8.15 refers to signaling and computational overhead. For E-OTD, OTDoA, and A-GPS this overhead depends on whether positioning is carried out in terminal-based or terminal-assisted mode and whether point-to-point or broadcast signaling is used. Generally, it can be stated that terminal-based positioning and broadcast signaling causes less signaling but more computational overhead (at the terminal) than terminal-assisted positioning and point-to-point signaling.

The costs listed in the table refer to investments an operator has to make to install the different positioning methods. As expected, Cell-Id is the cheapest solution as it requires only minor modifications on the network's infrastructure. The high investments for E-OTD, O-TDoA, and U-TDoA mainly result from the necessity to build an LMU overlay network of more or less density. Generally, there exists only less publicly available material that delivers objective or consistent statements about the required LMU density. The published LMU to base station ratio for E-OTD varies from 1:1 in (BWCS 2003) and (IST Locus Project 2001) up to 1:3 (3GPP TS 03.71) and 1:5 (IST Emily Project 2002) respectively, while for U-TDoA it is claimed by vendors that the required LMU density is less than for E-OTD (TruePosition 2003). Irrespective of these statements, it should be noted that the required density also varies for urban, suburban, and rural areas. Finally, A-GPS is a comparatively cheap solution as it works with a small number of reference stations, each of it serving several tens of kilometers.

9

Indoor Positioning

Traditionally, LBSs were designed for supporting typical outdoor applications, for example, fleet management and navigation. However, the potentials of location-related indoor applications were realized much earlier as LBSs entered the market. They were explored in conjunction with research on ubiquitous computing since the beginning of the 1990s. Typical application fields that were and still are under investigation address office environments and shopping malls, to name only a few. Though research and development considered outdoor and indoor LBSs strictly separated from each other for a long time, there is now an increasing demand for an integrated approach. From the users' point of view, it would be very convenient to invoke indoor and outdoor LBSs by using the same mobile devices, while service providers might be interested in reusing their infrastructures, interfaces, and protocols for both types of LBSs.

However, the major problem of an integrated approach is the absence of a common, universal positioning technology. Conventional GPS receivers do not work inside buildings, while cellular positioning methods generally fail to provide a satisfactory degree of accuracy. The delivered position fixes cannot even be used for determining whether a target person stays inside or outside a certain building, not to mention that it is by no means possible to locate it with the granularity of rooms or floors. Thus, there is a demand for stand-alone indoor solutions. This chapter gives an overview of the most promising technologies, which are WLAN fingerprinting, RFID positioning, and indoor positioning with GPS. Furthermore, an overview of the research prototypes is given, which are based on non radiolocation positioning with infrared and ultrasound.

9.1 WLAN Positioning

Wireless Local Area Networks (WLANs) are specified by IEEE in the 802.11 series and are available in different variants. A WLAN installation inside a building is basically a cellular network that comprises several cells, each served by a base station. In the IEEE 802.11 terminology, a base station is called *access point*, and the coverage area of an access point is referred to as *Basic Service Area* (BSA). The set of all terminals served by an access point is called *Basic Service Set* (BSS). Several BSSs, in turn, can be interconnected via a wired infrastructure to form an *Extended Service Set* (ESS). The infrastructure described here is needed for operating IEEE 802.11 in the so-called *infrastructure mode*. Alternatively, IEEE 802.11 enables direct communication between terminals without any infrastructure in between, which is known as the *ad hoc mode*. However, for positioning as described here, only the infrastructure mode is of relevance.

The IEEE 802.11 specifications comprise two layers, one representing the air interface, which is the *physical layer*, and the other for coordinating multiple access, which is the *medium access layer* (MAC). IEEE 802.11 is available in various versions with different frequencies, modulation, and multiple-access schemes, types of signals (radio or infrared), data rates, and bandwidth, ranges, and other facilities. Today, most installations are based on IEEE 802.11b and 802.11g. The former was developed in 1999. It operates in the license-free 2.4 GHz frequency range, provides data rates of up to 11 Mbps, and achieves ranges between several tens of meters and a few hundreds of meters, strongly depending on the surrounding environment. In 2003, networks based on IEEE 802.11g began to appear. They are backward compatible to IEEE 802.11b installations and enable increased data rates of up to 54 Mbps. Other IEEE 802.11 specifications deal with the provision of service guarantees with respect to classical quality of service parameters and the support of global roaming. It is not the intention of this book to further introduce IEEE 802.11 in detail, but only to describe positioning issues in these network. The interested reader can obtain all IEEE 802.11 documents from the web site of the related standard group (see (IEEE 802.11 Working Group Web site)).

WLAN installations are available in most public buildings today, and vendors are increasingly equipping PDAs and phones for WLAN connectivity. This makes WLAN very attractive for serving as a basis for indoor positioning. The next section identifies the main principles WLAN positioning relies on, followed by a more detailed overview of a special method called *fingerprinting*.

9.1.1 Principles of WLAN Positioning

Almost all WLAN positioning systems that have been developed as prototypes or are available as commercial products so far rely on measurements of the received signal strength (RSS), the received signal-to-noise ratio (SNR), or proximity sensing. Timing measurements are not an option, because precise timing synchronization is a very difficult issue in WLAN and time differences would be very hard to measure owing to the extremely short ranges the signals travel in indoor and local environments. RSS and SNR observations are based on the so-called *beacons*, which are broadcast either in the uplink or the downlink. Upon arrival of such a beacon, the receiver measures the RSS or SNR and makes it available to user-level applications, which is a standard feature of most WLAN equipment.

For measurements in the uplink, the mobile terminals must explicitly generate these beacons, which are then received by all access points that are in range. This is the basis for realizing network-based positioning methods. For measurements in the downlink, a standard feature of WLAN known as *passive scanning* can be utilized. Mobile terminals continuously perform passive scanning to detect nearby access points and select the best one for transmission. For this purpose, each access point periodically emits a beacon, which carries several parameters like a timestamp, supported data rates, and the access point's cell identifier, the so-called *Basic Service Set Identifier* (BSSI). The interval between two beacons can be dynamically configured and is typically in the range of several tens or hundreds of milliseconds. The terminal permanently listens to the possible channels for receiving beacons from nearby access points and records their parameters and measured RSS and SNR values. It then selects the access point with the best signal quality for transmission. This process is very similar to that performed by a GSM terminal for selecting a suitable base station. Alternatively, if the terminal does not receive a beacon during passive scanning in due time (maybe because the interval is configured to be too long), it can send a probe request, whereupon all access points in range respond with a beacon. This is known as *active scanning*. Thus, both active and passive scanning can serve as a basis for realizing terminal-based or terminal-assisted positioning.

The observation of beacons either in the downlink or uplink leads to the following three basic positioning methods:

- **Proximity sensing.** The position of the terminal is adopted from the position of the access point that the terminal has scanned with the best signal quality.

- **Lateration.** The position of the terminal is derived from lateration. The distances between access points and terminal are determined by the path loss a beacon experiences during transmission.

- **Fingerprinting.** The observed RSS patterns originating from or received at several access points are compared with a table of predetermined RSS patterns collected at various positions. The position for which this comparison fits best is then adopted as the terminal's position.

Not surprisingly, proximity sensing in WLAN, like for cellular systems, suffers from degraded accuracy, but is the simplest one to implement. It is supported by a database for mapping BSSIs onto room numbers. In an advanced approach, access points may broadcast the room numbers where they are located, which would make such a database unnecessary. Accuracy is in the range of several tens or hundreds of meters, depending on the transmitted signal strength and the density of access points in the building. In the worst case, such a system can only be used to detect whether a certain target person currently resides within a building or within a particular wing of a building. Even worse, it might be difficult to discriminate between different floors, which makes proximity sensing unusable for many applications. In the best case, it might be possible to derive location information with the granularity of rooms.

For applying lateration, it is required to accurately align the positions of access points inside the building. These positions can be expressed either by a local Cartesian coordinate system overlaid on the base of the building or by a global one such as ECEF. However, the former approach might be preferred, because the calculated position fixes can then be easily assigned to room numbers by taking into account the building's construction plans. Compared to cellular or satellite positioning, lateration in indoor environments generally suffer from negative impacts of multipath propagation if no line of sight exists. When traveling from the sender to the receiver, the beacon signals may be reflected and scattered at walls and ceilings several times, and each time experience a certain degree of attenuation that is hard to predict. This makes it nearly impossible to infer a realistic range from the path loss and imposes significant errors to the derived position fixes. Note that the path loss cannot be solely derived from the RSS; it is also necessary to know the transmitted signal strength, which requires dedicated signaling between sender and receiver.

Owing to the difficulties of conventional positioning methods in WLAN environments, a popular approach gaining more and more momentum in recent times is WLAN fingerprinting. The following section highlights its basic principles and gives an overview of implementations adopting this approach.

9.1.2 WLAN Fingerprinting

Remember from Chapter 6 that fingerprinting takes place in two phases. In the off-line phase, the system records RSS patterns for well-defined *reference positions* and stores them in a radio map. In a simple approach, this radio map consists of entries of the form $(p, RSS_1, \ldots, RSS_n)$, where p denotes the reference position and RSS_i for $i = 1, \ldots, n$ represents the RSS value of the ith access point. However, the RSS strongly depends on line-of-sight conditions on the spot, and these conditions, in turn, vary depending on the target's orientation. Therefore, most systems record RSS patterns for several directions d (e.g., north, south, west, east) at each reference position, resulting in entries of the form $(p, d, RSS_1, \ldots, RSS_n)$. Table 9.1 shows an example of a radio map generated in this way.

Table 9.1 Example of a radio map

Position	Direction	RSS/[dBm] from 00:02:2D:51:BD:1F	RSS/[dBm] from 00:02:2D:51:BC:78	RSS/[dBm] from 00:02:2D:65:96:92
p_1	0°	−59	−75	−71
	90°	−54	−73	−67
	180°	−49	−72	−69
	270°	−55	−73	−65
p_2	0°	−35	−64	−50
	90°	−27	−64	−43
	180°	−40	−65	−52
	270°	−30	−60	−46
p_3	0°	−69	−66	−73
	90°	−65	−60	−68
	180°	−63	−66	−70
	270°	−68	−62	−76

The reference positions may be selected equally spaced according to a regular grid or selected irregularly depending on accuracy demands and the building's structure. The latter is demonstrated in Figure 9.1. The reference positions might be specified by means of Cartesian coordinates, descriptive room numbers, or any other reference system. The off-line phase is also referred to as *training* or *calibration*. In the on-line phase, RSS patterns related to the target are recorded and compared with the RSS fields of the entries stored in the radio map. The position of the target is then extracted from the reference position of the

INDOOR POSITIONING

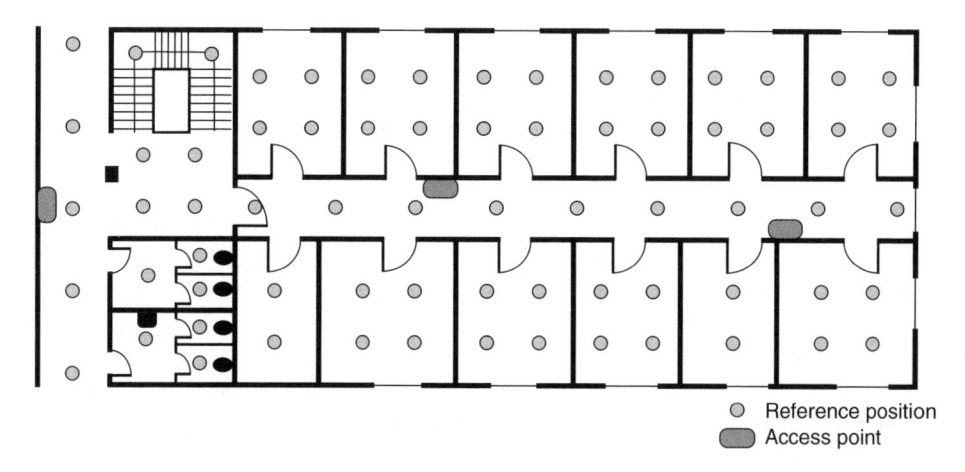

Figure 9.1 WLAN fingerprinting environment.

entry with the closest match. This matching can be done according to different approaches and algorithms, which will be explained later.

Fingerprinting can be performed terminal assisted, terminal based, and network based (see Figure 9.2). For terminal-assisted and terminal-based approaches, the radio map is derived from RSS downlink measurements made at the various reference positions in the off-line phase. For this purpose, a terminal observes the beacons emitted by the access points in the surrounding area from several directions and records the associated RSS values. The procedure during the on-line phase is then as follows: the target terminal permanently records RSS patterns in a similar fashion as in the off-line phase, but only in the direction of user movement. In the terminal-assisted approach, it sends regular measurement reports to a server in the network (see Figure 9.2(a)). This server maintains the radio map and performs the matching of RSS patterns for deriving the position. In the terminal-based approach, the radio map is maintained in the terminal, and the matching can be performed locally (see Figure 9.2(b)).

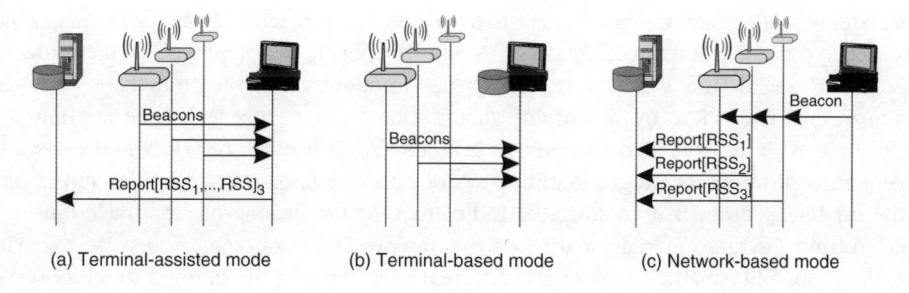

Figure 9.2 Modes for fingerprinting.

For network-based fingerprinting, the radio map is created from RSS measurements made in the uplink. In the off-line phase, the terminal periodically emits beacons from several directions at each reference position. The access points in the surrounding area receive these beacons and record the associated RSS. The measurement results are then centrally merged for composing the radio map. During the on-line phase, the target terminal must periodically emit beacons for measurements at the surrounding access points, which then transmit the resulting measurement reports to the server for position matching (see Figure 9.2(c)).

A significant drawback of fingerprinting is obviously the overhead for composing radio maps in the off-line phase. Performing measurements in a close grid of coordinates for completely covering an entire building is a time-consuming and cumbersome process. Even worse, this process has to be repeated whenever the configuration of access points changes, for example, if a new one is installed or an existing one is moved, removed, or replaced. An alternative is therefore to derive radio maps from a mathematical model that calculates the radio propagation conditions taking into consideration the positions of access points, their transmitted signal strengths, the free-space path loss, and obstacles reflecting or scattering signals like walls and furniture in the surrounding area. In this way, it is possible to conveniently and rapidly create updated radio maps for a close-meshed grid of reference positions whenever the configuration of access points changes and without performing any measurements at all. In the following text, this approach is referred to as the *modeling approach*, while the creation of radio maps from measurements is called *empirical approach*.

The empirical approach can be further subdivided into deterministic and probabilistic methods. For the former, several RSS samples are recorded for each reference position and direction, and the radio map is then created from the mean values of these samples. During the on-line phase, the matching between observed and recorded RSS patterns happens according to a metric. A common method is to compute the Euclidean distance $\sqrt{(RSS_{o,1} - RSS_{r,1})^2 + \ldots + (RSS_{o,n} - RSS_{r,n})^2}$, where $(RSS_{o,1}, \ldots, RSS_{o,n})$ is the observed RSS pattern and $(RSS_{r,1}, \ldots, RSS_{r,n})$ is the recorded RSS pattern at a particular reference position. From all reference positions stored in the radio map, the position with the smallest Euclidean distance is then assumed to be the target's current position. This method is also known as *Nearest Neighbor in Signal Space* (NNSS) and has been proposed by (Bahl and Padmanabhan 2000). However, other metrics can be used, too.

The drawback of deterministic methods is that for each position fix the entire radio map has to be searched and the matching is based on averaged RSS patterns only. The latter may cause significant accuracy degradations if the RSSs are subject to large variations, which may occur due to a number of reasons. A more sophisticated approach is therefore not to record averaged RSS values, but to describe variations of signal strengths experienced during the off-line phase by probability distributions. Often, this technique is applied in conjunction with the so-called *joint clustering*, see (Youssef et al. 2003). A *cluster* is a set of reference positions sharing a common set of access points covering them and is built in the off-line phase when creating the radio map. In the on-line phase, first a cluster is selected from the radio map depending on the observed (or observing) access points. After that, the probability distributions of the different access points are applied to the observed RSS pattern to find the most probable position. In this way, it is possible to reduce the computational overhead needed for searching the radio map and to refine accuracy of position estimation (both compared to deterministic methods).

In recent years, the research community has created a multitude of WLAN fingerprinting systems, which cannot be covered here completely and in detail. A selective overview

INDOOR POSITIONING

of some systems and their properties is presented in Table 9.2. The systems are compared against the observables they make use of, the delivered accuracy, the mode of measurements and calculations (ta = terminal-assisted, tb = terminal-based, nb = network-based), the method for creating radio maps, and the type of matching algorithm applied. The pioneer of all WLAN fingerprinting systems is the RADAR system developed by Microsoft Research in 1999 (see (Bahl and Padmanabhan 2000)). One of the few commercial systems that are available for purchase is the Ekahau system. Its developers have published a probabilistic approach for position matching in (Roos et al. 2002). The Nibble system is special in that it observes the SNR of incoming beacons instead of the RSS. The approach is presented in (Castro et al. 2001). Detailed information on the Horus system can be obtained from (Youssef et al. 2003), and on the WhereMops system from (Wallbaum and Wasch 2004).

Table 9.2 Overview of WLAN fingerprinting systems. Adapted from (Wallbaum and Diepolder 2005)

System	Observable	Accuracy	Mode			Radio Map		Matching	
			ta	tb	nb	emp.	mod.	det.	prob.
RADAR	RSS	2.1 m/50%		×		×		×	
Ekahau	RSS	3.1–4.6 m/90%	×			×			×
Horus	RSS	2.1 m/90%		×		×			×
Nibble	SNR	10 m/80%	×			×			×
WhereMops	RSS	1.5 m/50% 6.0 m/95%		×			×		×

The table clearly demonstrates that the achievable accuracy with WLAN fingerprinting is in the range of a few meters, which should be quite acceptable for most LBS indoor applications. In general, it can be concluded that, besides accuracy, the major advantage of WLAN fingerprinting is that it is based on existing WLAN installations, which most public buildings today are equipped with, and standard WLAN equipment in notebooks, PDAs, and mobile phones. It does not require any modifications on the hardware, but only special software for creating radio maps and making RSS observations in the on-line phase.

9.2 RFID Positioning

Radio Frequency Identification (RFID) is an emerging technology that is primarily used today for applications like asset management, access control, textile identification, collecting tolls, or factory automation. It is based on radio signals that are exchanged between an RFID reader and RFID tags (or transponders).

A reader consists of an antenna, a transceiver, a processor, power supply, and an interface for connecting it to a server, for example, by a serial port or via Ethernet. An RFID tag has an antenna, a transceiver, and a small computer and memory. There is a distinction between *active* and *passive tags*. The former is equipped with power supply in the form of a battery, while the latter extracts the required energy from the radio signals emitted by the readers. This has fundamental impacts on the communication range: active tags typically bridge distances of several tens of meters, while passive tags have a range between tens of centimeters and a few meters. Furthermore, active tags contain more memory and are equipped with more intelligence. They often contain additional sensors, for example, for checking temperature or humidity, and are able to store the history of sensor data and to calculate statistics from it. Passive tags, on the other hand, have a memory size of a few kilobytes only, and their functions are restricted to deliver an ID or other information stored in memory.

The RFID systems available at the market operate at different frequency ranges. Typically, they are categorized as high-frequency (850–950 MHz and 2.4–5 GHz), intermediate-frequency (10–15 MHz), and low frequency (100–500 kHz) systems. The frequency range has an impact on the communication range, the data rates, and the costs. Generally, systems operating at higher frequencies achieve larger ranges and faster data rates, but also suffer from increased acquisition costs.

RFID technology can be basically categorized in the positioning classification scheme as proximity sensing. However, there is no beacon broadcast in a one-way fashion as for other proximity methods. Rather, the RFID reader explicitly prompts the tags within its communication range to respond with some information from their memory. For positioning purposes, this might be a simple identifier that either represents the location of the tag or the target person to be located. This leads to the question of whether to deploy RFID in a network- or terminal-based approach. Basically, both are possible and are done in practice. In a network-based approach, the readers are mounted to walls, corridors, entrances and exits within a building and have to be connected to a location server for gathering location data. The tags are carried by the target persons and the respective identifier is transferred when a person is passing a reader. In a terminal-based approach, the reader is integrated into a mobile device like a PDA or mobile phone and catches location data from the tags in the infrastructure when passing them. Which mode to choose is a design choice and depends on a lot of circumstances like the demands of the respective application, privacy issues, and the ratio between the number of person to be tracked and the size of the area to be covered. The latter is due to the fact that RFID readers are much more expensive than tags, and hence their number should be limited.

9.3 Indoor Positioning with GPS

This section briefly describes indoor positioning with GPS, which is a new receiver technology and which is gaining more and more momentum recently.

Signals of GPS satellites arrive at a GPS receiver located in open-air environments with a strength of at least -130 dBm, which, when compared with other radio technologies, is rather weak. For example, WLAN and Bluetooth typically operate with RSSs of between -20 dBm and -80 dBm. In indoor environments, the situation is even worse: owing to shadowing and multipath GPS signals are additionally attenuated by approximately 30 dB compared to the RSS that is achievable outdoors. As a consequence, conventional GPS receivers located indoors cannot properly detect and interpret GPS signals and positioning is not possible. Thus, besides the long acquisition time for the first position fix (TTFF), the absence of GPS in indoor environments is the other major drawback of the system.

INDOOR POSITIONING

Remember from Chapter 7 that GPS receivers perform a technique known as *code phase ranging* for time measurements. Before that, the receiver first has to identify a sufficient number of visible satellites, which is done by building the autocorrelation between the received signals and all C/A-codes. This is executed by correlators, of which conventional GPS receivers have two to four dedicated to each satellite. For identifying the signals of a certain satellite, the correlators must detect the code delay between the local generated and the received codes, which, according to the number of chips per C/A code, is in the range of 0 to 1,023 chips. In addition, this procedure must be repeated for different deviations of the L1 carrier frequency resulting from Doppler shifts. To execute the acquisition phase in a reasonable amount of time, conventional GPS receivers spend one millisecond for checking one of 1,023 possible code delays at each frequency, which is also known as *dwell time*. If, after multiplying generated and received signal and then integrating the product, the correlator delivers a peak (see also Section 4.3.4), the receiver identifies a satellite and is synchronized to it.

The problem that occurs in indoor environments is that the signals are much too weak and can hardly be discriminated from background noise. However, it is possible to increase the sensitivity of GPS receivers by increasing the dwell time of correlation. Each additional millisecond delivers additional data for integration with previous results and improves the SNR ratio. For example, performing correlations for a duration of 10 ms instead of 1 ms leads to a sensitivity gain of 10 dB. However, the problem of this approach is that the required acquisition time would also dramatically increase.

Therefore, the idea behind indoor positioning with GPS is to equip GPS receivers with a high number of parallel working correlators, which enables the integration of hundreds of milliseconds of data for delivering the required sensitivity for indoor operations. The typical number of correlators is in the range of tens of thousands and the latest generation of GPS receiver technology even enables one-chip solutions on the basis of 200,000 correlators and A-GPS support. With these solutions, it is possible to hear signals with a strength of only -160 dBm and they significantly reduce acquisition times to a few seconds, thereby contributing to a fast TTFF. The reported accuracy is in the range of 10 to 20 m. For details about Indoor GPS, see (van Diggelen and Abraham 2001).

9.4 Non Radiolocation Systems

This section provides an overview of alternative positioning methods that make use of infrared and ultrasound signals. The presented systems are the results of research initiatives and are not available as commercial products or specifications.

9.4.1 Infrared-based Systems

A number of prototypes developed in research projects are based on infrared signals for realizing positioning methods based on proximity sensing. Unlike radio signals, infrared signals have the advantage in that their emitters have a short range of some meters only and do not penetrate walls. These properties predispose them for use in conjunction with proximity sensing for applications that require to resolve a target's position at least with the granularity of rooms inside a building. However, they are only slightly reflected and scattered at obstacles inside a room, and hence it may be necessary to have a line of sight between emitter and receiver. When the first infrared positioning systems were developed at the beginning of the 1990s, infrared was already a mature technology available in a broad range of products (e.g., remote controls of domestic appliances). Compared to radio, the required equipment was and still is comparatively cheap, straightforward, and shows only a low power consumption.

The pioneer not only of indoor positioning but also of location-based applications in general is the ActiveBadge system developed by Olivetti research at the beginning of the 1990s (see (Want et al. 1992)). It focuses on the localization of staff members within an office environment for automatically forwarding incoming telephone calls (selective routing) to the telephone set that is next to the current location of the callee. For this purpose, each staff member wears an infrared tag, the so-called *active badge*, which periodically emits a beacon. The beacon carries an ID that uniquely identifies the person wearing the badge. It is emitted every 15 s and lasts for a duration of one-tenth of a second. In this way, the batteries inside the badges are not exhausted too fast and last for about one year. The ratio between beacon duration and its periodicity also decreases the chance of collisions between the emissions of badges from different persons staying close by. In this way, it is not required to establish a complicated multiple-access scheme (the method applied is actually a kind of statistical multiple access).

The beacons are received by infrared sensors installed inside the building (see Figure 9.3(a)). For the prototype, they have been placed high up on walls or ceiling tiles of rooms and on the entrances and exits of corridors. The initial systems consisted of 128 sensors, which were controlled from RS232 ports of several workstations, which, in turn, were interconnected via Ethernet. At regular intervals, the sensors are polled by a central location server, which stores the gathered data as triplets of the form (ID, location, time). The location server can be interconnected with a switch for automatically forwarding telephone calls to the right place or it can be requested by a receptionist for manual switching. Thus, ActiveBadge follows a network-based approach for positioning. Want et al. (1992) also discuss the need for advanced control mechanisms for staff members and privacy issues.

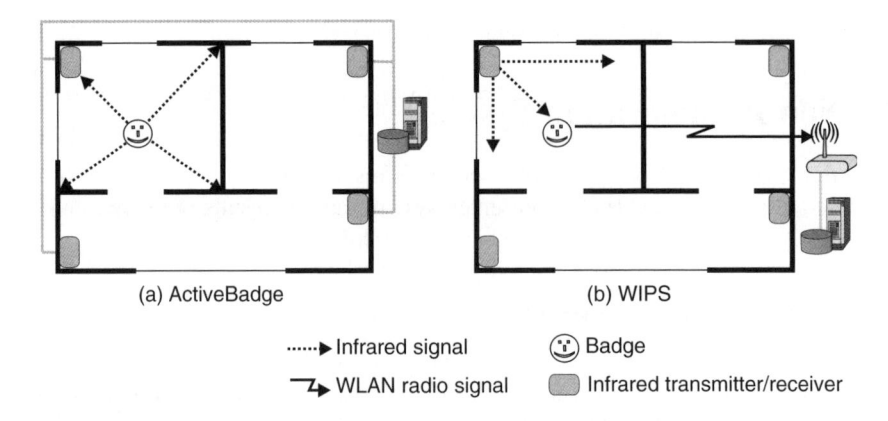

Figure 9.3 ActiveBadge and WIPS. Adapted from (Roth 2004).

An example for an infrared terminal-based approach is the *Wireless Indoor Positioning System* (WIPS) developed at the Royal Institute of Technology in Sweden (see (Royal

Institute of Technology, Sweden 2000)). In WIPS, infrared beacons are emitted by transmitters mounted in different rooms of a building (see Figure 9.3(b)). The beacons carry information from which location data can be derived and are caught by badges attached to persons when entering the coverage area of a transmitter. In order to make location data collected thus available to a distributed application, the badges forward it to a central database via WLAN. Thus, WIPS incorporates two wireless technologies, infrared for positioning and WLAN for connectivity, and hence the badges are more complex and power consuming than those used in the ActiveBadge system. WIPS comprises several location-related applications like the listing of all people connected to the system together with their locations, the listing of persons staying close by to a certain person, or the announcement of nearby peripherals.

9.4.2 Ultrasound-based Systems

An alternative to positioning via infrared and radio signals is the use of ultrasound. The major advantage of ultrasound signals is their propagation velocity of 1,243 km/h, which is very low when compared to that of infrared and radio signals of approximately 300,000 km/s. Consequently, ultrasound provides a very convenient basis for timing measurements and can thus be utilized for lateration without the need for complicated and expensive synchronization mechanisms. Ultrasound signals do not penetrate walls and do not require a line of sight between sender and receiver. Unfortunately, their propagation range is very limited, which makes them impracticable for localization in large coverage areas and thus as an alternative to cellular positioning methods. However, inside buildings ultrasound can achieve positioning accuracies in the range of centimeters, assuming that there exists a close-meshed network of transmitters and receivers respectively.

The working of ultrasound is again demonstrated by means of network-based and terminal-based approaches. An example for the former is the *ActiveBat* system, which was developed at the University of Cambridge and by Olivetti (see (Ward et al. 1997)). Each person to be located must carry a so-called *bat*, which consists of a hemispherical array of ultrasound transducers and a radio transceiver operating at 418 MHz. Each of these bats is assigned a unique address. The area inside a building to be covered for positioning must be equipped with a close mesh of ultrasound detectors, which are mounted on the ceiling (see Figure 9.4). The detectors are connected to a PC via a serial network. The prototype described in (Ward et al. 1997) consisted of a four-by-four square grid where the detectors were placed 1.2 m apart.

The positioning process is then controlled by radio signals. It is triggered by a broadcast message sent by the controlling PC on the 418 MHz radio link. This message carries the address of the bat to be located. At the same time as this message is emitted, the PC sends another one over the serial network for resetting the ultrasound detectors and for establishing a common time basis in this way. Upon arrival of the broadcast message, the addressed target bat emits an ultrasound impulse that propagates in a hemispherical pattern toward the detectors mounted at the ceiling. The arrival time of this impulse is recorded at each detector. Finally, the PC polls the timing information from the detectors and calculates the target bat's position by lateration. The time differences between the controlling broadcast message and the reset message sent over the serial network are very small compared to the slow propagation of ultrasound and can therefore be neglected. The reported accuracy is in

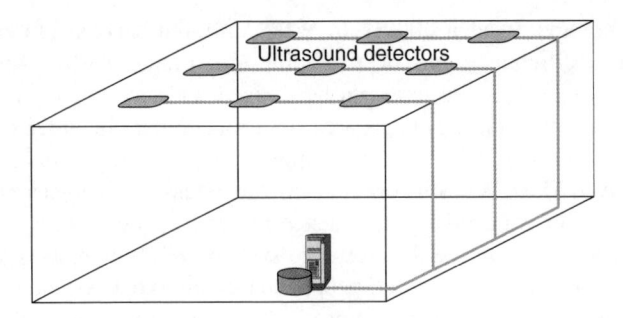

Figure 9.4 Ultrasound infrastructure.

the range of 10 cm for a 90% confidence level. ActiveBat also permits the determination of the orientation of the person wearing the bat.

The Cricket system implements a terminal-based approach for proximity detection with ultrasound. Again, the person to be located wears a kind of badge that is equipped with both a radio receiver and an ultrasound detector. Devices located in the rooms, which in this system are called *beacons* (in contrast to the signals this term is used for in other systems), emit a radio signal and an ultrasound impulse at the same time. Because the radio signal travels much faster, it arrives at the badge first and can thus be seen as a kind of announcement for indicating the impending arrival of the ultrasound impulse. This feature can be utilized for measuring the range between beacon and badge. If the badge is within range to at least three beacons, it would be possible to apply lateration in this way. However, the Cricket system uses range information only for determining the nearest beacon and for extracting the location information carried in the radio message from it. On the other hand, the advantage of this approach is that the installation overhead is less complex compared to ActiveBat, as a closed meshed grid of ultrasound devices is not required for proximity sensing.

9.5 Conclusion

This section has highlighted the most important technologies for indoor positioning. Although there already exist some commercial systems based on ultrasound positioning, for example, in the area of health care, non radiolocation methods in general are still in an experimental phase and therefore cannot be considered to be mature enough for entering the mass market, considering the prevailing circumstances. This is also due to the fact that they require a stand-alone infrastructure to be installed and special devices.

The latter argument obviously holds for RFID, too. However, this technology has experienced a massive push at the markets in recent years. It is not only of relevance for business applications in the field of asset tracking today, but is also predisposed for tagging goods and products in shopping malls, and hence end consumers will be faced with this new technology very soon. This trend is accompanied by the efforts of vendors of mobile devices and phones, who have already started benefiting with the integration of RFID readers into their products. If these RFID initiatives succeed, there will be a large potential to utilize this technology for indoor positioning also. The major advantages of WLAN fingerprinting are that it does not require stand-alone installations and equipment and that it delivers a very good accuracy. The major drawback is that the empirical generation of radio maps is a complicated and long-lasting process that must be repeated whenever the installation of access points is changed, while the modeling approach is still premature. From the point of view of integrating indoor and outdoor applications, indoor positioning by GPS is certainly the most attractive solution. Unfortunately, the delivered accuracy in the range of 10–20 m is rather moderate, when compared to WLAN fingerprinting, and does not allow a discrimination between different rooms and floors within a building.

Part III

LBS Operation

10

Interorganizational LBS Operation

Generally, there are many actors involved in the operation of an LBS. In the following text, an *actor* denotes an individual, organization, department, or enterprise that offers services to other actors, or consumes services from other actors, or does both of them. From a system's point, each actor autonomously operates and controls its own administrative technical domain, given, for example, by a network infrastructure, a server farm, or only a single mobile device. It is common practice in the telecommunications or Internet sector to classify the different actors participating in the operation of a service according to their *roles*, where a role represents a certain field of activity of an actor associated with a set of functions for realizing and controlling portions of a service as well as making it accessible to the end user. For this purpose, the roles have to interact with each other. A *reference point* covers the interaction between a pair of roles and is basically an abstract term representing communication links, interfaces, protocols, and transactions that are needed to exchange any kind of user and control data a service or portions of a service are based on. The classical roles in mobile communications are the roles of subscriber, network operator, application service provider, and content provider. A single actor may adopt several of these roles, for example, the roles of network operator and application service provider often coincide, because companies operating a cellular network do not want to restrict to connectivity services only, but want to offer application services, too. At the same time, there is a need to separate these roles in order to enable open, highly competitive service markets, where multiple actors participate in a *supply chain* for composing and offering sophisticated IT services.

In the particular case of LBSs, this supply chain comprises positioning, the refinement of position fixes to *location data*, relating location data either with geographic content or location data of other targets, and translating the result to *application data* suitable for presentation to the LBS user. This chapter deals with the interorganizational aspects of LBS supply chains. At first, it presents a general overview of the various actors participating in the realization of an LBS and of a number of scenarios covering different constellations of

---

Location-based Services: Fundamentals and Operation Axel Küpper
© 2005 John Wiley & Sons, Ltd

their interaction. The remainder of the chapter then deals with patterns for the dissemination of location information and with privacy aspects.

10.1 LBS Supply Chain

Figure 10.1 shows a general model for an LBS supply chain, which identifies the participating roles and the reference points in between. Note that this model is not part of any LBS or LBS-related standard or recommendation. It just provides a basis for the understanding of mechanisms and protocols that the LBS operation relies on and which will be introduced in subsequent chapters. Nevertheless, the roles identified in the supply chain have been selected in consensus with most LBS approaches, standards, and recommendations, although the terminology used there sometimes differs. The following list explains the functions of these roles:

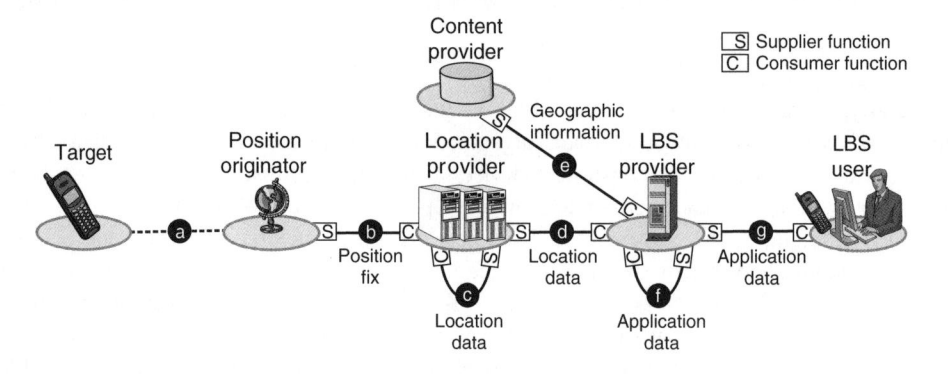

Figure 10.1 LBS supply chain.

- **Target.** As already mentioned in the previous chapters, a target is a mobile individual or object that is to be located, tracked, or sighted. For this purpose, it is equipped with a mobile terminal, for example, a cellular phone, PDA, GPS receiver, or badge, which is at least able to perform or to support measurements needed for positioning.

- **Position originator.** The position originator is the actor that calculates the position fix and thus represents the initial member of the LBS supply chain. For (network-assisted) terminal-based positioning, this role coincides with the target, while, for (terminal-assisted) network-based positioning, the role of the position originator is mostly adopted by an operator maintaining the positioning infrastructure.

- **Location provider.** The location provider is an intermediate role arranged between position originator and LBS provider. It triggers and controls positioning on behalf of an LBS provider, gathers position fixes of one or several targets from one or several position originators, refines them to more sophisticated location data, and returns it to the LBS provider. This service is referred to as *location service* (LCS).

- **LBS provider.** The LBS provider is the central role that offers the service logic for realizing an LBS, and maintains subscriptions with the LBS users. It collects location data from one or several targets, executes spatial analysis, combines it with other geographic content, and transfers the resulting application data to the LBS user.

- **Content provider.** A content provider may support an LBS provider in offering geographic content such as maps, routing data for navigation, or points of interest. For this purpose, it maintains a spatial database and GIS.

- **LBS user.** The LBS user is the actor that "consumes" an LBS. It usually requests it via a mobile device like a cellular phone or PDA or from a fixed terminal like a PC.

Note that these actors generate or consume different kinds of location information. It is distinguished between *position fixes*, *location data*, *geographic content*, and *application data*. As mentioned earlier, a position fix is an immediate result of positioning and in most cases represents the bare coordinates of the target. Location data, on the other hand, is more suitable for processing by the respective LBS application, particularly by an application, that is independent of the positioning method used. For this purpose, it contains a target's location in the format of the reference system required by the application as well as the target's identifier, quality data, and optionally other location-related data such as velocity and direction of motion. Geographic content is the description of stationary real-world entities and infrastructures (e.g., streets, buildings, and so on), as has been introduced in Chapter 3. Finally, application data is the result of an LBS function that is presented to the LBS user. An important example for this is the map that displays the locations of one or several targets. For the sake of simplicity, the term "location information" will be used in this chapter where the distinction between these kinds of locations is of less relevance.

Figure 10.1 also demonstrates the passing of location information along the supply chain. For this purpose, actors here adopt the function of a consumer, a supplier, or both of them. A *consumer* requests location information from a *supplier*, which collects, combines, and refines it, and passes it to one or several consumers. The interaction between consumer and supplier may be based on synchronous or asynchronous communication. In the former case, the consumer requests for location information of a particular target and immediately receives a response, while in asynchronous communication, the consumer subscribes for certain events and receives location information only when the event occurs. As will be explained later, the former category focuses on the realization of reactive LBSs, while the latter is primarily used for proactive LBSs.

The supply chain starts with positioning, which is represented by reference point (a) in Figure 10.1 and which happens between the target to be located and the position originator that calculates the position fix. This reference point is the only one without explicit supplier/consumer interaction. It is strongly dependent on the positioning method used and comprises the measurements of observables and signaling messages for exchanging measurement data between target and position originator. Once measurement data has been derived at the position originator, it generates the position fix, which represents the actual raw material of an LBS. The position originator acts as a supplier toward the location provider. The position fix is passed between both roles via reference point (b). The location provider then refines the position fix to location data and offers it to an LBS provider. Thus, the location provider adopts both functions, that of a consumer toward the position originator and that of a supplier toward the LBS provider. Furthermore, it is possible that multiple location providers incorporate into a federation, which is represented by reference point (c).

The LBS provider may have two subscriber functions, one for the interconnection with a location provider (d) for receiving location data of one or several targets, which is mandatory, and the other for receiving geographic content from a content provider (e), which is optional. The latter may be used to get map material or the coordinates of certain points of interest that are to be matched against location data of a target. Again, multiple LBS providers may incorporate into a federation, which is realized by reference point (f). Finally, the application data that has been generated by the LBS provider based on the locations of one or several targets and possibly after taking into consideration other geographic content is passed to the LBS user via reference point (g).

10.2 Scenarios of the LBS Supply Chain

The supply chain presented in Figure 10.1 does not match reality in that each role is adopted by another actor. Rather, as has been mentioned earlier, actors typically act in several roles during the operation of an LBS. The conjunction between roles and actors depends on a lot of circumstances like the technical capabilities of the terminal and positioning infrastructure, the yield of a positioning method, the type of LBS (e.g., reactive versus proactive, and client/server versus peer-to-peer service), and finally the business model that describes LBS operation from a market perspective. As a result, there exist many different configurations for the teamwork between actors during LBS operation, which have a significant impact on its realization. To give an overview of this complex matter, this section highlights five different scenarios of the LBS supply chain.

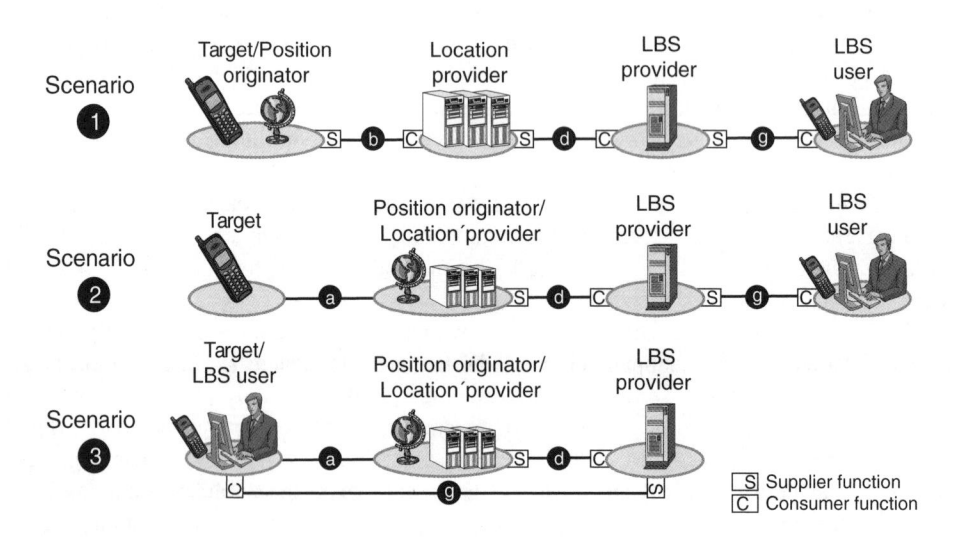

Figure 10.2 Supply chain scenarios 1–3.

Figure 10.2 covers its typical configuration in cases of cellular positioning. Scenario (1) represents an approach with terminal-based positioning like E-OTD or A-GPS. The position fix is calculated at the target's terminal, and hence the roles of target and position originator coincide here. As a consequence, reference point (a) is not visible as it is included inside the terminal. Scenario (2) differs from Scenario (1) in that positioning is done by a network-based approach like Cell-Id or U-TDoA. The position originator represents the network where the position fix is calculated, and adopts also the role of the location provider. Reference point (b) is realized within this network. Finally, Scenario (3) is a modification of Scenario (2) in that target and LBS user are the same actor here. Certainly, LBS user and target may also coincide in Scenario (1), which is, however, not explicitly shown here.

In these scenarios, cellular network operator and location provider coincide. Position fixes are refined to location data and offered to other actors via a *Gateway Mobile Location Center* (GMLC), which will be covered in Chapter 11. In Scenarios (2) and (3), the network operator acts also as position originator, as the position fix is calculated in the SMLC based on the measurements delivered by one or several base stations. Typical applications that can be realized by Scenarios (1) and (2) are community services and mobile gaming, that is, all applications where multiple users share their locations. Scenario (3) covers self-centric applications where the user's own location is processed and no external user's locations are involved. This is typical of mobile marketing and information retrieval services.

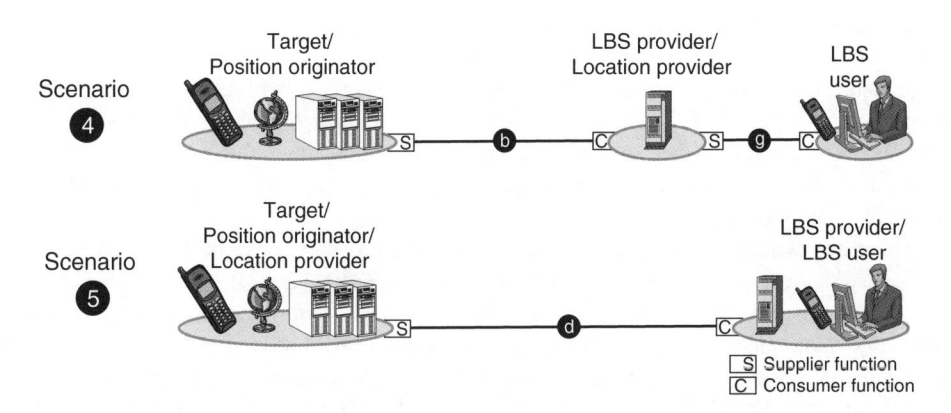

Figure 10.3 Supply chain scenarios 4 and 5.

Scenarios (4) and (5), which are shown in Figure 10.3, are rather related to LBSs that are based on noncellular positioning and where the targets are attached to the supply chain via connectivity services like WLAN or GPRS. In Scenario (4), target and position originator functions reside in the terminal. This scenario appears if a terminal is able to autonomously perform positioning, for example, by GPS or indoor beacons. The terminal calculates the position fix, and sends it to an actor adopting the roles of location and LBS provider. Therefore, reference point (b) is realized within the terminal and (d) within the domain of the location/LBS provider. Scenario (5) covers a peer-to-peer LBS. This time, target, position originator, and location provider functions coincide in the target's terminal, and the functions of the LBS provider are adopted by the LBS user itself.

Among other applications, Scenarios (4) and (5) cover LBS applications outside a cellular network. This means that positioning happens through another approach, but it does not exclude that the target and the user may be attached via connectivity services of a cellular network. Scenario (4) may represent an indoor positioning system for tracking persons inside a building, while Scenario (5) may cover fleet management applications, where the trucks of a fleet are equipped with GPS-enabled OBUs. The GPS receiver calculates the position fix, and the OBU translates it to a highway number and milestone, which is then directly transferred over a public cellular network to the fleet management center.

Note that the scenarios presented here are just examples. Basically, any other configuration of roles may appear in reality, too. From the point of view of software reuse, it would be desirable that a given reference point uses the same protocols and mechanisms independent of a particular scenario. Unfortunately, many of today's systems and services have been tailored to a particular scenario, infrastructure, or, even worse, a particular application, thereby making software reuse and rapid service creating and deployment a difficult matter.

10.3 Supplier/Consumer Patterns for Location Dissemination

The previous sections have clearly demonstrated the need for exchanging location information along the supply chain, which in the following text is referred to as *location dissemination*. Although it must be distinguished between position fixes, location, and application data, and this data is accordingly exchanged between actors of different roles, location dissemination always follows similar patterns, which are termed *supplier/consumer patterns* here and which refer to the functions of supplier and consumer the different actors may adopt. This section explains these patterns. In particular, they can be applied to reference points (b), (c), and (d).

There are two basic patterns, *reporting* and *querying*, which can be further subdivided into several subcategories (see also Leonhardi and Rothermel (2002)). In addition, it is possible to dynamically parameterize them on the basis of the requirements of the application. The deployment of a certain pattern and its configuration may have a significant impact on the frequency with which location information is exchanged. The aim is generally to limit the overhead caused by the dissemination, while simultaneously meeting the requirements of the respective LBS. Depending on these requirements, a suitable strategy must be selected and appropriately configured.

10.3.1 Querying

Querying is based on synchronous communication between consumer and supplier and thus follows the traditional RPC (*Remote Procedure Call*) approach. The consumer explicitly requests location information from the supplier and is immediately served. The corresponding sequence diagram is depicted in Figure 10.4. In a simple approach, the consumer invokes a location request whenever it needs to process the location information of a particular target or if it has been requested by an external actor to acquire it. The location request contains at least the target's identity. The supplier then obtains the location information of the specified target and returns it to the consumer.

An advanced approach would be querying in combination with caching, where the consumer holds a copy of a target's location information and reuses it until it expires. The decision whether or not a copy has expired may be determined by a timer or it may

INTERORGANIZATIONAL LBS OPERATION

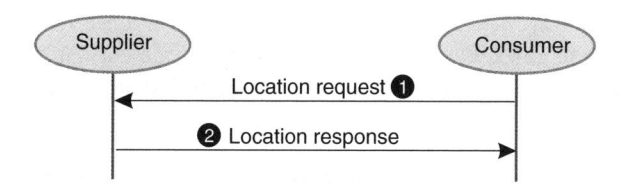

Figure 10.4 Sequence diagram for querying.

be based on an estimation that, for example, takes the target's velocity into account, if available. Different strategies exist for such an estimation (see (Leonhardi and Rothermel 2002) for an overview). If the copy expires, the consumer may immediately request for fresh location information at the supplier or it may defer this request until there is a demand for processing it.

10.3.2 Reporting

As opposed to querying, reporting is based on asynchronous communication. In terms of design patterns, it is also known as the *observable pattern* (see (Gamma et al. 1997)), or as *publish/subscribe paradigm* (see (Eugster et al. 2003)). The basic character of this pattern is that it is event based, that is, reporting is initialized by the supplier when a trigger condition becomes true. In most cases, the consumer has to subscribe first to the supplier for receiving reports and specify the trigger conditions. The general procedure of reporting is illustrated in Figure 10.5.

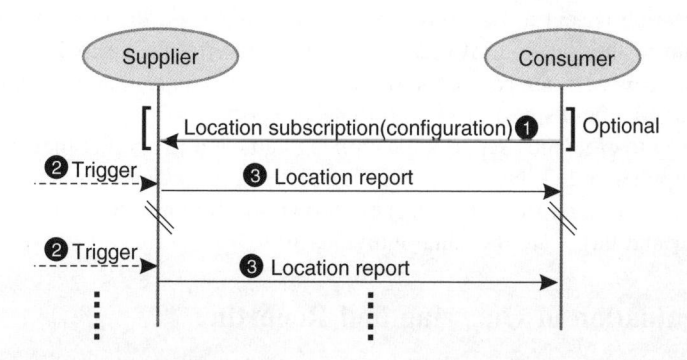

Figure 10.5 Sequence diagram for reporting.

Basically, a trigger condition acts as a filter that is applied on the location information gathered by the supplier. Each time new location data becomes available, it is checked against the filter, and only if the condition is true, it is sent to the consumer. Reporting can be subdivided into the following categories:

- **Immediate reporting.** Each time new location information arrives at the supplier, it is reported to the consumer. Thus, for this particular category, no trigger condition needs to be specified.

- **Periodic reporting.** The trigger condition is given by a timer value that is applied on a timer at the supplier. Each time the timer expires, the current location information is sent to the consumer and the timer is reinitialized, that is, location information is reported periodically.

- **Distance-based reporting.** Each time new location information becomes available, the supplier determines the distance between this location and the last reported one. If this distance exceeds a threshold prescribed by the trigger condition, the current location information is reported to the consumer.

- **Zone-based reporting.** A report occurs if the target has entered or left a predefined zone. A zone, similar to the definition of spatial objects in GIS, represents any physical location and may be defined as a single point, a circle or ellipse of a certain center and axis, a line, or a polygon. For example, a circle can be used to model the borders of a city, a line may represent a street or river, and a polygon can be used to specify the outlines of a building. The trigger condition defines the zone and whether the report is to be sent on entering or leaving the zone.

Another, more advanced reporting strategy is based on dead reckoning (see Section 6.2.4 for an introduction), where the trigger condition is given by a threshold that refers to the deviation of a position estimate from the actual position fix. To apply this strategy, the dead reckoning algorithm must be performed on both sites, the supplier and the consumer. The supplier calculates a position estimate based on the position last reported to the recipient as well as the target's direction and speed of motion. The result is then compared with the current position, and if the difference between them exceeds the threshold, a report with the current position is sent to the consumer. The consumer, on the other hand, applies the same algorithm to the location information last received from the supplier whenever it is requested to process between two reports. This, however, requires that the supplier attaches the target's direction and speed of motion to each report.

Another way to efficiently report a location is to attach it to another message exchanged between the supplier and the consumer. This is called *piggybacking*. It is a preferred strategy if an LBS user invokes an LBS and attaches its own location information to the invocation (i.e., LBS user and target are the same individual).

10.3.3 Evaluation of Querying and Reporting

Location dissemination causes a certain amount of network load that is intended to be reduced by careful deployment and combination of the various supplier/consumer patterns. Especially, this concerns the air interface, which is always the most valuable resource in a wireless network, and which is affected by location dissemination for all scenarios where suppliers reside in the terminal, that is, for all terminal-based positioning methods (Scenarios (1), (4), and (5)). Note, however, that network-based positioning also burdens the air interface for each position fix, but this overhead is not caused by dissemination but by positioning itself and is therefore not treated here.

Another evaluation criteria in cases the supplier resides in the terminal is charging. In Scenarios (1) and (3), the dissemination of position fixes and location data belongs to the explicit tasks of a cellular network operator and is thus processed by control channels at the air interface and the SS7 network in the fixed infrastructure, both within the operator's domain. Generally, the subscriber to be located is not charged for using these resources. Rather, the LBS user may be charged for each location data delivered by the GMLC for his service session. However, if the infrastructure of the network operator is bypassed (see Scenarios (4) and (5)), the target has to access connectivity services such as GPRS for transferring location information, which is commonly charged according to a time- or volume-based tariff. In this case, each query or report must be paid by the target, which is thus another motivation for reducing the frequency of querying or reporting.

Taking into consideration these criteria, it can be generally stated that querying is a prerequisited for any kind of reactive services, which are initiated by the LBS user, for example, in order to receive a list of nearby points of interest like restaurants (for enquiry and information services, see Section 1.2.1.1) or nearby buddies (for community services, see Section 1.2.1.2). The former example can also be managed by reporting in conjunction with piggybacking, as user and target are the same individual here. Querying has the advantage that location information is as up to date as possible, but it suffers from a delay caused by the request, processing at the supplier, and response. On average, this delay may be shorter for a caching strategy, which may also reduce network load. However, on the other hand, caching may deliver inaccurate or outdated location information if the target has changed its position since the last query.

Reporting, on the other hand, is useful whenever a target needs to be tracked, which especially holds for proactive services that are automatically executed or perform any actions if the target triggers an event, for example, entering, approaching, or leaving a certain point of interest or approaching, meeting, or leaving another target. For example, consider a tourist guide service that tracks the user in order to check whether he stays in close vicinity to certain landmarks, and if so, notifies him and delivers useful information about their history. In this particular case, tracking may be realized by zone-based reporting, where the zones represent the landmarks of interest, and a report is triggered on entering a zone. Other examples are mobile gaming or community services, where a number of users share their current locations among each other by displaying them on a digital map at their mobile terminal. To efficiently disseminate location information from and to the users, a periodic or distance-based reporting strategy can be deployed. Thus, the variety of reporting strategies allow to send reports only if it is relevant to the particular LBS application and to save valuable resources at the air interface and in the network respectively.

10.4 Privacy Protection

Users of IT services are always exposed to the risk that their personal information and data collected and processed during service usage may be misused by unauthorized parties or by the service provider itself. The motivation behind this misuse is often to observe and analyze the user's behavior, attitudes, and social situation in order to tailor special offers or advertisements for him, but sometimes it may also be with criminal intentions. An example of the former motivation is the observation of shopping habits and product preferences, while the spying on credit card information almost always belongs to the latter category. Therefore, the acceptance of IT services strongly depends on the existence of technical mechanisms for protecting the user's privacy. Many countries also certify the right of data protection by law and prohibit the utilization of personal information, and especially its dissemination to third parties, for purposes other than those for which it was initially collected.

While privacy is thus a hot topic for IT services in general, it is even more sensitive with regard to location dissemination. This section highlights the background of privacy protection in LBSs, shows how the common definition of privacy applies to LBSs, and explains the basic mechanisms of LBSs for protecting privacy.

10.4.1 Characteristics of Privacy Protection for LBSs

Privacy is always an issue whenever a target's location information is transferred from one actor to another. While privacy protection could easily be achieved if location information were to be passed from the target to the LBS user directly, as indicated by the peer-to-peer approach in Scenario (5), the multistep location dissemination with multiple intermediates as sketched in the last section makes it rather complicated. To put it simply, a target must be able to control the dissemination of its location information in that it specifies to whom, when, and in which form it is made available to other actors.

In the following text, it is assumed that the target that is subject to positioning is an individual person, and must be distinguished from locating mobile artifacts. For example, consider a service for tracking stolen cars. It assumes that the car has a GPS receiver on board and is attached to a cellular network for transferring location information. In this scenario, the target is an artifact, that is, the car, and privacy protection is not necessary. Strictly speaking, it is even not desired as it may prevent the tracking of the stolen car. However, if the same car with the same positioning equipment is navigated by the legitimate driver, and the driver is registered with a traffic telematics service, the target is an individual, that is, the driver, for which privacy rules must be strictly kept.

As compared to conventional IT services, LBSs impose much higher requirements on mechanisms for saving privacy, which is due to the following reasons:

- A target's location information passes many different actors along the LBS supply chain. Each additional actor increases the potential risk of misuse of location information. Also, the target can hardly understand the conglomeration of actors, and it is often even desired that this complexity is hidden from it. This is in contrast to many IT services, where users directly interact with the service provider and explicitly enter into a trusted relationship with it.

- The target is passive in that it is automatically tracked by an LBS provider and related actors during its everyday activities, and it is often not aware of this fact permanently. This is in contrast to the typical IT service user, who sits in front of a PC and explicitly discloses personal data.

- Location information is often regarded as belonging to a category of high-level information that is desired to be saved more than other personal information, for example, address, gender, and age. This becomes clear when imagining that not only a target's current location becomes observable by location dissemination but that it is even possible to record its personal location information for a longer period of time in order to generate traces reflecting the visits of the targets to political and religious groups, medical doctors, or nightclubs, to name only a few very personal locations.

Basically, the dilemma in privacy protection for LBSs is that positioning and tracking represent inherent key functions without which LBSs will not work and even make any sense at all. On the other hand, the same functions represent a potential source for misuse and are therefore the reason LBSs are often exposed to distrust in public, which may prevent the success of LBSs in general.

This contradiction between the desired positioning functions on the one hand and the risk of their misuse on the other is in close analogy to location management in cellular networks, where the network operator tracks mobile subscribers on the basis of location and routing areas (or even radio cells) in order to deliver network-originated calls or data to him. Without location management, reachability would simply not be possible, and cellular networks would never have gained the popularity they have today. Interestingly, in contrast to LBSs, privacy protection with regard to location management is not really a topic that is publicly talked about, and most subscribers of cellular services are even not aware of the fact that they are permanently tracked by their operators. It is also common practice that operators record location management data and store it for several days or even weeks.

Privacy protection in location management is primarily achieved by authentication and encryption mechanisms between the terminal and the BSS. And, apart from a few exceptions, transferring a subscriber's true identity (i.e., its MSISDN or IMSI) over the air interface is avoided by using the TMSI instead. The TMSI is dynamically assigned by the network and used during authentication, for example, when making calls or performing location updates (see Section 5.5). Eavesdroppers can hardly derive a subscriber's true identity from a TMSI they have caught from listening to the air interface. However, privacy protection for LBSs is much more complicated than for location management, which is due to the following reasons:

- Data collected for location management, for example, a subscriber's TMSI and current location area, is exclusively processed by the network operator itself and, in contrast to geographic location information, not passed to other actors.

- The signaling infrastructure used for location management is a closed network accessible only by the operator itself, while location information for LBSs may pass the public Internet, which by its very nature cannot be considered safe.

- Geographic location information may be much more accurate than location management data, which merely allows tracking of a target on the basis of location and routing areas or cells.

Before introducing the basic mechanisms of privacy protection, it is at first necessary to give a general definition of privacy and show how it can be applied to LBSs.

10.4.2 Definition of Privacy

There are many definitions, which often simply equate privacy with confidentiality or anonymity. A more comprehensive and often cited definition stems from Westin (1970): "Privacy is the claim of individuals, groups, and institutions to determine for themselves, when, how, and to what extent information about them is communicated to others." Westin (1970) further defines the following four states of privacy:

- **Anonymity** permits the engagement and interaction with others without being identifiable to them.

- **Solitude** is the right of being alone and secure from intrusion, interruption, and observation.

- **Intimacy** is the right to decide with whom, how much, and when to interact.

- **Reserve** is the freedom to withhold a personal information or the option to choose when to express it.

Applying these states to the area of LBSs leads to the following examples. Anonymity is enabled if location information is disclosed without the target's true identity or without an identity at all. In the former case, the true identity, which in most cases is given by an MSISDN or IMSI, might be replaced by a pseudonym or a temporary identifier comparable to the TMSI used for location management. Such a "false identity" is then attached to any kind of location information like position fixes or location data, which makes it very difficult for official LBS actors or for eavesdroppers to observe a target for purposes other than for the intended LBS operation, assuming that the mapping between true and false identity is kept secret. This technique is known as *identifier abstraction* and will be explained in Section 10.4.3.3 in detail. On the other hand, disclosing location information without any identification features might only be useful for statistical LBSs, which, for example, focuses on the prediction of traffic jams.

Solitude is given if the target is protected from unwanted positioning and disclosing of its location information. Thus, a target must be able to determine by itself when to subscribe for positioning and passing of location information to other actors and when to cancel this subscription. Furthermore, solitude means that an LBS target is protected from unwanted messages or any other interruptions that result from matching its location with other geographic information. This is of special importance for mobile advertising services, which inform the user about nearby shopping opportunities and may thus significantly disturb him.

Intimacy focuses on mechanisms for controlling the group of actors to which the location information is passed, to what extent, and at which time. A target must be able to compose a well-defined list of users that are authorized to receive its location via an LBS, and must also be able to control the intermediate actors processing its location information. This can be achieved in that the target specifies privacy policies that represent rules on how an actor has to deal with the target's personal location information. This will be introduced in Section 10.4.3.2. Furthermore, it must be possible to intentionally falsify the accuracy of position data, or to express a position in terms of a reference system that can only be interpreted by the target and the intended LBS user, for example, "at home", "in office" and so on. Also, a target may decide that location information is only disseminated with a delay. The intentional falsification of location and time data is known as *information content abstraction* and may be used to support anonymization, which will be introduced in Section 10.4.3.3.

Finally, reserve means that each positioning attempt must be explicitly authorized by the target.

10.4.3 Concepts and Mechanisms for Privacy Protection

For the sake of simplicity, location information here is considered to be represented by a tuple of the form (*target identity, location*), although it may comprise additional data items such as timestamps, accuracy estimations, and the spatial reference system the location is based on. Thus, the combination of the target's identifier and its location is the entity to be protected from misuse by regular actors of the value chain as well as by unauthorized parties like eavesdroppers and intruders. The underlying mechanisms for privacy protection are secure communications, the specification and enforcement of privacy policies, and anonymization. Figure 10.6 gives an overview of these mechanisms and their subcategories, which will be introduced in the subsequent sections.

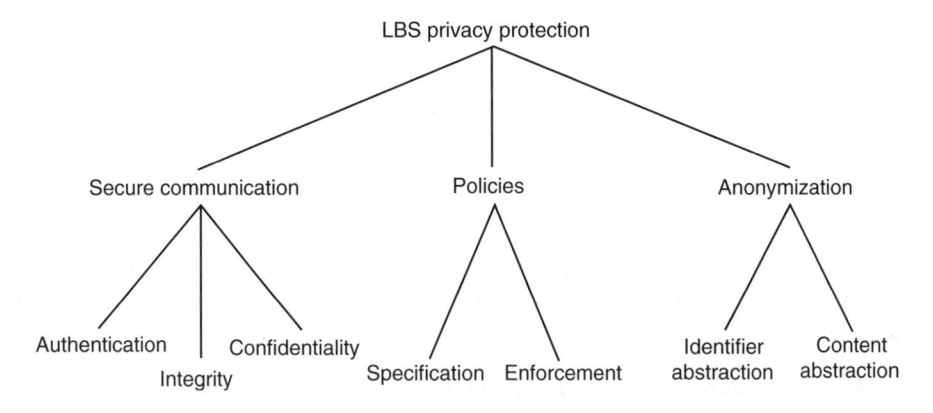

Figure 10.6 Overview of LBS privacy protection.

10.4.3.1 Secure Communications

A secure communication protocol is deployed not only for location dissemination, but also for negotiating and enforcing privacy policies, as well as for managing anonymity and identifier abstraction. It is strongly recommended to apply for location dissemination between actors, which in many cases happens over the public Internet, as well as for the transfer over the air interface. Location dissemination within an actor's domain, however, mostly happens in a closed network, for example, within an operator's SS7 signaling network as has been demonstrated in Scenarios (1)–(3) (i.e., between SMLC and GMLC), and hence a secure communication protocol is not necessary here.

In communications, security is usually associated with the three areas of *confidentiality*, *integrity*, and *authentication*, which are also known by the acronym *CIA*. Confidentiality enables that information is not accessed by unauthorized parties, integrity means that information is not allowed to be altered by unauthorized parties in such a way that it is not detectable by the authorized recipient, and authentication ensures that the parties involved in a communication session are really the parties they claim to be.

In this way, CIA prevents location information from being caught by eavesdroppers during transmission. Furthermore, it prevents intruders from pretending to be a legitimate consumer to the supplier of location information or vice versa, that is, intruders from pretending to be a legitimate supplier to the consumer. Finally, it is not possible for an intruder to falsify location information of a target or to fake a target's identifier. The underlying mechanisms of CIA are symmetric or asymmetric encryption for ciphering messages exchanged between supplier and consumer. Authentication may be supported by the challenge-response method or by user certificates, which are generated and stored by a trusted certification authority and which contain the user's public key and his signature that has been ciphered with the user's private key. There are various transport protocols offering these mechanisms. The most prominent one is obviously HTTPS, which is based on *Secure Socket Layer* (SSL) and the Internet standard of SSL known as *Transport Layer Security* (TLS). As it is not the intent of this book to cover secure communication protocols in detail, the interested reader can find comprehensive introductions in (Stallings 2002b) and (Schäfer 2004).

10.4.3.2 Privacy Policies

A secure communication protocol protects the privacy of targets in that unauthorized eavesdroppers and intruders get no access to location information or can fake it. However, a target must also have full control on how location information is treated by legitimate actors, especially to whom these actors grant access to location information or forward it. This control can be achieved by the specification of a privacy policy.

According to Cuellar (2002), a privacy policy is an assertion that a certain amount of information (identity or identifier plus location) may be released to a certain entity (or group of entities) under a certain set of *constraints*. Examples of privacy policies in colloquial language are:

- *"My wife is allowed to know the city in which I'm currently staying."*

- *"My superior is allowed to automatically track me in terms of exact coordinates if and only if*

    - *it happens during the working hours, and*
    - *I'm staying on the company's premises, or*
    - *I'm visiting clients."*

- *"Outside working hours, my superior is only allowed to track me on explicit request."*

- *"Members of my community are allowed to receive my location via the buddy finder service on weekends."*

These examples clearly demonstrate that there exist various types of constraints a privacy policy may consist of. For example, it must be possible to specify a set of LBS users and intermediate actors that are granted access to a target's location information, the time period during which access is granted, the granularity of location data passed to them, and the services that are allowed to receive and process a target's location information. The following list, although not complete, provides an overview of obvious constraints for location privacy policies (see also (Myles et al. 2003)):

- **Actor constraints.** It must be possible to restrict access to location information to a limited set of actors. For example, the target must be able to identify LBS users and providers that have access to its location information, possibly under consideration of other constraints.

- **Service constraints.** The target may identify a set of LBSs or types of LBSs for which it either grants or denies access to its location information and allows or does not allow to process it for service operation. For example, it may allow requesting and processing of its location information by navigation services, but may deny it for buddy finder services.

- **Time constraints.** A target must be able to restrict positioning and location information access to a certain period of time. This may be accomplished either by specifying predefined time periods, or by explicitly activating and deactivating positioning and location information access.

- **Location constraints.** Furthermore, it must be possible to limit positioning and location information access to predefined locations. For example, a target may allow unlimited positioning on the company's premises or the campus of the university, but forbid it if it stays outside these locations.

- **Notification constraints.** With these constraints, the target can specify whether or not it wishes to be informed about positioning attempts or attempts to access its location information. Upon arrival of such a notification, it can authorize or deny positioning and access respectively.

- **Accuracy constraints.** By specifying accuracy constraints, a target can intentionally degrade the accuracy of location information passed to another actor. For example, while GPS positioning achieves accuracy in the range of 15 m on a 95% confidence level, the target may decide to falsify its GPS position passed to another actor such that it matches only 1 km accuracy. Accuracy degradation is one means to anonymize location data, which will be introduced in Section 10.4.3.3 in detail.

- **Identity constraints.** Using identity constraints, the target can determine to pass location information to other actors either by using a pseudonym instead of its true identity or without any identity at all. Thus, these constraints enable to enforce another type of anonymization, and this will be introduced in Section 10.4.3.3.

Considering the manifoldness of these constraints, it is of particular concern to provide a convenient, easy, and flexible means for specifying them. The main problem results from the fact that different constraints may be combined by the target and therefore may interact with each other when matching them against a location request. For example, an employee wants to be tracked by a navigation service all the time, but simultaneously restricts positioning to weekdays because he does not want his principal to locate him in his free time. While this example is still easily manageable, it becomes increasingly complex when additional constraints are to be considered. Thus, although the interaction between constraints may be explicitly desired in order to provide flexibility, it may easily result in contradictions, conflicts, and unwanted decisions about whether to accept a location request or to decline it.

In order to cope with these problems, the enforcement of a privacy policy must be organized according to well-defined rules that prescribe priorities, sequences, and relations between different constraints that are applied when checking a location request. These rules must also be reflected by the policy language or the user interface that is accessed by the target in order to specify its policy.

At the time of writing this book, practicable and system-independent privacy policy rules and languages for LBSs are still missing. The 3GPP does not specify a genuine language, but at least considers the specification of the so-called *privacy options*, which will be treated in Chapter 11. However, compared to the different requirements and constraints presented here, the 3GPP concept has so far been rather rudimentary and only envisages the specification of actor and notification constraints. Another approach stems from the *World Wide Web Consortium* (W3C), which standardizes the *Platform for Privacy Preferences* (P3P), see (W3C 2002). P3P enables web sites to express their privacy policies and compare them with the user's privacy preferences. The policies are transferred to the user's browser and then matched to his personal preferences there. However, as stated in (Cuellar 2002), P3P has been specified for more general private data, and has not been tailored to the special requirements of location information and positioning. Therefore, it is not possible to specify some of the constraints presented earlier, for example, identity or accuracy constraints. Furthermore, the matching of constraints is always executed in the web browser, which is not necessarily practicable for all the LBS scenarios presented earlier. In the research community, various approaches for privacy policies are under investigation (see, for example, (Myles et al. 2003) and (Langheinrich 2002)), but their breakthrough into the market is yet to happen.

Another concern is where to enforce a target's privacy policy during LBS operation. The general problem is that many actors may be involved in location dissemination and thus each actor adopting a supplier function must be aware of a target's privacy policy. For example, the LBS provider has to consider actor constraints in that it passes location information to only those LBS users that have been authorized by the target, and it must enforce service constraints, too. The location provider, which is responsible for controlling positioning and which plays the major role in location dissemination, may adopt the enforcement of all constraints mentioned earlier to a greater or lesser extent. However, the position originator may care for enforcing some of them too, for example, accuracy, time, and identity constraints, but certainly not all of them. The problem of where to enforce policies also raises the question of where to locate the so-called *policy repository*, that is, the component needed for specifying and storing them.

Obviously, a solution for these problems is strongly dependent on the scenario considered. Basically, it can be stated that the lesser the number of actors involved in LBS operation the simpler is the enforcement and management of privacy policies. For example, in Scenario (5) there is no intermediate actor between target and LBS user, and hence privacy policies could simply be stored and enforced in the target's terminal. As opposed to that, Scenarios (1) and (2) have many actors involved, making policy management and enforcement a very complicated matter.

To cope with these problems, it is possible to follow a distributed or a central approach. In the former, the target has to fix its constraints at all actors involved in LBS operation. It may fix the LBS users authorized for receiving its location information as well as service constraints at the LBS provider, may advice the location provider to forward location information to a particular LBS provider only, and may determine accuracy and time constraints of position fixes at the position originator. In this way, each of these actors needs to maintain a policy repository. As a consequence, policy specification is very inconvenient and needs to be carefully prepared in advance. It is very cumbersome to subscribe for new services at LBS providers not specified before and to use their services in an ad hoc manner. Moreover, if the target roams in foreign networks and the standard position originator changes, it needs to redefine policies if it wants to make further use of its LBSs or wants to be still trackable for another LBS user.

In a central approach, there is only one policy repository, which stores all privacy constraints of the target and which is located at a particular actor. This actor is then responsible not only for policy enforcement with regard to its own tasks in location dissemination and processing, but also to give instructions to other actors for saving the target's privacy interests. In most approaches presented in Chapter 11, the central policy repository is maintained by the location provider. If an LBS provider requests for location information, it has to provide the location provider with all relevant information in the location request that may affect the privacy policy of the specified target, especially the (type of) services and the LBS users the location information is to be processed for. If this information does not match the target's policy, the location provider must decline the request. Otherwise, it obtains and returns the location information, but only after taking into consideration the other constraints.

A more general approach would be to arrange the additional role of a *policy holder*, which maintains the policy repository and which is connected to each supplier of location information. This is illustrated in Figure 10.7. Like for other roles, the role of a policy holder may be adopted by a dedicated actor or it may coincide with other roles. A similar approach is followed by the Geopriv architecture, which will also be covered in Chapter 11.

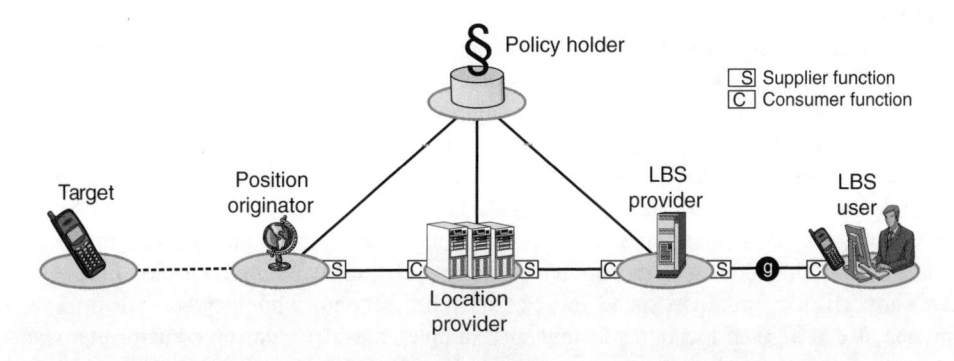

Figure 10.7 Integration of policy holder into the LBS supply chain.

10.4.3.3 Anonymization

Policies may be an efficient means for protecting privacy if all actors along the LBS supply chain are trustworthy. However, they do not prevent that an actor having access to a target's location information may violate its privacy policy, that is, that an actor may "talk about the target behind its back". For example, even if a location provider follows privacy policies and only passes location data to an LBS provider that has been authorized by the target, basically nothing could prevent this LBS provider from forwarding the location data to other, unauthorized parties (that even do not officially participate in the supply chain). However, such a violation does not necessarily occur on purpose. Cuellar (2002) identifies four possible reasons:

- misbehavior or negligence of an actor,
- attacks from hackers or other outsiders,
- unauthorized access from insiders, and
- technical or human errors.

A means for protecting a target against these violations is the anonymization of location information, for which *identifier abstraction* and *information content abstraction* are two basic mechanisms. According to (Cuellar 2002), identifier abstraction means the changing of an identifier by another, more anonymous one, while information content abstraction refers to reducing the information content of location information, for example, by intentionally degrading the temporal or spatial resolution of location information. Before introducing them, it is important to note that these mechanisms so far have represented rather theoretical approaches that have not been proven in reality or in large-scale systems.

Identifier abstraction. For identifier abstraction, an identifier unambiguously referring to a target on a global basis, for example, an MSISDN or IMSI, is replaced by a pseudonym that hides the target's true identity from other actors, for example, from the LBS provider and other LBS users. Of particular concern is whether this pseudonym is assigned permanently or on a temporary basis.

A permanent pseudonym has the advantage that a target remains identifiable toward other LBS users for several service sessions. In this way, it can be compared to the pseudonyms used for chat or instant messaging applications. The users here refer each other by their permanent pseudonyms that are assigned when registering with the service for the first time and that remain valid until they are explicitly changed. However, the general drawback of permanent pseudonyms is that according to Beresford and Stajano (2003) it is easy to de-anonymize a target by having some background information about it (for example, the address of its residence) and correlating two simple checks: first, where does any given pseudonym spend most of its time? second, who spends more time than anyone else at a given location? To give an example, consider a corrupted actor that wants to get the traces of a particular person and for this purpose at first seeks out the address of this person's residence, for example, by using a phone book. For de-anonymization, it then filters the amount of location information it has access to with regard to the coordinates belonging to this address. In a next step, it scans the remaining data set for a pseudonym with frequent and long-lasting visits there, which then, in all likelihood, is the pseudonym used by the person. Once having de-anonymized it in this way, the actor can scan the amount of location information again in view of the pseudonym's visits to religious or political groups, medical doctors, or nightclubs, or any other location.

Therefore, it is highly recommended to frequently change pseudonyms, as it is, for example, done in cellular location management by dynamically assigning a TMSI to the subscriber after each location update. However, this approach entails the risk that an actor can link old and new pseudonyms assuming that the spatial and temporal resolution of location traces is sufficiently high. To cope with this problem, Beresford and Stajano (2003) introduces the concept of *mix zones*, which is based on *mix networks* for anonymous communications as proposed by Chaum (1981). A mix network is a store-and-forward network consisting of several mix nodes. Each mix node collects a number of equal-length packets as input and reorders them by some metric before passing them to the next node, thereby avoiding that outgoing messages are attributed to incoming ones.

Applying this principle to geographical locations, a mix zone is a connected geographical region where targets do not divulge their location information. An *application area*, on the other hand, is a region where targets make their location information public, either by responding to a query from or reporting to a consumer. An arrangement of mix and application zones is depicted in Figure 10.8. Since no actors receive location information from targets residing in a mix zone, their identities can be mixed. Each time a target enters a mix zone, it changes to another pseudonym. The next time the target enters an application area, this new pseudonym is then attached to the location information, which is then passed to one or several actors of the supply chain. From the point of view of these actors, a particular target thus disappears when entering a mix zone, and it cannot be reidentified when it leaves the mix zone again to enter another application area. The drawback of this approach is that mix zones must be carefully dimensioned. If they are too large, the respective LBS application could be negatively affected or make no sense at all. If they are too small, they might accommodate only a very small number of targets or even only a single one, which would significantly ease the linking of old and new pseudonyms and thus the de-anonymization. The approach of mix zones and its evaluation is presented in (Beresford and Stajano 2003).

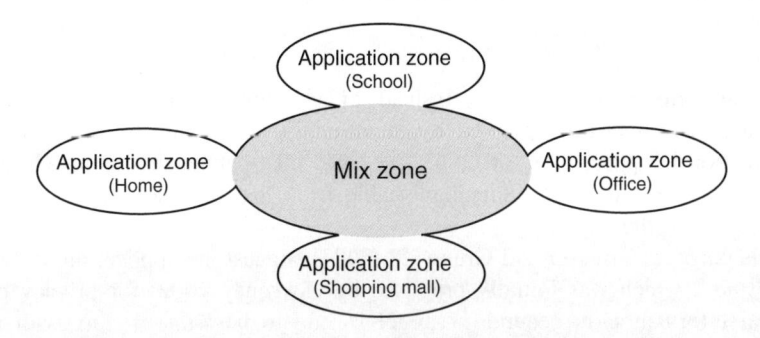

Figure 10.8 Arrangement of mix and application zones.

It must be emphasized that identifier abstraction raises a number of unsolved questions and is not necessarily suited to each type of LBS application. First, knowing the network address of the target's terminal (which is attached to IP packets or known from signaling connections used for transferring location information) increases the risk of de-anonymization, unless operators support anonymous connectivity services. This may be achieved by deploying complicated technical mechanisms like mix nodes or by selling prepaid cards. However, technical mechanisms come along with a lot of overhead, while the anonymous subscription with prepaid cards is prohibited by law in many countries.

Therefore, one has to act with the assumption that at least the network operator the target is attached to represents a security hole with regard to anonymization. However, if location information passes the operator's location server, that is, the cellular network operator is also the location provider as in Scenarios (1) and (2), the terminal's network address is replaced by the server's address and is therefore not available to other actors.

Second, usage of a pseudonym is only an appropriate means if the LBS application is stateless with regard to the identities of targets. This is the case, for example, if users want to receive a list of nearby shopping opportunities, or, more generally, if user and target are the same person. Other applications, however, focus on the mutual localization of several users including their tracking over a longer period of time (e.g., community services), that is, they need to keep a state with regard to the identities of participating targets in order to make sense at all. Often, these applications may be personalized by the users in that they may want to be notified when another well-known user stays close by. It is hard to imagine how such an application would work if all users disappear to and emerge from mix zones, each time having new pseudonyms.

From the point of view of privacy and anonymization, one might therefore prefer to realize this kind of applications by a peer-to-peer approach as covered by Scenario (5), thereby exchanging location information directly between the LBS users and without the detour over an application server. A peer-to-peer location-based community service, for example, is proposed by (Amir et al. 2004). However, compared to a central server-based approach, peer-to-peer communication requires enhanced service logic in the mobile terminals and also increased overhead of location dissemination. Another way to cope with functional drawbacks due to changing pseudonyms is to use protocols known from e-cash as proposed by (Cuellar 2002). These protocols enable anonymous payment transactions over the Internet, which require similar mechanisms for anonymization like location dissemination.

Information content abstraction. Instead of changing pseudonyms in mix zones, anonymization can also be supported by intentionally degrading the resolution of location information in space, time, or both. The basic idea behind this approach is to make a certain target's location data indistinguishable from that of a number of other persons staying close by to the target.

For this purpose, Gruteser and Grunwald (2003) suggest the application of the concept of *k-anonymity*, which was initially proposed by (Sweeney 2002) for privacy protection of personal information in general. A target is said to be *k-anonymous* with regard to location information if this data is indistinguishable from the location information of at least $k - 1$ other persons. Thus, instead of representing location information in the form of exact coordinates and attaching the exact timestamp when the target was there, it is here given by the following tuple of uncertainty ranges: $([x_1, x_2], [y_1, y_2], [t_1, t_2])$. The intervals $[x_1, x_2]$ and $[y_1, y_2]$ represent a two-dimensional area where the target is located, and $[t_1, t_2]$ describe a time period during which the target was present in that area. Thus, an actor receiving such a tuple only knows that the target was at a certain position within this area at a certain point of time within the temporal interval. Represented by such a location tuple, a target is *k*-anonymous if at least $k - 1$ other persons are present in the area and the time period described by the tuple. The larger the anonymity set k is, the higher is the degree of anonymity. This is illustrated in Figure 10.9.

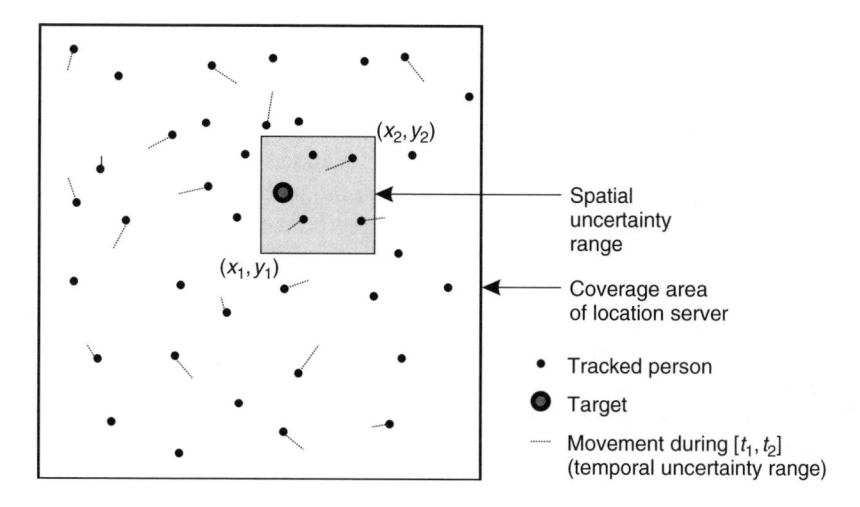

Figure 10.9 k-anonymity in the spatial and temporal domain for $k = 5$.

Gruteser and Grunwald (2003) also propose a cloaking algorithm for calculating the spatial and temporal uncertainty ranges. Because the positions of other individuals have to be taken into account, the algorithm cannot be executed at the target's terminal, but at an independent location provider and is therefore applicable to Scenarios (1)–(3) only. For the space domain, the algorithm takes as input the minimum acceptable anonymity set k_{min}, the geographic region covered by the location server, and the positions of the target as well as of all other persons staying in this area, and calculates the quadrant ($[x_1, x_2], [y_1, y_2]$), which meets the requirements of k_{min}. For the temporal domain, the location request of an actor is delayed until at least k_{min} other persons have visited the quadrant, that is, t_2 is set to the current time, and t_1 is set to the time of request minus a random value.

It must be stressed that this approach is of course contradictory to the efforts made for introducing high accuracy positioning methods such as A-GPS, E-OTD, or U-TDoA. Also, accuracy of disseminated location information may significantly vary depending on the population density on the spot, that is, the smaller this density, the larger the quadrants calculated by the cloaking algorithm and vice versa. Whether to deploy information content abstraction in this way may be decided taking into consideration the applications a target has subscribed for and the sensitivity of its location information. General mechanisms for privacy protection, which may also be applicable to location information, are also under consideration in the database community. As it is not intended to cover them in this book, the interested reader can find an overview of this in (Adam and Worthmann 1989).

10.5 Conclusion

This chapter has covered the interorganizational aspects of LBS operation. It has identified the roles the participating actors adopt and presented different scenarios of a supply chain. Furthermore, the chapter has explained the different interaction patterns for exchanging location information, which can be applied for realizing different types of LBSs (e.g., reactive and proactive). Of great concern and an essential for the success and acceptance of LBSs is privacy protection. This chapter has therefore identified the basic mechanisms like privacy policies and anonymization.

Besides the functional aspects that have been covered in this chapter, an interorganizational operation of services also requires that management issues are taken into consideration. In general, management of networks and services is a very complex field of activity that comprises various dimensions, such as the classical subdivision of management tasks into fault, configuration, accounting, performance, and security management (FCAPS). Though the management of conventional networks, services, and applications has been a well-established discipline for many years, there has so far been no efforts to cover the management needs of LBSs. This concerns, for example, *Service Level Agreements* (SLAs) that need to be closed between the LBS actors of the supply chain. According to (Hegering et al. 1999), an SLA contains an exact description of what is offered in a service and defines which costs are applied when a customer uses the service. Accordingly, this requires fixing several conditions of service operation, for example, functional specifications, technical requirements, operating and maintenance times, service levels, system faults and penalties, as well as costs. While many of the proven concepts and mechanisms used for conventional services might also serve very well for LBSs and related subservices, there is a strong need to develop solutions tailored specifically to LBSs. Among other things, it must be possible to negotiate quality of location information, for example, in terms of accuracy and up-to-dateness, and to relate the costs of deriving and delivering it to its quality. Furthermore, there is a need for flexible charging mechanisms that allow to charge either the LBS user, the LBS target, the LBS provider, or another actor for performing positioning and delivering location data, both depending on the respective LBS application and the business model it relies on.

11

Architectures and Protocols for Location Services

A very important subservice needed for building an LBS is the so-called *location service* (LCS). While an LBS takes location data in order to compile, filter, or select any information or provide any other added value to the user depending on location data, an LCS is primarily concerned with the mere delivery of location data. Compared to position fixes, which only contain a position and perhaps the target's identity or the positioning method used, location data contains additional high-level information that is more appropriate for processing by an LBS application. Usually, location data is composed by a location provider if a location request arrives from a consumer, for example, an LBS provider. This request specifies at least the target location for which the data is needed, the desired format of representation, and the desired quality. In terms of the supply chain introduced in the last chapter, an LCS is based on reference point (b) for controlling positioning and generating location data, as well as on reference points (c) and (d) for making it available to other actors.

The elements of location data can be derived from Table 11.1. Most of them have already been covered in previous chapters. The type of location refers to the up-to-dateness of location data. "Current location" means that the target has been located on request and the delivered location is the most recent one. "Initial location" indicates that the location data has been derived at the point of time the target has initialized the service session for which location data needs to be generated and processed. Finally, the "last known location" means that the target has not been explicitly located for the respective request, but that the consumer is supplied with cached location data.

In order to compose location data and provide the LCS, the location provider has to adopt the following tasks:

- **Selection and control of positioning method.** If several methods exist in parallel for locating a particular target, an LCS must select an appropriate positioning method and control positioning. The selection must be made taking into consideration the quality parameters like accuracy, which are usually defined by the LBS provider requesting for the target's location. However, the capabilities of the target's terminal

---

Location-based Services: Fundamentals and Operation   Axel Küpper
© 2005 John Wiley & Sons, Ltd

Table 11.1 Elements of location data

Element	Description
Location	Represents the target's location, but not necessarily in the original format resulting from positioning.
Type of location	Indicates whether the location is the current, initial, or last known location.
Format of representation	Specifies the spatial reference system the location is based on.
Quality	Contains quality parameters such as accuracy of location and the time when it has been generated.
Identity	Specifies the target's identity and the identity type. Examples are MSISDN, IMSI, IP address, name or a pseudonym
Direction	Denotes the direction of the target's motion.
Speed	Denotes the speed of the target's motion.

and of the positioning infrastructure must be taken into account, too. Depending on the particular scenario, positioning control may include the allocation of resources, the determination of an appropriate set of base stations for measurements if necessary, and the triggering of positioning.

- **Conversion to another reference system.** The position of a target might be transferred into the format of another spatial or descriptive reference system, which also depends on the requirements of the requesting LBS provider. For example, most positioning methods generate spatial coordinates in WGS-84, and some applications may require UTM coordinates. In particular, this step is necessary if the position originator delivers location in a proprietary, operator-specific format, for example, cell identifiers.

- **Tagging.** The LCS must attach the target's identity to location data. Typical identifiers are an MSISDN, IMSI, IMEI, IP address, or a pseudonym. In most cases, this identifier is also used to address the target location the information is needed for.

- **Quality indication.** The LCS must estimate the quality of the derived position, above all the vertical and horizontal accuracy, and indicate it.

- **Dissemination.** The LCS is responsible for passing location data to other actors via its supplier function. It should offer interfaces for querying and reporting.

- **Protecting privacy.** Tagging and dissemination must happen under consideration of the target's privacy interests. The LCS must guarantee that these privacy interests are kept.

- **Accounting.** It must be possible to charge actors for consuming the functions described here.

ARCHITECTURES AND PROTOCOLS FOR LOCATION SERVICES 273

This chapter shows the realization of different types of LCSs in GSM/UMTS networks. The following section identifies the components involved in their operation and explains the related signaling flows between them. An important application area that makes use of LCSs are emergency services, which are covered in a subsequent section by taking E-911 as a showcase. Finally, the chapter introduces different location protocols for making location data available to external actors.

11.1 GSM and UMTS Location Services

In terms of the supply chain introduced in the last chapter, the LCS specifications focus on the realization of reference point (b) for controlling positioning and generating location data. Reference points (c) and (d) are not subject to standardization by 3GPP, but are covered by other standardization authorities. These approaches will be presented in subsequent sections.

The overall description of GSM/UMTS LCSs is given in (3GPP TS 22.071). At a high level, an LCS is organized between the three logical entities depicted in Figure 11.1. The LCS client is the entity requesting and receiving location data from an LCS server, which coordinates positioning for a target specified in the request, collects its location data, and delivers it to the LCS client. Note that the figure only shows a logical view, that is, the entities do not represent physical components. The LCS client may be an application server, the terminal of an LBS user, or another component. It may be part of an operator's network that also performs positioning, or it may be an external entity located in the domain of an LBS provider. The LCS server is the placeholder for the set of all network components needed for positioning and transferring location data, for example, access networks, MSCs and SGSNs, as well as HLRs and VLRs. The target is always represented by the terminal of the person to be located.

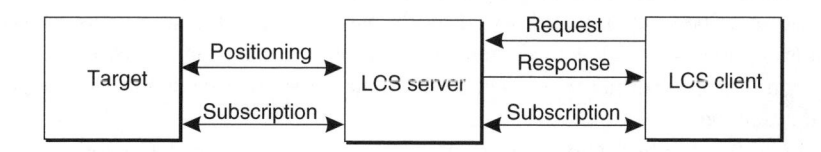

Figure 11.1 LCS logical reference model.

There is a distinction between *mobile terminating* and *originating location requests*. Mobile-terminating LCSs are initialized by an entity that is not the target's terminal (for example, an LCS client), while mobile-originating requests refer to location requests that are initialized by the target's terminal itself in order to perform self-positioning. Mobile-terminating requests appear in two variants:

- **Immediate location request.** The immediate location request follows the querying pattern presented in Section 10.3.1. The request of an LCS client must be processed immediately, and the result must be returned to the LCS client within a predefined time period. Only a single response to this type of request is returned.
- **Deferred location request.** The deferred location request realizes some of the reporting patterns as presented in Section 10.3.2. The response is not returned immediately,

but after the fulfillment of a trigger condition specified by the LCS client. The trigger conditions supported so far allow for periodic reporting or when the target switches on its terminal and registers with the network.

For each request, the LCS client may specify QoS parameters in terms of horizontal and vertical accuracy as well as response times. Also, it is possible to assign different levels of priority to each location request. A request with higher priority is served faster and with more reliable and accurate location data than one with lower priority. This is useful, for example, for efficiently and reliably serving emergency services or other sensitive services like child tracking.

Mobile originating location requests are available with the following options:

- **Basic self-location.** The terminal explicitly requests the network for each separate positioning procedure.

- **Autonomous self-location.** The terminal initializes positioning at the network and is then tracked by the network over a predetermined period of time.

- **Transfer to third party.** The location data of the target is transferred by request of the terminal to a specified LCS client.

As can be derived from Figure 11.1, there is another relationship that addresses the subscriptions of a target and an LCS client at the LCS server. The target subscription encompasses the target's privacy options, which is the 3GPP term for privacy constraints as introduced in the last chapter. These options are defined in the form of a privacy exception list, which at least contains a list of LCS clients allowed to request the target's location data (i.e., actor constraints) and the target subscriber notification settings indicating whether positioning must be explicitly authorized by the target (i.e., notification constraints). Note that this exception list is subject to extensions in future releases that focus on the consideration of additional constraint types. These additional types will be shortly presented at the end of this section.

The LCS client subscription is only necessary if the LCS client is an external entity. When subscribing, the LCS client negotiates terms of LCS usage with the LCS server, for example, the default range of QoS and the LCS features (e.g., immediate or deferred) it is allowed to request. The results of these negotiations are then stored in the client's subscription profile at the LCS server. Each time the LCS server receives a request from an LCS client, it checks it against the conditions of this profile as well as against the target's privacy options in order to decide whether to process or to decline it. The LCS client subscription also contains a *Privacy Override Indicator* (POI). It indicates whether location requests originating from the LCS client must be processed, although the circumstances of these requests offend against the privacy options of the respective targets. Thus, an LCS client possessing this override capability is allowed to track targets irrespective of their privacy interests. The override capability is to be assigned only to emergency centers or investigating authorities (e.g., policy, justice) for enabling location-based emergency services and lawful interception.

Starting with an overview of the network architecture, the following sections show how the LCSs described here and their features are realized.

ARCHITECTURES AND PROTOCOLS FOR LOCATION SERVICES 275

11.1.1 LCS Network Architecture

As outlined in Chapter 5, a smooth migration path from 2G toward 3G has been prepared in the long term in that GERAN and UTRAN access networks coexist under an integrated core network architecture that offers common infrastructure components, protocols, and management mechanisms. This integrative approach has also been adopted for LCSs, which comprise the different positioning methods developed under consideration of the radio links in both types of access networks on the one hand (see Chapter 8) and a common part situated in the core network on the other, which controls positioning irrespective of any particularities of the respective access network. Figure 11.2 gives an overview of the LCS architecture and its most important components.

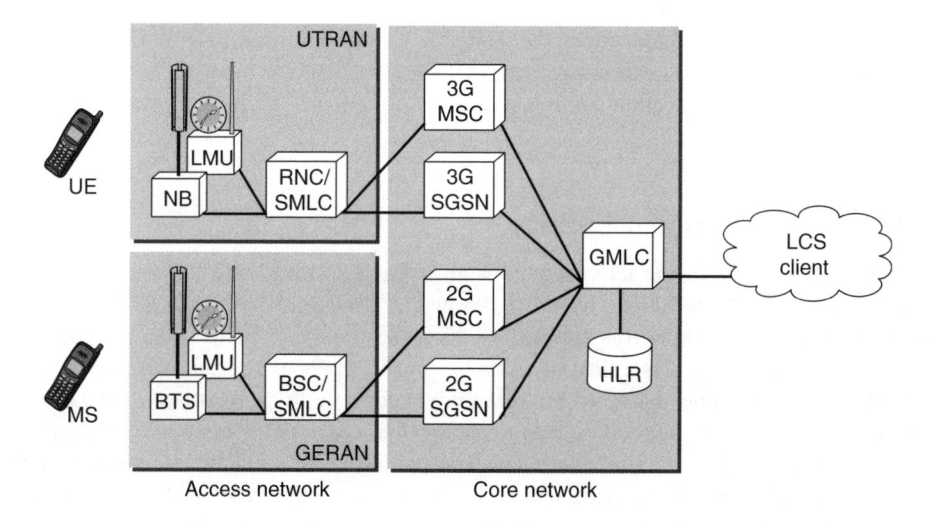

Figure 11.2 Overview of 3GPP LCS architecture.

The main component of interest is the *Gateway Mobile Location Center* (GMLC), which represents the interface between the SMLCs in the various access networks and an LCS client (in the figure, the SMLCs are assumed to be integrated into the BSCs and RNCs respectively). An operator may maintain one or several GMLCs, depending on the size of its network, and GMLCs of different operators may be interconnected to support positioning of roaming subscribers. The GMLC receives location requests from an LCS client, coordinates the entire position process, and finally returns location data to the client. It is connected to the SMLCs via MSCs or SGSNs, depending on whether positioning is carried out in the CS or PS domain.

Figure 11.2 also shows the interfaces interconnecting these components. Each of them is realized by a certain signaling protocol that is needed for exchanging messages for controlling the LCS. Because it is very cumbersome to cover these protocols in detail, Table 11.2 presents an overview of the official interface terminology, the names of protocols used to realize them, and the associated 3GPP specification documents. Thus, the interested reader can then obtain additional information from the related specifications.

Table 11.2 Overview of interfaces for LCSs

Interface	Signaling protocol	3GPP spec.
A	Base Station System Application Part (BSSAP)	3GPP TS 48.008
	Base Station System Application Part LCS Extension (BSSAP-LE)	3GPP TS 49.031
Gb	Base Station System GPRS Protocol (BSSGP)	3GPP TS 48.018
	Base Station System Application Part LCS Extension (BSSAP-LE)	3GPP TS 49.031
Iu	Radio Access Network Application Part (RANAP)	3GPP TS 25.413
Lg	Mobile Application Part (MAP)	3GPP TS 29.002
Lh	Mobile Application Part (MAP)	3GPP TS 29.002
External	Mobile Location Protocol (MLP)	
	WAP Location Protocols	
	Parlay	

11.1.2 LCS Functional Entities

It is common practice in the telecommunications domain to organize the various network functions into well-defined units, which are known as *functional entities* or *building blocks*. The main motivation behind this approach is to separate service logic from network functions, thereby achieving reusability of software and independence from different types of networks and components delivered by different vendors. In this way, it is possible to introduce new services rapidly without having to change the functionality of network components (Magedanz and Popescu–Zeletin 1996). This approach leads to a functional architecture that is, at first, independent of the particular network architecture. Both are then combined so that functional entities are mapped onto network components. This section identifies the functional entities needed for LCSs and shows their mapping onto the components depicted in Figure 11.2. See also (3GPP TS 23.271) for a general description, as well as (3GPP TS 43.059) and (3GPP TS 25.305) for the particularities of GERAN and UTRAN respectively.

The LCS functional architecture is depicted in Figure 11.3. It consists of an LCS client, which requests location data, and an LCS server, which delivers it. The LCS client encompasses the *client handling component*, which, in turn, bundles a number of *Location Client Functions* (LCFs). LCFs request and receive location data for one or more target terminals within a specified QoS. There may exist several of these functions in parallel, each representing a particular protocol for obtaining location data. An LCF is not necessarily an external component, but may also be located on different internal network components if the operator offers internal LBS applications. Also, the LCF may be used for supporting non-LCS-related functions, for example, the location-assisted handover. In this case, the LCF resides in the access network and is denoted as *internal LCF*.

The actual complexity of the functional architecture resides in the LCS server. It is subdivided into four different components which are introduced in the following text:

Client handling component. The client handling component comprises a number of functional entities for managing and coordinating location requests from clients. The *Location*

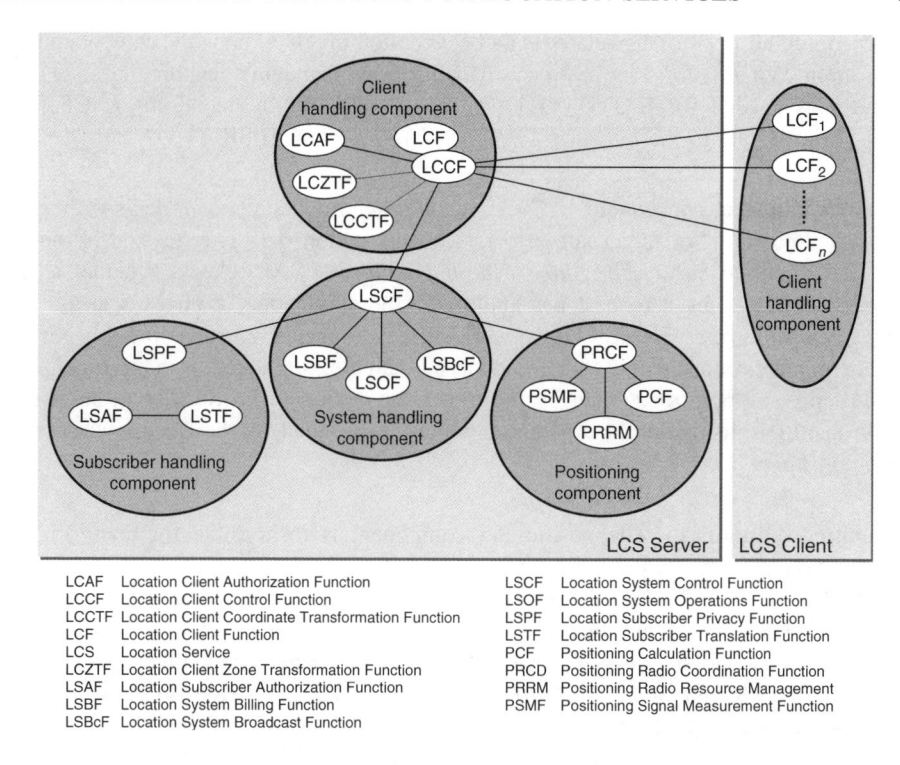

ARCHITECTURES AND PROTOCOLS FOR LOCATION SERVICES 277

Figure 11.3 LCS functional architecture.

Client Control Function (LCCF) identifies the LCS client by requesting client verification and authorization through interaction with the *Location Client Authorization Function* (LCAF). The LCAF performs a number of checks, for example, whether the LCS client is registered and authorized to use the specified LCS request type, and whether it is allowed to request location data for the target subscriber specified in the request. Furthermore, the LCCF is responsible for passing a request to the MSC or SGSN the target subscriber is currently attached to. After receiving the position fix, the LCCF checks whether it corresponds to the requested QoS, and may ask the *Location Client Coordinate Transformation Function* (LCCTF) for translating the coordinates to another spatial reference system requested by the client. The *Location Client Zone Transformation Function* (LCZTF) is used in the United States for translating the coordinates of a position fix into the corresponding emergency service zone. The latter will be explained in detail in Section 11.2.

System handling component. The system handling component coordinates location requests via the *Location System Control Function* (LSCF). Basically, this entity acts as an intermediate between the LCCF, from which it receives location requests and returns location data, and the *Positioning Coordination Function* (PRCF), which controls positioning. In this way, the LSCF triggers the allocation of network and radio resources for positioning under consideration of the capabilities of the network and the terminal as well as various other parameters. For this, it makes use of the *Location System Operation Function* (LSOF) for exchanging all positioning-related data between the involved network components, and the *Location System Broadcast Function* (LSBcF) for managing the broadcast of assistance data as needed for E-OTD, OTDoA, and A-GPS. Furthermore, the LSCF passes charging-related data to the *Location System Billing Function* (LSBF).

Subscriber handling component. The subscriber handling component deals with aspects of privacy protection for target subscribers whose location data is requested by an LCS client. The *Location Subscriber Authorization Function* (LSAF) checks whether a target subscriber addressed by a request has authorized the application of an LCS at all. If so, the *Location Subscriber Privacy Function* (LSPF) then checks whether to apply the LCS under the given circumstances, which is done by matching the location request against the individual privacy options defined by the target subscriber. From Release 6, anonymization by identifier abstraction is envisaged, which is managed by the *Location Subscriber Translation Function* (LSTF).

Positioning component. The positioning component is responsible for controlling the entire positioning process. Its main component is the *Position Radio Coordination Function* (PRCF), which determines the positioning method to be used taking into consideration the requested QoS, the capabilities of the access network and the target terminal. It is also responsible for exchanging measurement and position data between the other entities of this component. The *Position Signal Measurement Function* (PSMF) performs uplink and downlink measurements, depending on the respective positioning method used, and gathers the resulting data. The *Position Radio Resource Management* (PRRM) controls the effects of positioning on the overall performance of the radio network, for example, in order to ensure that the PSMF does not degrade the quality of other air interface connections. In UTRAN, it coordinates different RNCs in order to ensure that idle periods of different neighboring base stations do not overlap, which is required for proper working of OTDoA-IPDL. Finally, the *Positioning Calculation Function* (PCF) computes the terminal's position fix by taking into account various measurement results delivered by the PSMF and the coordinates of the base stations and LMUs involved in the measurements.

The mapping of functional entities onto network components is shown in Table 11.3. In most cases, the LCF either resides in an external LCS client, for example, the LBS application server, or in the terminal. For offering internal applications, the operator may also equip SGSNs and MSCs with an LCF. As mentioned earlier, an internal LCF may support non-LCS-related functions in the respective access network. All functions of the client handling component are located in the GMLC, while the entities of the system handling component are distributed over various network components. Note that the LSBcF, which is not listed in the table, resides either in the CBC (for GSM) or the RNC (for UMTS), which controls and manages the broadcast of messages for each cell separately (see also (3GPP TS 23.041)). Subscriber handling is organized for the entire network in the HLR, and for the various subnetworks in the respective SGSNs or MSCs. Finally, positioning is handled in the access networks and in the terminals, depending on whether it is terminal or network based. In the former case, terminals must be equipped with the PSMF and PCF for performing measurements and position calculation. For network-based positioning, on the other hand, it is not required to equip the terminals with these entities.

Table 11.3 Mapping of LCS functional entities onto network components

	MT/UE	BSS/RAN	GMLC	SGSN	MSC	HLR	Client
LCS client							
LCF	×			×	×		×
LCF int.		×					
Client handling component							
LCCTF			×				
LCCF			×				
LCAF			×				
LCZTF			×				
System handling component							
LSCF		×		×	×		
LSBF			×	×	×		
LSOF	×	×	×	×	×		
Subscriber handling							
LSAF				×	×		
LSPF				×	×	×	
Positioning component							
PRCF			×				
PCF	×		×				
PSMF	×		×				
PRCF			×				

11.1.3 Location Procedures

The procedures carried out in the network for executing mobile terminating and originating location requests can be subdivided into three general steps:

- **Location preparation procedure.** The procedure is responsible for checking the location request against the target's privacy policies, reserving network resources, communicating with the target terminal, and determining the positioning method used under consideration of the desired QoS, terminal, and network capabilities.

- **Positioning measurement establishment procedure.** This procedure aims to perform measurements, including the exchange of measurement data between the components involved, that is, SMLC, LMUs, and the terminal. It is dependent on the positioning method used.

- **Location calculation and release procedure.** After successful measurements, this procedure is responsible for calculating the terminal's position, either in the terminal or in the network and for releasing all network and terminal resources that have been involved.

The procedures are defined separately for CS and PS domains, but are very similar to each other and mainly differ only in the components (i.e., MSC and SGSN) involved. Therefore, only the procedures of the CS domain are introduced in the following text. For an overview of request procedures in the PS domain, see (3GPP TS 25.305). Note also that the message flows described subsequently are realized by the signaling protocols listed in Table 11.2.

11.1.3.1 Mobile Terminating Location Request

For mobile terminating procedures, it must be distinguished between whether the location request is immediate or deferred. Figure 11.4 illustrates the immediate location request.

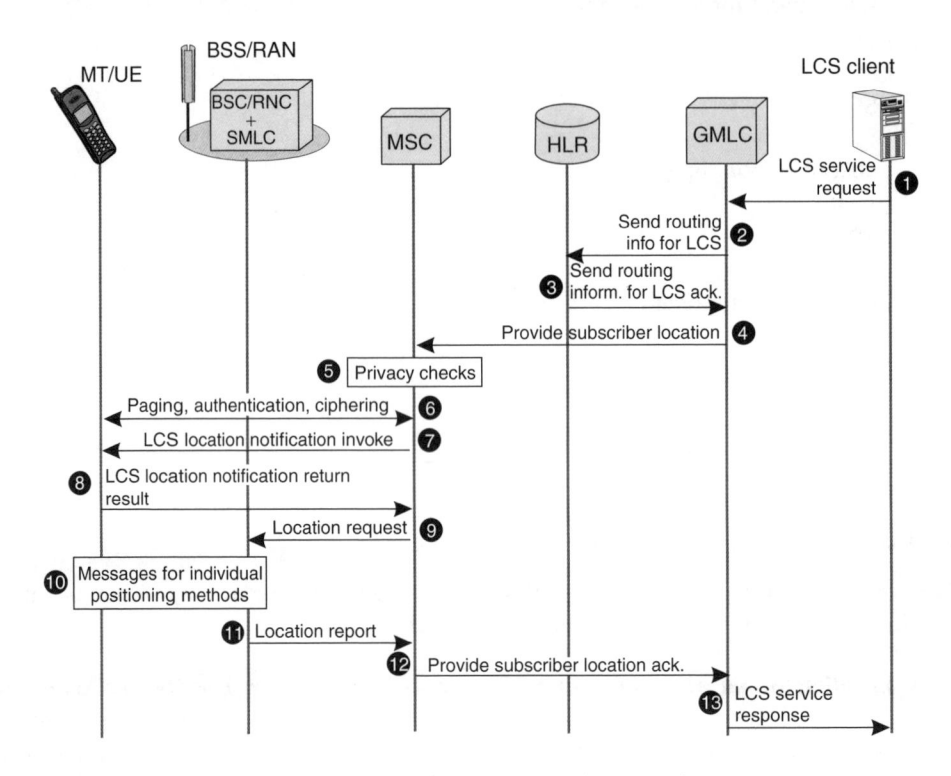

Figure 11.4 Mobile terminating immediate location request.

In this figure, steps (1)–(9) belong to the location preparation procedure introduced earlier. In the first step, the LCS client sends a location request to the GMLC (1), which contains the MSISDN or IMSI of the target subscriber to be located as well as the desired QoS. Note that, as mentioned earlier, this procedure is not part of the 3GPP specifications, but is covered by other standardization authorities. In order to process the request, the GMLC at first has to find out the access network the target subscriber is currently attached to; that is why it requests the HLR to send the routing information to the corresponding MSC (2). If for any reason the access network is already known, steps (2) and (3) can be skipped. After having received this information (3), the GMLC advices the MSC for further coordinating the location request (4). For this purpose, it passes the identity of the requesting LCS client to the MSC, which is needed there for matching the request against the subscriber's privacy options (5). This will be explained in detail in Section 11.1.4. If the location request is accepted, the MSC proceeds with checking whether the target terminal is idle. If so, the subscriber's cell is unknown and the terminal must first be paged (6). Otherwise, that is, if the terminal is busy, the subscriber's serving base station is already known and Step (6) can be skipped. If the notification settings of the subscriber's privacy options indicate that positioning must be explicitly authorized, the MSC informs the subscriber about the positioning attempt (7), which he can then accept or decline (8). In the latter case, the location request would be canceled, and the MSC would proceed with a negative acknowledgement carried by the message returned in step (12). However, if it is accepted, the MSC passes the location request, together with QoS requirements and terminal capabilities, to the respective access network (9), that is, strictly speaking, to the SMLC. This step terminates the location preparation procedure.

The positioning measurement establishment procedure is covered by step (10). Depending on the requested QoS, and the capabilities of the terminal and the access network, the SMLC selects an appropriate positioning method and triggers the associated measurements. These procedures strongly depend on the positioning method used, and have already been covered in Chapter 8.

After all measurement results have been gathered, either in the terminal or the SMLC, the target's position is calculated, which is the first step of the location calculation and release procedure (and which also depends on the positioning method used; hence it is not shown in the figure). In case of successful positioning, the position fix is sent to the MSC in a location report, which also indicates the used positioning method (11). Otherwise, an error notification is returned, indicating that positioning could not be successfully performed. The MSC then forwards the position fix to the GMLC or, if positioning has failed, the subscriber's last known position (if available) (12). Optionally, the MSC may also record billing information. Finally, the GMLC may translate the position in another spatial reference system or in a descriptive location as desired by the LCS client, and subsequently return the location data to the LCS client.

Figure 11.5 shows the procedure of the deferred location request. The main difference between the deferred location request and the immediate request is that positioning does not follow the location preparation procedure. Steps (1)–(5) are the same as for the immediate request. However, after finishing with the privacy checks, the MSC acknowledges the request (6), whereupon the GMLC passes a response to the LCS client (7). As opposed to the immediate request, this first response does not carry any location data, but only indicates to the LCS client that deferred positioning has been activated.

Positioning is not performed before the trigger condition of the deferred location request becomes true. As mentioned earlier, this may happen as soon as the subscriber registers with the network (8), or if a timer expires, that is, if the periodic reporting strategy is applied. As will be discussed in the following text, later releases of the 3GPP specifications will cover a zone-based strategy, too. If the trigger condition is fulfilled, the MSC initializes positioning, and the remaining steps are analogous to the immediate request.

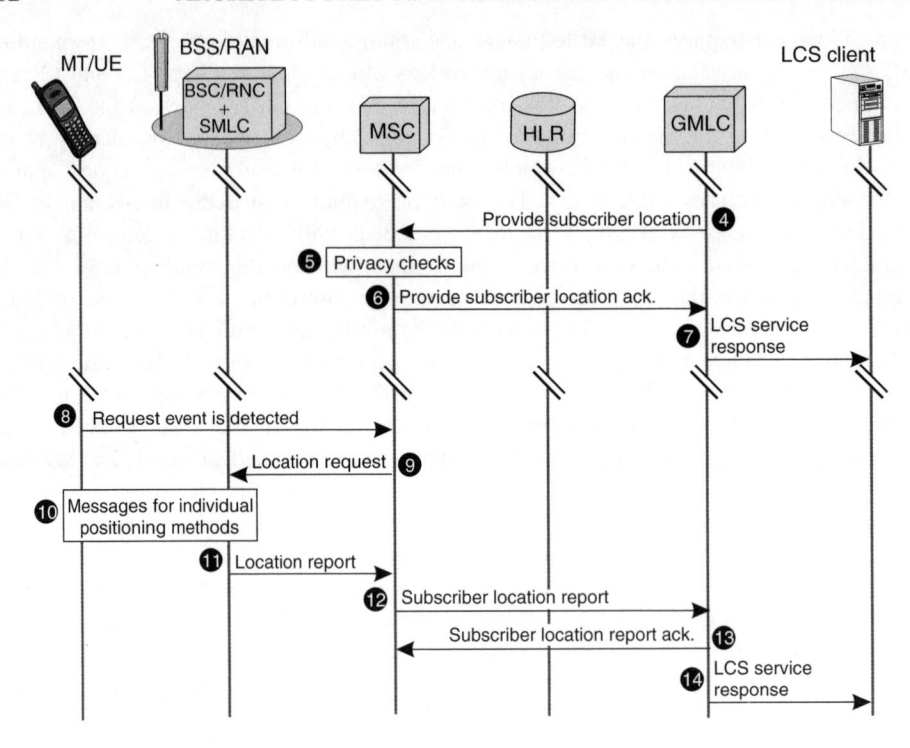

Figure 11.5 Mobile terminating deferred location request.

An open issue not addressed in Figure 11.5 is the case where the target subscriber changes the MSC area when the MSC waits for the trigger condition to become true. In this case, the old MSC sends a location report to the GMLC, indicating that the deferred request must be reinitiated against the new MSC, whereupon steps (4)–(6) must be repeated for this new MSC.

11.1.3.2 Mobile Originating Location Request

Mobile originated location requests are always initiated by the target terminal itself; hence the general procedure of a location request differs from that of mobile terminated requests. It is depicted in Figure 11.6.

The location preparation procedure encompasses Steps (1)–(5). In the first step, the terminal sends a service request to the access network (1), which is forwarded to the MSC (2). The MSC acknowledges the request by returning a service acceptance to the terminal (3). If the terminal is in idle mode, authentication and ciphering is necessary before sending the accepted message. After that, the terminal invokes the actual location service (4) and specifies the type of LCS (i.e., basic self-location, autonomous self-location, transfer to third party). This invocation contains a number of parameters, for example, the terminal's positioning capabilities, the desired QoS, or, in case of transfer to third party, the identity of the LCS client. Upon arrival of the invocation, the MSC checks whether the target subscriber has permission at all to request the specified LCS, and if so, it delegates positioning to the access network.

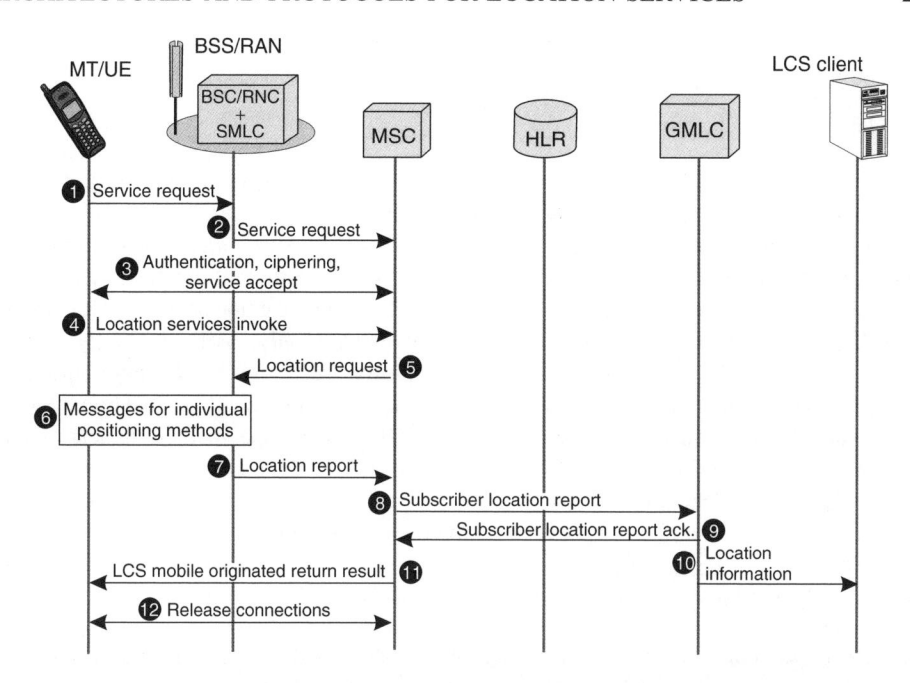

ARCHITECTURES AND PROTOCOLS FOR LOCATION SERVICES 283

Figure 11.6 Mobile originating location request.

Steps (5)–(7) then correspond to Steps (9)–(11) of Figures 11.4 and 11.5 respectively. After positioning, the position fix is sent to the GMLC (8,9), where it may be translated to another reference system and forwarded to an LCS client. Steps (8)–(10) are only executed if the target subscriber has requested for transfer to third party. Finally, a return result is passed to the terminal (11), which includes the position fix or a notification that location data has been passed to the specified LCS client. Finally, all resources involved in positioning are released (12).

11.1.4 Privacy Options

From the constraint types that a privacy policy may consist of (see Chapter 10), only two are supported in Releases 99 and 4 of the 3GPP specifications, that is, notification and actor constraints (options). A subscriber's privacy options are contained in its *Subscriber LCS Privacy Profile* (SLPP), which is stored in the HLR and transferred to the VLR or SGSN (either of the home network or of any visited network) the subscriber is currently attached to.

The notification options allow specification of whether the target subscriber wishes to be explicitly notified about each positioning attempt. They can be specified in the following gradations of strictness:

- Positioning allowed without notifying the target subscriber.

- Positioning allowed only with notification to the target subscriber.

- Positioning requires notification and verification by the target subscriber; it is allowed only if granted by the target subscriber or if there is no response to the notification.

- Positioning requires notification and verification by the target subscriber; it is allowed only if granted by the target subscriber.

- Positioning not allowed.

Thus, positioning without notifying the target subscriber represents the loosest degree and the general refusal of positioning the strictest one.

The actor options are subdivided into GMLC and LCS client options. With the GMLC options, the target subscriber can determine a set of GMLCs that are authorized to trigger positioning and to process its location data. It is possible to grant access in general for all GMLCs, for GMLCs belonging to the home network only, or to explicitly identify one or several GMLCs from which location requests are allowed. The latter might be applied by subscribers roaming in visited networks in order to allow location requests from trusted GMLCs only. In this case, the GMLCs must be specified by means of their E.164 addresses (see (ITU–T Recommendation E.164)). The LCS client options are used to authorize external LCS clients to request and receive location data of the target subscriber. Like for GMLC options, the clients are identified by E.164 addresses.

The options can be specified for different *privacy classes*, and within each class interrelated in a hierarchical fashion. The following classes are distinguished:

- **Universal class.** If the target subscriber is subscribed to the universal class, positioning is not restricted, and location data can be requested and received by any internal or external LCS client. The privacy options of other privacy classes are not considered.

- **Call/session-related class.** This privacy class contains options that are to be applied in the context of a call (in the CS domain) or a session (in the PS domain), assuming that the call or session has been established between the target subscriber and the external LCS client requesting for location data.

- **Call/session-unrelated class.** The privacy options of this class are applied if an external LCS client requesting for location data is not involved in a call or session with the target subscriber.

- **PLMN operator class.** The privacy options of this class refer to location requests from internal LCS clients located in the home or visited networks, which are officially denoted as home or visited *Public Land Mobile Network* (PLMN).

The relation between privacy classes and the different types of options is shown in Figure 11.7. In the universal class, no options are possible at all. If the target subscriber activates this class, it allows unrestricted positioning regardless of which LCS client has requested for it. In the call/session-related and unrelated classes, the target subscriber determines notification options that are to be applied for location requests in general. Optionally, it can define an exception list by using the LCS client options. This list contains the identities of all LCS clients that are not subject to the general notification options, but for which dedicated options are to be applied. With these dedicated options it is possible to fix the GMLCs that are allowed to transfer location requests and responses between this LCS client and the target subscriber as well as to define special notification options for this LCS

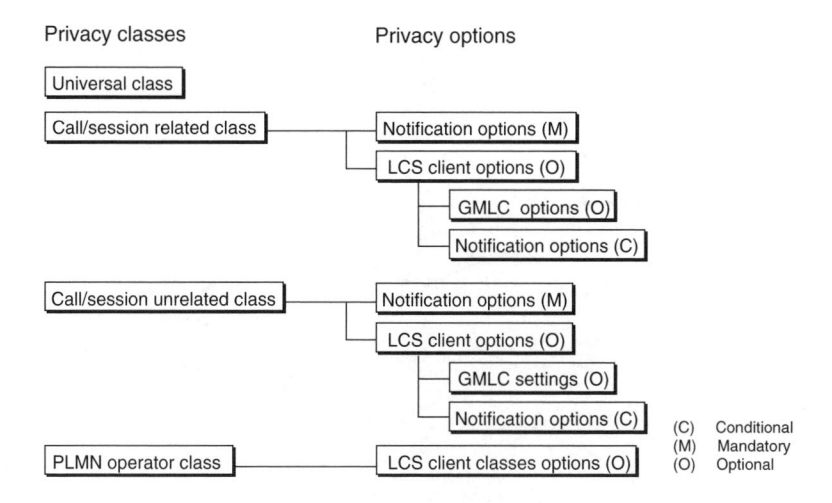

Figure 11.7 Relation between privacy classes and options.

```
Privacy classes              Privacy options

Universal class

Call/session related class ──── Notification options (M)
                            └── LCS client options (O)
                                           ├── GMLC options (O)
                                           └── Notification options (C)

Call/session unrelated class ── Notification options (M)
                            └── LCS client options (O)
                                           ├── GMLC settings (O)
                                           └── Notification options (C)

PLMN operator class ─────────── LCS client classes options (O)

(C) Conditional
(M) Mandatory
(O) Optional
```

client. For example, a target subscriber may generally forbid positioning in the call/session-unrelated class, but it may explicitly exclude the LBS provider from these options that offer the community service it has subscribed for. For this LBS provider, it can then grant access to location data for all GMLCs of its home network and he can allow positioning without notification.

The PLMN operator class contains the privacy options to be applied for location requests from internal LCS clients. This class does not enable the definition of notification options. Instead, access to location data can only be restricted to the following classes of LCS clients, for which positioning is then allowed without notification:

- LCS clients providing a location-related broadcast service,

- LCS clients being part of the home or visited network's operation and maintenance center (for example, in order to perform non-LCS-related functions like the location-assisted handover),

- LCS clients recording anonymous location data, and

- LCS clients providing supplementary services or other services the target subscriber has subscribed for.

Figure 11.8 shows a flow diagram for selecting the relevant privacy class if the target subscriber has activated more than one of them. The procedure given by this diagram is used during the privacy check at the MSC or SGSN when a location request arrives (see Step (5) in Figures 11.5 and 11.4).

At first it is checked whether the universal class is activated (1), and if so, location requests are allowed in general and further checks are not necessary (2). If not, it is subsequently checked whether the request belongs to the PLMN operator, the call/session-related or unrelated class (3, 4, and 5). If one of the class criteria is met, it is checked whether the subscriber has defined options for this class (6), and if so, the location request

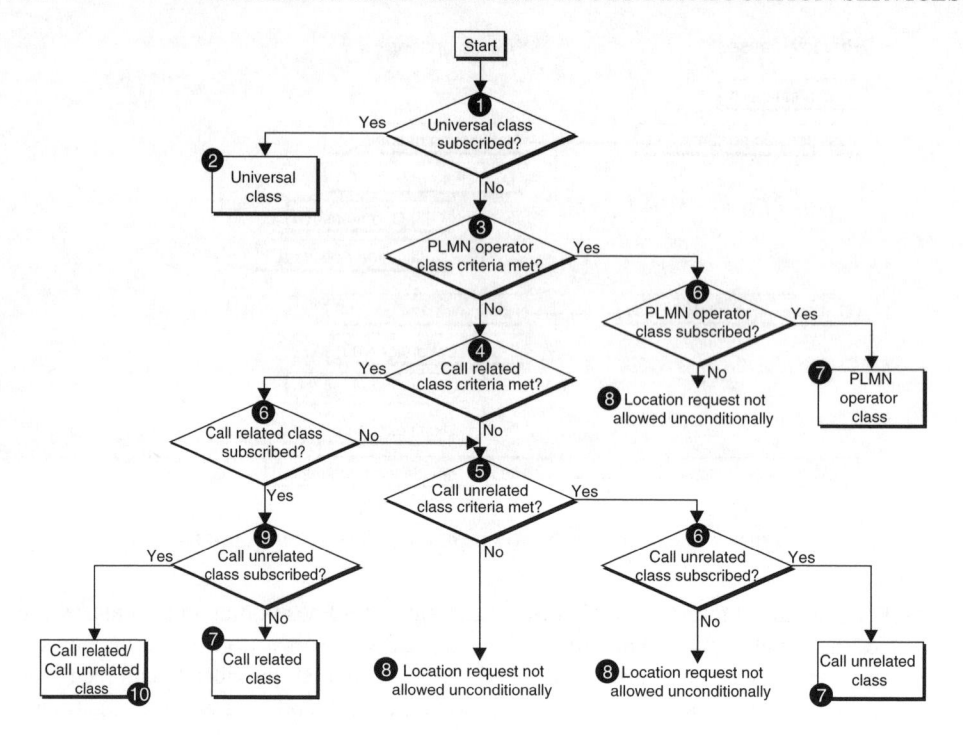

Figure 11.8 Privacy class selection rule (3GPP TS 23.271).

is then checked against them (7). If not, the location request is not allowed (8). In the particular case of a call/session-related location request where the subscriber has defined policies for both the call/session-related and unrelated classes (9), the loser privacy options are selected (10).

11.1.5 Outlook to Future Releases

In this chapter, 3GPP LCSs have been described from the point of view of Releases 99 and 4. For future releases, positioning and LCSs are subject to some extensions. Table 11.4 gives an overview of the features of each release.

The majority of new features concentrate on privacy aspects. From Release 5, it is envisaged to support the specification of what has been denoted as service constraints in Section 10.4.3.2. In this way, it will be possible to limit location data access to certain types of LBSs. In order to make service-type options interpretable by various actors, for example, visited networks the target subscriber is roaming to, Release 5 and subsequent releases envisage to categorize LBSs according to the scheme presented in Table 11.5. Note that 3GPP does not standardize interfaces or behavior of these services, but only their naming or identifier scheme according to which they are referenced in the privacy options.

Furthermore, from Release 5 it will be possible to indicate the identity of the LBS user, called the *requestor*, who has triggered the location request to the target subscriber. Optionally, target and LBS user may agree on a code word, which has to be delivered

Table 11.4 Schedule for the introduction of new LCS and positioning features

Release	New positioning and LCS features
99	Positioning: Cell-Id combined with TA/RTT, E-OTD, IPDL-OTDoA, A-GPS LCS: Immediate LCS, deferred LCS with periodic reporting and reporting upon subscriber registers with network Privacy: Notification, GMLC, LCS client options
4	–
5	Privacy: Service type options, requestor code word, and notifications including requestor identity
6[1]	Positioning: UL-TDoA LCS: Support of roaming subscribers, deferred LCS with zone-based reporting (change of area) Privacy: Anonymization by identifier abstraction, implementation of Pseudonym Mediation Device (PMD) and Privacy Profile Register (PPR)

[1] not frozen at the time of writing

together with the requestor's identity. The checking of this code word may be obliged either to the target subscriber itself or to the LCS server. In this way, the target subscriber is protected against location requests from actors that "have stolen" the identity of a requestor but do not know the appropriate code word. Anonymization by permanent or temporary pseudonyms will be covered not before Release 6. For this purpose, it is intended to implement a new component, the so-called *Pseudonym Mediation Device* (PMD), which care for the mapping between pseudonyms and true identities (MSISDN or IMSI). Also, from Release 6, privacy options will not be contained in the HLR, but will be managed by a dedicated component, the so-called *Privacy Profile Register* (PPR).

Furthermore, it is planned to extend the deferred location request by a so-called *change-of-area event*, which basically corresponds to zone-based reporting. The change-of-area is an event where the target enters or leaves a predefined geographical area. The geographical area is defined by the LCS client and is mapped onto the identifiers of one or more radio cells, location or routing areas by the LCS server. Their identifiers are then transferred to the target's terminal, and if the terminal recognizes that the target enters or leaves one of these predefined areas, it notifies the network that a change-of-area has occurred. This corresponds to Step (8) in Figure 11.5. Furthermore, Release 6 also provides improved roaming support in that it is possible to interconnect GMLCs of different operators and pass location requests and reports between them. Finally, as already mentioned in Chapter 8, Release 6 introduces the additional positioning method U-TDoA.

Details of the new features can be found in the respective release versions of (3GPP TS 22.071) and (3GPP TS 23.271). Note that at the time of writing the functions of Release 6 had not been frozen yet.

Table 11.5 Standardized LBS types according to releases 5 and beyond of (3GPP TS 22.071)

LBS categories	Standardized LBS types
Public safety services	Emergency services
	Emergency alert services
Location sensitive charging	
Tracking services	Person tracking
	Fleet management
	Asset management
Traffic monitoring	Traffic congestion reporting
Enhanced call routing	Roadside assistance
	Routing to nearest commercial enterprise
Location-based information services	Traffic and public transportation information
	City sightseeing
	Localized advertising
	Mobile yellow pages
	Weather
	Asset and service finding
Entertainment and community services	Gaming
	Find your friend
	Dating
	Chatting
	Route finding
	Where-am-I
Provider specific services	

11.2 Enhanced Emergency Services

As outlined in Chapter 1, emergency services represent a very obvious and reasonable application area for the deployment of location services. However, compared to other applications, location-based emergency services, which are commonly referred to as *enhanced emergency services*, impose very special requirements on the underlying networks and systems and therefore need tailored solutions, which are presented in this section. The development of enhanced emergency services has specifically been fostered in the United States (known as the E-911 initiative there), but in the meantime is considered in other countries too. For example, the European Union has launched activities for E-112, and in Japan, enhanced emergency services based on GPS-enabled mobile phones are expected to go into operation in 2007. However, this section specifically covers the situation in the United States, as the introduction of enhanced emergency services has progressed to a great extent there.

Basically, an enhanced emergency service is characterized by the following three service features:

ARCHITECTURES AND PROTOCOLS FOR LOCATION SERVICES 289

- Delivering and displaying the telephone number of the emergency caller. This feature is called *Automatic Number Identification* (ANI).

- Routing of an emergency call to the *Public Safety Answering Point* (PSAP) that serves the geographical area the emergency call originates from. This feature is called *selective routing*.

- Determining the location (in terms of a street address) of an emergency caller. This feature is called *Automatic Location Identification* (ALI).

In many countries, the geographic area covered by an emergency service is subdivided into several *emergency service zones* in that there is no overlapping between them and each point of the geographic area is covered by a zone. Each zone is assigned a PSAP as well as different emergency response agencies like fire, police, and ambulance departments that are connected to a PSAP. A PSAP is a network component that is responsible for receiving emergency calls and forwarding them either to another, secondary PSAP or to the emergency response agency addressed by the call. According to the *National Emergency Number Association* (NENA), the territory of the United States is covered by approximately 6,000 PSAPs (NENA Web site), while in Europe there exist between 14 and 1,060 stage 1 PSAPs and numerous stage 2 PSAPs in each country (Salmon 2004).

11.2.1 Wired Enhanced Emergency Services

In order to understand the difficulties of enhanced emergency services for wireless networks, it is useful to first get an overview of their principal working in fixed telephony. Figure 11.9 depicts the basic configuration of an emergency service network as it exists, with more or less modifications, in many countries.

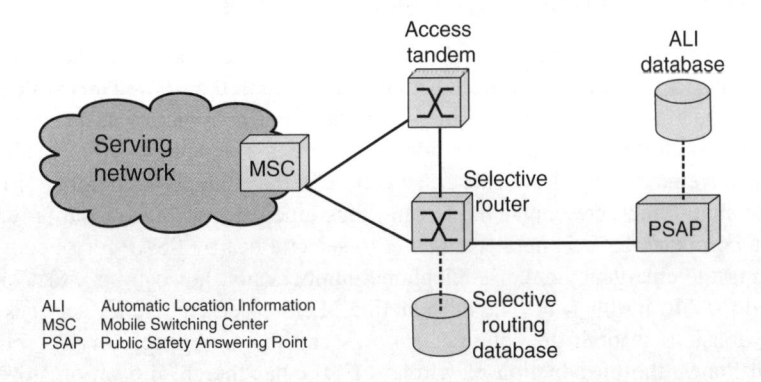

Figure 11.9 Basic configuration of emergency services network.

The provision of enhanced emergency services in wired networks is essentially based on the ANI feature, that is, on the network's capability to deliver the caller's telephone number to the recipient. While this represented a barrier in the former analogous telephony systems, ANI is a standard feature in modern digital telecommunications, which is realized by *call-associated signaling* (CAS) over a separate signaling infrastructure like SS7. CAS is used to set up, maintain, and release calls, as well as for controlling service features associated with a call, for example, reverse charging, call forwarding, or ANI. The counterpart is *non-call-associated signaling* (NCAS), which happens independent of any call and is used, for example, to track subscribers in the context of location management in cellular networks as has been explained in Chapter 5.

ANI is the key function for selective routing and ALI, as it allows for resolving a caller's location from his telephone number. This is supported by two databases, the *Selective Routing Database* and the *ALI database*. The former is connected to a selective router, which connects several PSAPs to the networks of different operators either directly or via an access tandem that concentrates the trunks from and to several networks. Upon arrival of an emergency call, the selective router extracts the caller's telephone number delivered via CAS and requests the Selective Routing Database for the PSAP that is responsible for the emergency service zone the call originates from. For this purpose, the telephone number is mapped onto an emergency service number representing this emergency service zone and the associated PSAP, to which the call is then forwarded. When arriving there, the caller's street address may be obtained by using the ALI database in order to display it to the staff of the emergency response agency. Similar to the Selective Routing Database, the street address is resolved there from the caller's telephone number.

The ALI and Selective Routing Databases may be physically collocated, but are shown as separate components in Figure 11.9 because the queries to them are quite different (3GPP2 N.S0030). The ALI database is composed from the subscription data of the wired telephone companies, which are usually obliged by law to record the identity of each subscriber together with its telephone number and the street address where the wired telephone line is terminated. As the location of wired telephone lines change rarely, there is no need to update the ALI database in real time. Instead, it is considered to be sufficient to update it every 24 h.

11.2.2 Wireless Enhanced Emergency Services

Mobility support of subscribers in cellular networks makes the realization of enhanced emergency services a very complicated matter. As opposed to wired networks, where a telephone number is permanently assigned to a wired line terminating at a certain location, the MSISDN in cellular networks is location independent. The main intention behind cellular location management is to allow subscribers to keep their telephone numbers irrespective of the base station they are connected to (and thus independent of the location where they stay). As a consequence, it is neither possible to select a nearby PSAP for selective routing by analyzing the emergency caller's telephone number only, nor can his exact location be determined for ALI in this way. The introduction of enhanced emergency services therefore requires extensions in both the cellular networks and the emergency service networks. In the United States, the introduction of wireless E-911 has therefore been organized in two phases.

11.2.2.1 Phase I

Remember from Chapter 1 that the first phase of E-911 requires to derive a caller's location only from the coordinates of the serving base station and to enable ANI. For this purpose, a set of pseudo telephone numbers is assigned to each base station. In the United States, this pseudonumber is known as *Emergency Service Routing Digits* (ESRD). The MSC knows

ARCHITECTURES AND PROTOCOLS FOR LOCATION SERVICES 291

the pseudo numbers of all base stations in its coverage area. The numbers are also known by the Selective Routing and ALI databases. The former provides the mapping between pseudo number and responsible PSAP, while the latter delivers the corresponding base station location to the emergency response agencies. When establishing an emergency call, the MSC attaches the pseudo telephone number of the base station serving the emergency caller to the CAS messages needed for call setup and forwards them to the selective router. The remaining steps are the same as for wired networks (see the last section). Note that the pseudo telephone number is for internal use only. It is not dialable and does not replace the caller's original telephone number. This has to be forwarded in addition in order to enable callback if the call is unintentionally interrupted.

The advantage of this approach is that it requires only minor modifications on the existing network infrastructure. The MSCs need to be upgraded for detecting emergency calls and attaching the pseudo numbers to CAS messages, while Selective Routing and ALI databases simply need to be extended by additional entries for resolving the pseudo telephone numbers. However, the drawback is that location information obtained in this way is much too inaccurate, especially when compared to the exact street addresses delivered for emergency callers attached to wired networks. Basically, the positioning method used here is Cell-Id, and its poor accuracy performance has already been discussed in Chapter 8. Another concern is that an emergency caller may change his position in the course of a call once or several times, and it is not possible to track him by this approach.

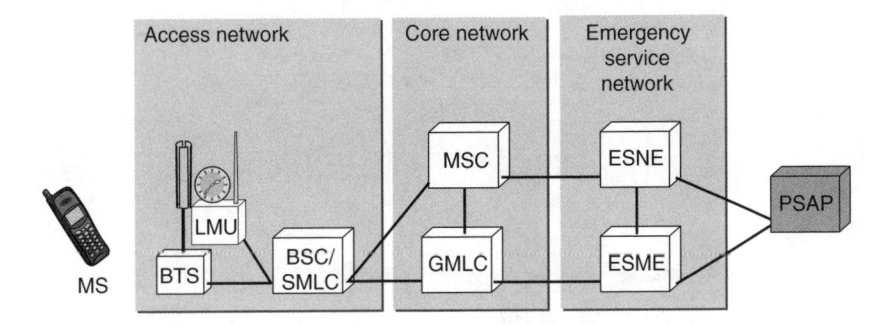

Figure 11.10 Network reference model for Phase II emergency services.

11.2.2.2 Phase II

As explained in Chapter 1, Phase II imposes to locate an emergency caller with a comparatively high accuracy and to deliver its location in terms of latitude and longitude instead of using any other location format. Furthermore, it is also desired to provide updated location data to the emergency response agency in the course of a call. Thus, these features require the deployment of advanced positioning methods and significant modifications and extensions to be made in the networks. One of the most significant changes concerns signaling for exchanging location data, which is now required to be available in two distinct modes:

- **CAS push.** CAS push represents the traditional way to deliver location information. It is always linked to an emergency call and automatically delivers, that is, pushes the initial position of an emergency caller at call setup.

- **NCAS pull.** Using NCAS pull, the position is requested, that is, pulled over signaling channels that are not related to the emergency call. It can be independently invoked regardless of the state of the call and is therefore useful for providing the emergency response agency with updates of position data.

The distinction between these signaling modes gives rise to a reorganization of the emergency services network, because a selective router cannot adopt the processing of NCAS messages. As can be derived from Figure 11.10, the new components are the *Emergency Services Network Entity* (ESNE) and the *Emergency Services Message Entity* (ESME). The former is responsible for routing emergency calls to the right PSAP and is therefore connected to an MSC. The ESME requests initial or updated position data of an emergency caller from a GMLC and forwards it to the PSAP. In addition, ESNE and ESME are interconnected in order to correlate emergency calls and related NCAS messages. For this purpose, each emergency call is assigned an *Emergency Services Routing Key* (ESRK) by the serving MSCs, which is used to correlate calls and related NCAS messages and to route the call through the network.

Note that ESNE and ESME are rather logical components than physical ones, because they might be realized by the existing but extended network components known from wired emergency services. For example, the ESNE may accommodate a selective router, and the ESME may include the ALI database. However, there exist many different configuration options that can hardly be described here. The interested reader can receive additional information from (3GPP2 N.S0030).

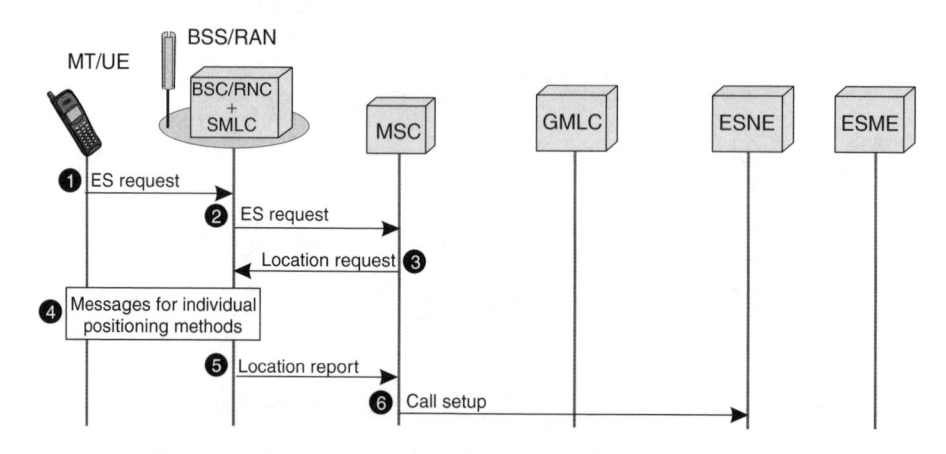

Figure 11.11 Emergency LCS with CAS push.

Figure 11.11 depicts the call setup of an emergency service and related positioning with CAS push. Like for any other call, a service request is initialized by the terminal (1) and forwarded to the serving MSC. From all calls arriving there, the MSCs filter out the emergency calls and delay them in order to determine the caller's position first. For this purpose, it returns a location request to the access network, whereupon positioning is performed there (either terminal- or network-based) (4). The resulting position is then returned to the MSC (5). The MSC assigns ESRD and ESRK identifiers to the call and

ARCHITECTURES AND PROTOCOLS FOR LOCATION SERVICES 293

sends them together with the caller's MSISDN and his position to the GMLC (6). These parameters are used later for correlating NCAS messages to the correct emergency call if updated position data is requested. Alternatively, the ESRK may also be assigned by the GMLC. After the GMLC has acknowledged the reception of this data (7), the MSC proceeds with the call setup and includes the identifiers mentioned before as well as the caller's position in the standard messages for call setup and transfers them to the ESME (8), which then establishes the call between the terminal and the PSAP.

Even though an emergency call might be processed with higher efficiency if the position data is immediately delivered together with the call to the emergency response agency, this approach may lack from the delay imposed on call processing during positioning in Steps (3)–(7). Depending on the concrete emergency situation, this delay, which might be in the range of several seconds, may be experienced very negatively by the emergency caller. To cope with this problem, the MSC initializes a timer when imposing the location request in Step (3). If this timer expires and position data has not arrived yet, the MSC passes the call without position data to the ESNE. Selective routing is not affected by that since the MSC includes the ESRD, which designates the cell site the call originates from. For ALI, the position might then be requested subsequently by using NCAS pull.

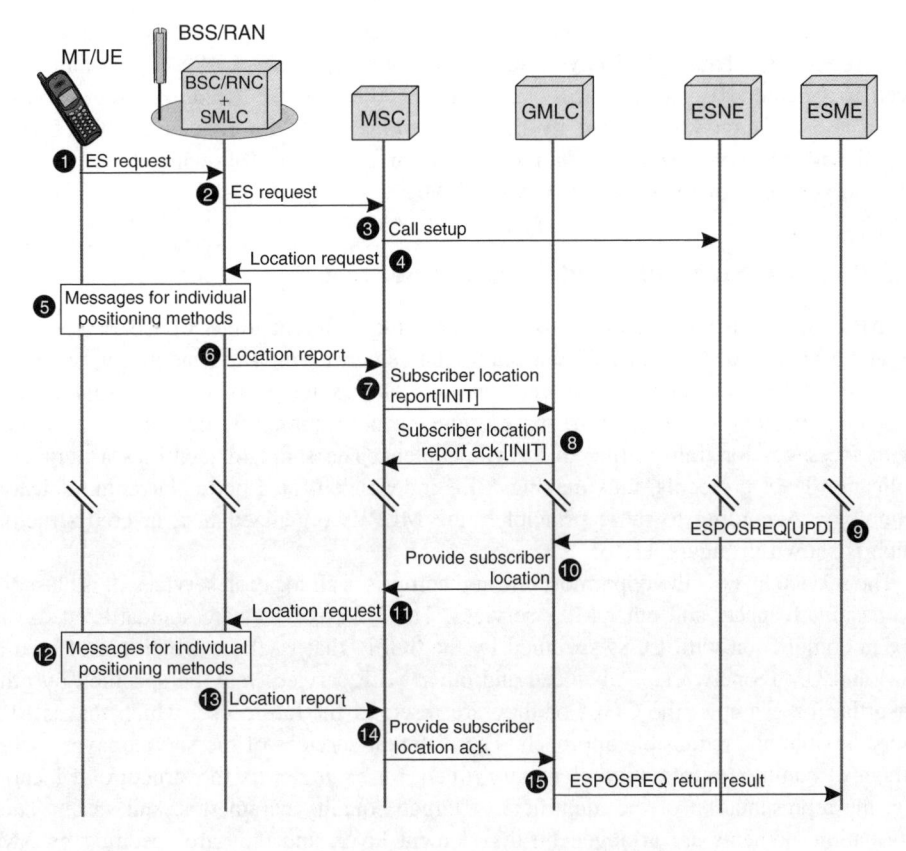

Figure 11.12 Emergency LCS with NCAS pull.

Alternatively, it is possible to abandon CAS push and to exclusively determine the position by NCAS pull, which is depicted in Figure 11.12. In this scenario, the emergency call is passed to the ESNE independent of positioning (1–3). After the call setup has been processed by the MSC, it triggers positioning (4–5) and receives the so-called *initial position*, which is forwarded to the GMLC together with the caller's MSISDN, as well as the ESRD and ESRK that have been assigned in Step 3. The initial position might then be requested explicitly by the ESME (not depicted in the figure). Furthermore, at any time in the course of the call an *updated position* may be requested by the ESME (9), whereupon the entire positioning process is repeated (10–15). It is also possible to request for the emergency caller's position after the call has been released, which is then denoted as the *last known position*.

A more detailed description of Enhanced Wireless 911 Phase II can be found in (3GPP2 N.S0030), while (3GPP TS 23.271) describes the realization of emergency services from the point of view of GSM/UMTS.

11.3 Mobile Location Protocol

The *Mobile Location Protocol* (MLP) is a location transport protocol for exchanging location data between an LCS client and a location server, for example, as specified by 3GPP (see last section). Thus, MLP covers reference point (c) of the LBS supply chain introduced in Chapter 10. The protocol has been specified by the *Location Interoperability Forum* (LIF), which was created by Ericsson, Nokia, and Motorola in 2000 and which merged with the *Open Mobile Alliance* (OMA) in 2002. The following descriptions are based on version 3 of MLP (see (OMA LOC MLP)).

11.3.1 MLP Structure and Location Service

The MLP defines a number of LCSs for supporting different kinds of LBSs (e.g., reactive and proactive ones) with different requirements concerning the availability, delay, and reliability of location data and its transmission. The exchange of content between LCS client and location server is XML-based, and each location service comprises a number of messages for transporting this XML content. The standard identifies a number of application-level protocols, thus making MLP independent of a particular communication technology. According to these principles, the MLP is organized in a layered structure, which is shown in Figure 11.13.

The service layer is the uppermost one and comprises all location services. It is classified into basic, advanced, and other MLP services. The basic services are standardized for use only in conjunction with LCSs specified by the 3GPP, that is, those which are offered by GSM and UMTS networks. Advanced and other MLP services may be specified by other consortia, for example, the OGC, or they are reserved for future use, which makes MLP a very flexible and extensible approach. The different services of the service layer share a number of common information elements, which, for example, fix the structure of location data, the representation of the identifiers of targets, quality parameters, and so on. These information elements are arranged on the element layer, and they are specified by XML *Document Type Definitions* (DTD).

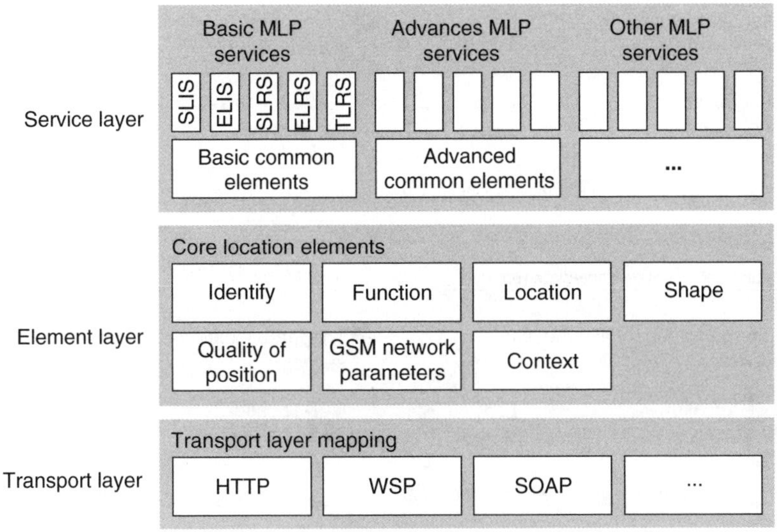

ARCHITECTURES AND PROTOCOLS FOR LOCATION SERVICES

Figure 11.13 MLP structure.

Finally, the transport layer represents the application-level protocol used to transfer the XML content that has been composed by the upper layers. Although, the MLP standard identifies a number of alternative protocols (for example, the *Simple Object Access Protocol* (SOAP) or the *Wireless Session Protocol* (WSP)), the only mapping available so far is that for the HTTP. The MLP standard fixes two socket ports for operation, namely, 9210 for insecure and 9211 for secure connections. An insecure connection can be used if location server and LCS client are located in a secure, private domain, while for operation over the public Internet encrypted communication is strongly recommended in order to safeguard security and privacy of location data. Encryption, data integrity, and authentication are based on the SSL/TLS service here.

Figure 11.14 gives an overview of the different basic MLP services and their messages. The following list presents a short description of each service:

- **Standard location immediate request (SLIR).** The SLIR realizes the querying of location data and is used whenever an LCS client requires the location data of one or several targets immediately, that is, within a predefined period of time. It is available in a synchronous and asynchronous mode. In the former one, the LCS client requests the location of one or several targets, whereupon the location server collects all location data and returns it with the answer to the request. A drawback of this approach emerges if location data of a higher number of targets are bundled in a single request. The answer will not be returned then before the network has determined all locations, which may take an unacceptable amount of time. In this case, the asynchronous mode might be preferred, where the location server creates dedicated reports for a target or a subgroup of targets specified in the request as soon as their location data is available. This guarantees that location data is passed without delays to the LCS client and is not outdated when it arrives there.

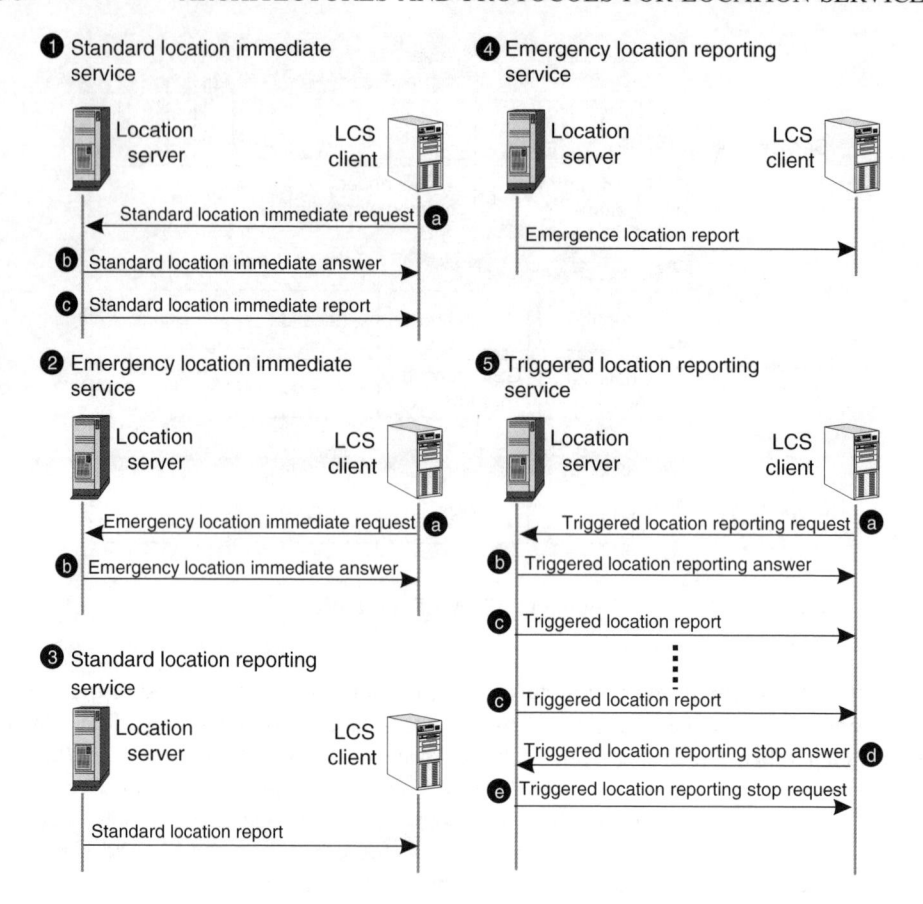

Figure 11.14 MLP location services overview.

- **Emergency Location Immediate Request (ELIS).** The ELIS is used for requesting the location data of a target that has initiated an emergency call. The answer to this request must be returned immediately, that is, within a predefined period of time to the LCS client, which is typically located at an emergency center.

- **Standard Location Reporting Service (SLRS).** The SLRS is to be used in conjunction with the mobile originating location service of 3GPP. It is explicitly triggered either by the target itself or by the location server in order to provide a certain LCS client with location data. Correspondingly, this client needs to be specified by the initiating entity.

- **Emergency Location Reporting Service (ELRS).** This is the counterpart to the SLRS for emergency cases. It is used when the network per default locates a target that have initiated an emergency call in order to send it location data to a nearby emergency center.

- **Triggered Location Reporting Service (TLRS).** The TLRS is used to track a target by an LCS client and thus corresponds to the reporting patterns described in

Chapter 10. The service is invoked at the location server by a request, which instructs the trigger when to send a location report. In its current version, MLP supports only periodic updating and updating when the target registers with the network. Advanced updating mechanisms, like distance or zone-based reporting, are not supported in the latest version. The tracking is performed until the LCS client sends a stop request to the location server.

The XML content exchanged in the messages of these services is specified by seven XML DTDs on the element layer. The following list provides an overview of these DTDs (their usage will then be demonstrated in a subsequent example):

- **Identity element definitions.** Specify several alternatives for identifying a target in a request, for example, by using MSISDN, IMSI, IPv4, or IPv6 addresses.

- **Function element definitions.** Determine a number of parameters for configuring messages, like timestamps, priorities, triggers, and push addresses.

- **Location element definitions.** Provide a means for classifying location data according to different shapes (points, circular areas, polygons, etc.), as well as for specifying related parameters like spatial reference systems, direction, and speed.

- **Shape element definitions.** Fix the concrete data format of shapes determined by the location elements definition.

- **Quality of position elements definitions.** Identify parameters for expressing the desired quality of location data in a request, for example, vertical accuracy, latitude/longitude accuracy, and maximum location age.

- **Network parameters element definitions.** Allow to express location data in terms of GSM/UMTS network parameters such as cell identifiers or timing advance values (if needed).

- **Context element definitions.** Used for authenticating the LCS client at the location server and to indicate the LBS user requesting a target's position (requestor).

The following section demonstrates the usage of these DTDs by means of the TLRS.

11.3.2 Example

An MLP message typically consists of a header and a body part. For all requests initiated by an LCS client, the header is mandatory, while for answers or reports from the location server it is only optional. In the former case, it is used to authenticate the LCS client at the location server, and to inform the location server about the identity of the LBS user who is requesting location data. In the context of the MLP, this user is denoted as *requestor*. The header of a request message from the LCS client to the location server may then look like this:

```
<hdr ver="3.0.0">
    <client>
        <id>TheFriendFinderServiceCompany</id>
        <pwd>ak451lbs</pwd>
```

```
            <serviceid>4711</serviceid>
        </client>
        <requestor>
            <id>4917226912499</id>
        </requestor>
</hdr>
```

The elements of the header are standardized by the Context Element Definitions. The `<client>` element contains a number of subelements needed for authentication at the location server, including elements carrying the identifier and the password that has been assigned to the LCS client. The `<serviceid>` element can be used to distinguish between different LBS applications. The `<requestor>` element denotes the identity of the requestor requesting location data. It is used at the location server to check whether this user is authorized at all to receive location data for the target specified in the subsequent body part of the message. Once an LCS client is authorized at the location server, subsequent messages may carry a `<sessionid>` element that replaces the `<client>` element and that remains valid as long the session between LCS client and location server lasts.

The XML content exchanged in the messages of the various location services differ from each other more or less significantly, but is composed according to the same principle. In the following, the composition of message content is demonstrated for the TLRS. A triggered location-reporting request may look like this:

```
<tlrr ver="3.0.0">
    <msids>
        <msid type="MSISDN">4917959896301</msid>
    </msids>
    <interval>00003000</interval>
    <start_time utc_off="-0100">20050330154500</start_time>
    <stop_time utc_off="-0100">20050330184500</stop_time>
    <qop>
        <hor_acc>100</hor_acc>
    </qop>
    <geo_info>
        <CoordinateReferenceSystem>
            <Identifier>
                <code>4326</code>
                <codeSpace>EPSG</codeSpace>
                <edition>6.1</edition>
            </Identifier>
        </CoordinateReferenceSystem>
    </geo_info>
    <pushaddr>
        <url>http://www.locationbasedservices.org/</url>
    </pushaddr>
    <loc_type type="CURRENT" />
    <prio type="HIGH" />
</tlrr>
```

The `<msids>` element specifies one or several targets for which location data is needed. MLP provides for various naming or numbering schemes, such as MSISDN, IMSI, IPv4 and IPv6 addresses. Several targets can be identified by a list of `<msid>` or, alternatively, as a range of consecutive numbers of a particular scheme (which is useful, for example, if the trucks of a fleet have been assigned consecutive MSISDN numbers by an operator). In the example, the request configures the location server to send periodic reports with a frequency specified by `<interval>` and for the time frame given by `<start_time>` and `<stop_time>`.

In the next steps, the request contains a closer specification of the desired quality and format of location data. The quality is defined by the `<qop>` element, and, in this example, contains only the horizontal accuracy in meters. However, MLP provides for the specification of various other quality parameters like latitude and longitude accuracy in seconds, the maximum allowable age of location data, or accuracy of altitude in meters. The spatial reference system that the location data refers to is defined in the `<geo_info>` element. In its current version, MLP has adopted the EPSG code space. The identifier 4326 refers to WGS-84.

Finally, `<pushaddr>` contains the address of the host to which the reports should be sent. The element `<loc_type>` specifies whether the location server should execute positioning in order to obtain the current position of the target (CURRENT), or whether the location data that has been cached in the location server (LAST should be returned). The `<prio>` element defines whether the request should be processed with normal or high priority.

Upon arrival of a triggered location reporting request, the location server generates a request identifier and returns it to the LCS client in a triggered location reporting answer:

```
<tlra ver="3.0.0">
    <req_id>12345</req_id>
</tlra>
```

Note that the request identifier must not be confused with the session identifier assigned after establishing a session between LCS client and location server. The latter represents a trusted relationship between these two entities, while the request identifier is used to discriminate between different request/report pairs exchanged within a trusted session.

According to the request in our example, the location server periodically reports the target's position starting at the time and with the frequency specified in the request. Such a report may then appear as follows:

```
<tlrep ver="3.0.0">
    <req_id>12345</req_id>
    <trl_pos trl_trigger="PERIODIC">
        <msid>4917959896301</msid>
        <pd>
            <time utc_off="-0100">20050330161500</time>
            <shape>
                <CircularArea srsName="www.epsg.org#4326">
                    <coord>
                        <X>48 08 57.200N</X>
                        <Y>11 35 47.600E</Y>
                    </coord>
                    <radius>15</radius>
```

```
            </CircularArea>
          </shape>
       </pd>
    </trl_pos>
</tlrep>
```

In the beginning, the report identifies the request this report refers to, the type of trigger that has activated this report, and the target whose location data is contained in the report. The location data is then encapsulated in the <pd> element, starting with a timestamp of the position fix. For expressing the actual position, the GML specification (version 2.1.1) has been adopted for MLP. Consequently, a position may be expressed as a number of different shapes like points, line strings, polygons, or, as shown in the example, circular areas.

Each request message depicted in Figure 11.14 is sent to the location server via a HTTP POST request, while answers from the server to the client are returned in a HTTP response. Reports, on the other hand, which in most cases are asynchronous and deferred, are issued in that the location server sends them in a HTTP POST request to the LCS client. Thus, both LCS client and location server must be equipped with HTTP servers for processing HTTP POST requests. Thus, MLP is not well suited for scenarios where the location server resides in mobile terminals as in Scenarios 4 and 5 of Chapter 10.

11.3.3 Outlook to Future Releases

The OMA is continuously working on extensions of MLP (see (OMA Location Working Group)). This concerns enhancements of MLP itself, as well as a number of auxiliary protocols. The goal is to define an entire protocol suite for location services, the so-called *Location Enabler Release* (LER), which comprises MLP, the *Privacy Checking Protocol* (PCP), and the *Roaming Location Protocol* (RLP). The LER is specifically tailored to comply with Release 6 of 3GPP LCSs. Since work on it is in progress at the time of writing, the extensions are only covered briefly here.

The PCP is used to check a target's privacy options if a location request arrives at the location server. It defines the message exchange needed for this purpose between the location server and an (external) *Privacy Checking Entity* (PCE), which is the OMA term for the PPR of 3GPP. It also cares for anonymization in that it defines procedures for the mapping between pseudonyms and the target's true identity. The RLP focuses on the interconnection of several location servers, for example, GMLCs, in order to pass location requests and responses between them, which is especially required to make location data of targets roaming in foreign networks accessible. Furthermore, MLP itself is extended and modified in order to meet requirements of Release 6 LCSs. This includes, for example, extended XML DTDs for defining "change-of-area" events for the TLRS.

Finally, the OMA has also launched activities for controlling and coordinating positioning methods. As was extensively described in this and previous chapters, positioning has so far been controlled by a number of signaling protocols defined by 3GPP. The OMA now follows another approach in which it defines IP-based control protocols. This approach is called *Secure User Plane* (SUPL) architecture. Its main advantage is that messages for assistance and measurement data can be easily transferred over IP, and thus the introduction of new positioning methods does not require expensive and cumbersome extensions on the network infrastructure. In this way, SUPL also meets the concerns of the ongoing fixed/mobile convergence, which envisages the disappearing of circuit switched technology and its replacement by IP, as intended for future releases of UMTS. In its first version, SUPL primarily focuses on the support of A-GPS.

11.4 WAP Location Framework

The *Wireless Application Protocol* (WAP) is an architecture for offering Internet-like information services in cellular networks. It has specifically been designed to cope with the limitations of wireless data transmission like reduced bandwidth and increased error rates, as well as with the reduced capabilities of mobile devices like low computational resources, small displays, and numerical keypads. The WAP architecture has its origins in the WAP Forum, which was an industry consortia founded by Ericsson, Nokia, Motorola, and Unwired Planet in 1997. Like the LIF, the WAP Forum has joined the OMA in the meantime and no longer exists as an independent organization. The WAP architecture also contains a protocol for exchanging location data in the context of a WAP application. Before introducing this protocol, a general overview of WAP is given in the following section.

11.4.1 WAP Overview

The WAP framework consists of a protocol suite covering layers 3–7 of the OSI (Open Systems Interconnection) reference model as well as markup and script languages for representing content. The protocols can be used in conjunction with CS bearer services, SMS, or GPRS, even though WAP usage is almost always based on the last one. Initially, the WAP framework defined its own protocols, for example, the *Wireless Transaction Protocol* (WTP) and the *Wireless Session Protocol* (WSP) as counterparts for HTTP. However, later releases of WAP also allow the usage of special configurations of HTTP and TCP. Content is represented either in the *Wireless Markup Language* (WML), which is an XML dialect, or in xHTML, which is a subset of HTML. WML or xHTML pages can be stored on conventional WWW servers in the Internet, and the user can access them via a microbrowser of his mobile device, which is the so-called *WAP client*. This is demonstrated in Figure 11.15.

The mediation between the WAP and the Internet protocol stack is managed by proxies that are located at the border between a cellular network and the Internet. WAP supports pull and push services. Pull services represent the classical, user-initiated WWW browsing and are supported by a *pull proxy*, which forwards requests from the WAP client to the application server and responses in the opposite direction. The proxy converts messages between the WAP and the Internet protocol stack. The application-level protocol between WAP client and pull proxy is WSP or the wireless profile of HTTP, while interactions between pull proxy and application server are based on conventional HTTP.

It is also possible to push content from the application server to the WAP client, that is, the session is here initiated by the server. Usually, the user has to subscribe for receiving personalized content, which may be delivered, for example, if it becomes available, or on the occurrence of predefined events. WAP push services can also be used for realizing proactive LBSs.

In general, the problem of mobile push services is that the application server is often unaware of the terminal's valid IP address. Remember, for example, that in GPRS this IP address is mostly assigned on a temporary basis only, while for CS bearer service

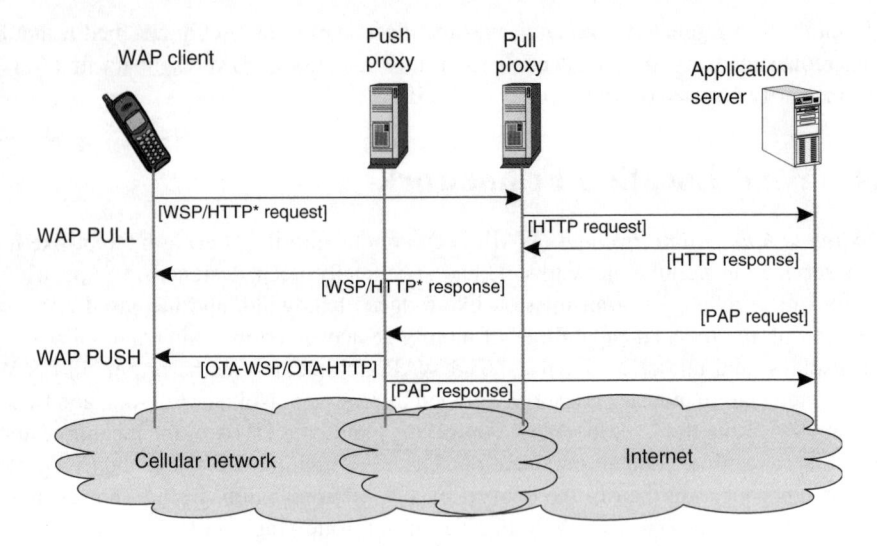

Figure 11.15 WAP architecture (WAP Forum 2001c).

the idle terminal is assigned no IP address at all. Therefore, mobile push services require alternative addressing and thus special protocols to cope with this problem. WAP envisages two protocols for this, the *Push Access Protocol* (PAP) between application server and a push proxy and the Push *Over the Air* (OTA) protocol between this proxy and the WAP client. The PAP is used by the application server to inform the push proxy that new content is available, about the URL where it can be requested, and the subscriber for whom this content is intended. The subscriber is usually identified by its MSISDN. The push proxy then converts this information to a message of the Push OTA protocol, which in most cases uses SMS as a bearer for transferring this message to the WAP client. Upon arrival of the SMS notification, the WAP client then initializes a WAP session to the URL contained in the notification to get the content. Thus, apart from the notification by SMS, the push service is actually based on conventional pull mechanisms.

A more detailed description of WAP push services can be found in (WAP Forum 2001e). For an overview of the entire WAP architecture, see (WAP Forum 2001a).

11.4.2 WAP Location Services

Similar to MLP, the WAP Location Framework is an approach for exchanging location data between a location server and an LBS application server and thus corresponds to reference point (d) of the LBS supply chain. However, it differs from MLP in that it has been specifically tailored to fit into the WAP architecture and for use in conjunction with WAP-based applications. Although not formally defined, the WAP Location Framework can be organized in a layered architecture very similar to that known from MLP (see Figure 11.13). It defines three location services, each composed by a number of messages carrying XML content according to XML DTDs. The application-level protocol for transporting these messages is preferably HTTP, although WSP is possible, too. For some services and some configurations, it may also be required to use PAP, which will be pointed out in detail later.

ARCHITECTURES AND PROTOCOLS FOR LOCATION SERVICES

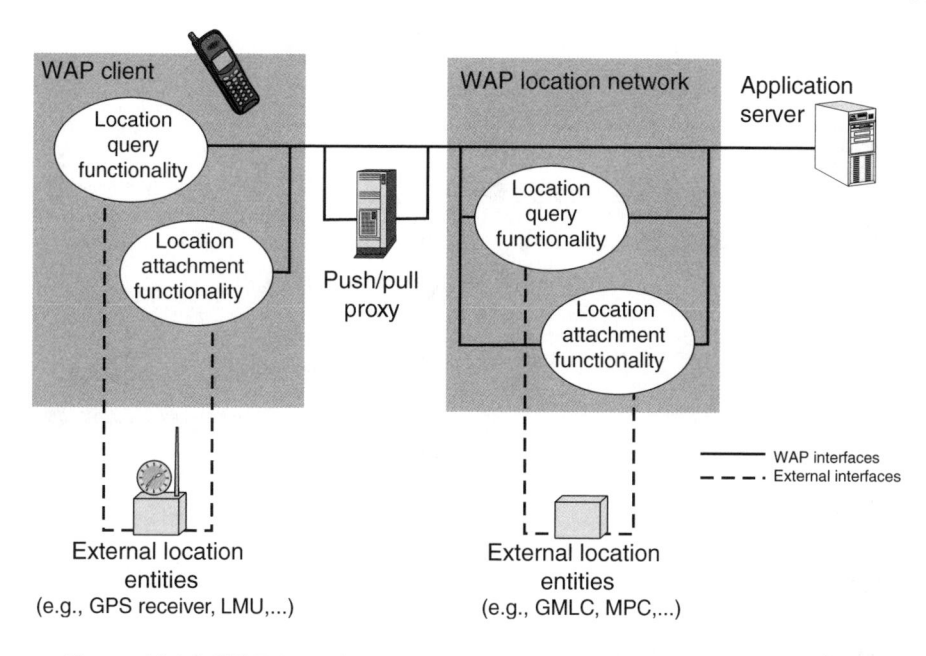

Figure 11.16 WAP Location Framework architecture (WAP Forum 2001c).

The framework provides for two functional building blocks, which offer location services and which are called *WAP Location Query Functionality* and *WAP Location Attachment Functionality*. Both of these building blocks hide positioning from the application and application server respectively. This implies that both of them can be located in the WAP client (i.e., the mobile terminal) as well as somewhere in the network domain, for example, in a WAP proxy, in a GMLC, or somewhere else. In this way, it is possible to realize scenarios with a dedicated location server as covered by Scenarios (1)–(3) in Chapter 10, as well as those where the location server resides in the target's terminal as represented by Scenarios (4) and (5). Figure 11.16 shows the principal architecture of the WAP Location Framework. The dotted lines represent interfaces not defined by it (but, for example, by 3GPP), while the solid lines indicate relationships considered by the framework.

The following location services are supported (see also Figure 11.17):

- **Immediate query service.** This service is used if an application server wants to receive location data of one or several targets immediately. Thus, it is very similar to the SLIR of MLP, however, differs from it in that only synchronous operation is supported. The service is provided by the WAP Location Query Functionality.

- **Deferred query service.** This service is used for tracking one or several targets in that their location data is periodically reported to the application server. Thus, the deferred query service is the WAP counterpart of the TLRS of MLP. Unfortunately, only periodic reporting is supported so far, and zone- or distance-based approaches are missing. This service is also realized by the WAP Location Query Functionality.

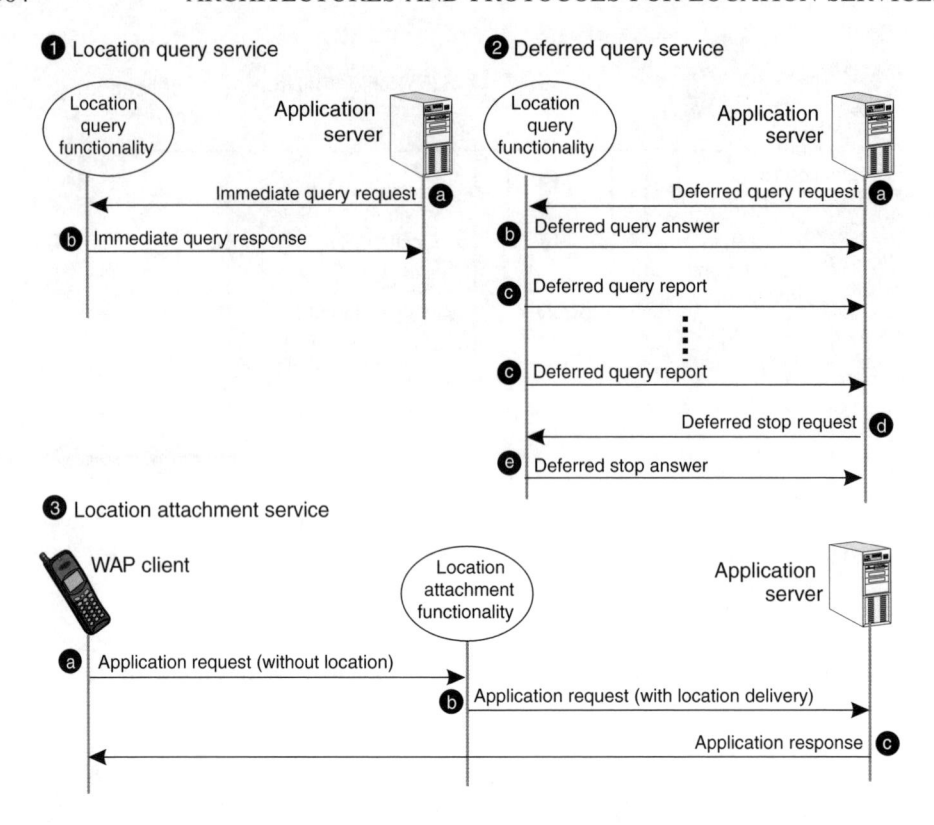

Figure 11.17 WAP location services overview (WAP Forum 2001c).

- **Attachment service.** The attachment service realizes a piggybacking approach, that is, it allows to attach the location data of a target WAP client to the client request sent to the application server. If, for example, a user wants to receive a list of nearby restaurants, he invokes the corresponding service via his WAP client. The WAP Location Attachment Functionality either of the terminal or of the location server in the network then adds the user's location data to the request and forwards it to the application server, which then compiles the list of restaurants and returns it to the WAP client. Thus, the entire LBS supply chain can be realized within one request/response cycle, which saves valuable network resources and causes only low delays.

The DTDs, according to which the messages content for each service is composed, look very similar to those that have been specified for MLP, and also, to a large extent, use the same identifiers for elements and parameters. Therefore, they are not explicitly treated here. The interested reader can find the detailed specification in (WAP Forum 2001b).

The WAP Location Framework also fixes the mappings for HTTP and WSP. Again, the mapping for HTTP is similar to that defined for MLP. The main difference between HTTP and WSP is that XML content is transferred in plain text by HTTP, while for WSP it has to be converted into *Wireless Binary XML* (WBXML) prior to transmission, and vice versa afterward.

ARCHITECTURES AND PROTOCOLS FOR LOCATION SERVICES 305

A special case is given if the WAP Location Querying Functionality resides in the WAP client and the LBS application server wants to invoke the deferred query service in order to periodically receive the target's location data (determined by terminal-based positioning). Against all practice, a WAP session is not initiated here by the WAP client, but the application server issues the request in order to push some content (i.e., the deferred query request) to the client. This is a typical scenario of a WAP push service, for which a particular mapping for location services has been specified in (WAP Forum 2001d). The procedure of a push-based deferred query service request is depicted in Figure 11.18.

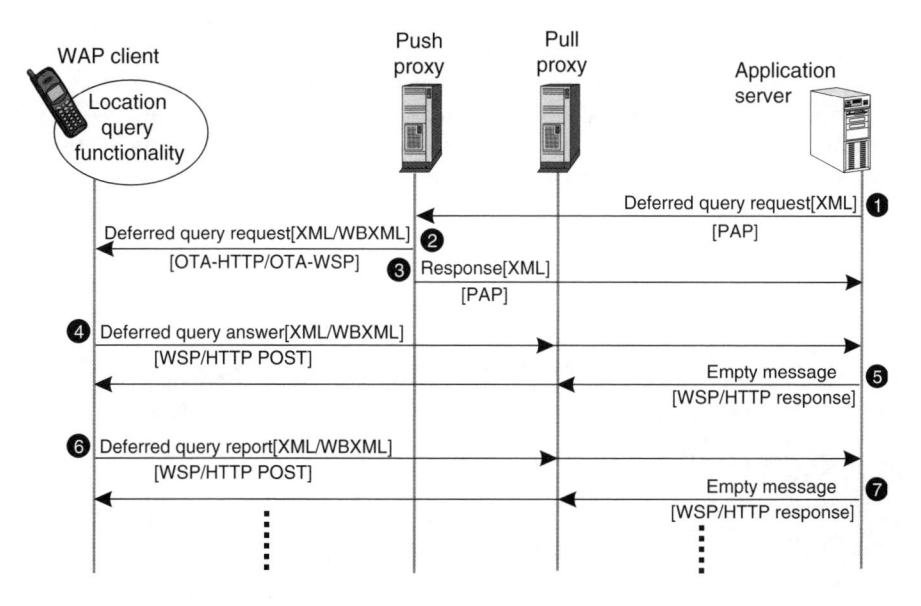

Figure 11.18 Deferred query service with content push.

At first, the deferred query service request is passed to the push proxy (1) using PAP. The push proxy includes the request in a message of the Push OTA protocol and sends it to the WAP client (2). The delivery of the content is confirmed to the application server by returning a response (3). Upon reception of the request, the Location Query Functionality composes a deferred query answer and sends it to the application server, either via HTTP POST or WSP (4), which confirms the reception by an WSP/HTTP response (5). In subsequent steps, the Location Query Functionality then periodically provides the application server with location reports (6 and 7).

11.5 Parlay/OSA

Another approach that covers reference point (d) of the supply chain stems from the Parlay Group, which, similar to OMA, is an industry consortium consisting of multiple vendors, service developers, and providers, and operators. The group was founded in 1998 with the intention of developing open, technology-independent APIs for making network and service functions of fixed and mobile telecommunication networks accessible to external, third-party service and application providers. In this way, it provides an abstract view of the network's core functions, such as (multimedia) call control, data session handling, messaging, charging, account management, presence, policy management, and user location (Karlich et al. 2004). Thus, it basically provides an API for accessing and controlling functions of the Intelligent Network from outside the network domain. Work on Parlay specifications has been done in close cooperation between 3GPP, 3GPP2, and ETSI. The 3GPP has adopted these specifications for GSM/UMTS under the term *Open Service Access* (OSA).

Figure 11.19 gives an overview of the basic Parlay/OSA configuration of a cellular network. A Parlay/OSA gateway works as the interface between the network functions and an application server. The application server is connected to the gateway over the Internet or dedicated IP links. In the network domain, the gateway is connected to various network components. The network used for this interconnection and the protocols applied in between are not covered by Parlay specifications, but one preferred method is to use the SS7 signaling infrastructure and related protocols.

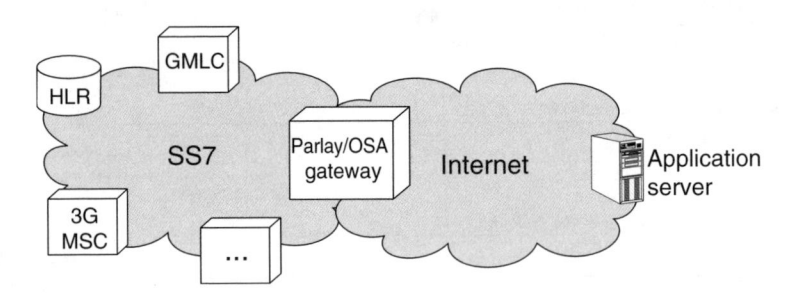

Figure 11.19 Parlay/OSA gateway.

The Parlay framework provides the so-called *User Location services* that are very similar to those of the MLP and the WAP Location Framework, but not only includes location data of and from mobile terminals but also provides the locations of fixed telephones and stationary Internet hosts in order to support emergency services also for fixed telephone networks and VoIP. In contrast to MLP and WAP, Parlay/OSA actually does not specify a protocol. Rather, as mentioned earlier, it fixes a set of APIs together with mappings for different representations, especially for the *Interface Description Language* (IDL), the *Web Services Definition Lang* (WSDL), and Java. The communication between application server and gateway is then realized by the protocols of CORBA, SOAP, or Java, all of which provide standard mechanisms for synchronous and asynchronous communication. Furthermore, in Parlay/OSA, the application server does not directly communicate with the location server and GMLC respectively, but over the gateway. For this purpose, it contains a building block referred to as the Mobility *Service Capability Feature* (SCF). It is defined in (ETSI ES 202 915-6).

The following location services are defined (and instead of explaining them again, only the corresponding MLP counterpart is identified):

- Interactive request (corresponds to MLP SLIR),

- Periodic request (corresponds to MLP TLRS),
- Triggered request (no MLP counterpart),
- Interactive emergency location request (corresponds to MLP ELIS), and
- Network induced emergency location report (corresponds to MLP ELRS).

Each request for location data contains a set of parameters that are very similar to MLP and WAP. For example, it is possible to address one or several targets in each request, to define the desired level of quality of location data, and also to specify the preferred positioning method. The returned location data is expressed in latitude/longitude (and optionally altitude) of WGS-84, and accuracy is given by an uncertainty shape in terms of circle/ellipse, circle/ellipse sector, or circle/ellipse arc stripe. Unlike MLP and WAP, the triggered request does not provide for periodically reporting a target's position (this is done by the periodic request), but implements a zone-based approach where a report is sent if the target enters or leaves a predefined area, which, however, must always be defined as an ellipse (other shapes are not supported). This feature must be supported by the positioning infrastructure.

Parlay/OSA provides a set of standard security mechanisms for secure communication between application server and gateway, but does not care for privacy mechanisms specifically tailored to the needs of protecting location data.

11.6 Geopriv

All approaches, including the emergency LCSs, have in common that they originate from the telecommunications sector and as such are specified by related authorities like 3GPP, 3GPP2, or OMA. Although many of them adopt IP-based protocols such as HTTP or SOAP or IP-related technologies such as XML for location dissemination, their basic intention is to provide for interoperability issues between cellular networks, that is, the classical telecommunications domain, and the Internet. Another approach for an LCS stems from the Internet community, strictly speaking, from the IETF, and is called *Geopriv*.

Geopriv is not a stand-alone approach. Rather, it is only one initiative in the efforts of establishing an overall signaling framework for creating, modifying, and terminating Internet sessions with one or more participants. These sessions include Internet telephone calls, multimedia distribution, and multimedia conferences. The framework is known as *Session Initiation Protocol* (SIP) and is basically the counterpart of the Intelligent Networks in the telecommunications sector. In the meantime, SIP and SIP-related approaches and protocols are covered by numerous IETF documents, but their essentials are summarized in (IETF RFC 3261 2002). The interested reader can also receive an overview from (Sinnreich and Johnston 2001). A particular type of applications covered by SIP is instant messaging, which requires the exchange of presence information between the participants of a session. Presence information indicates a participant's state in terms of "on-line", "off-line", "busy", or "away". The IETF standardizes the *Common Profile for Presence* (CPP), which comprises a set of subscription and notification operations for the delivery of presence information, and the *Presence Information Data Format* (PIDF), which is an XML-based representation of it (see (IETF RFC 3859 2004) and (IETF RFC 3863 2004)). As both dissemination and representation of presence information is very similar to what is needed for location dissemination, Geopriv envisages to reuse the presence architectures and protocols for the purposes of Internet-based LBSs and to continue with an integrated approach.

The main focus of Geopriv is to provide for the secure gathering and dissemination of location data and the protection of privacy of all individuals involved. The Geopriv working group was generated in 2001 and concluded in November 2004. At the time of writing, most contributions of this group were merely in the stage of an Internet draft, except for the document fixing the Geopriv requirements (see (IETF RFC 3693 2004)). Therefore, not all aspects of Geopriv can be presented in the following section, and the concepts described below may be subject to modifications.

11.6.1 Geopriv Entities

The core of Geopriv is a so-called *location object*, which carries location data and which is exchanged between different actors. Geopriv distinguishes between a *location generator*, a *location server*, and a *location recipient*, which correspond to the roles of position originator, location provider, and LBS provider of the LBS supply chain. These entities are depicted in Figure 11.20. The interfaces in between are the *publication* and *notification interfaces*, which realize reference points (b) and (d) of the supply chain. In addition, Geopriv envisages the role of a rule handler, which maintains the privacy constraints, which are officially termed *rules* in Geopriv, of different targets and makes them available for the location server. Like the other roles, the role of rule holder may be solely adopted by a special, trustworthy actor or it may be adopted by an actor in conjunction with other roles. It is connected to a location server via the *rule interface*.

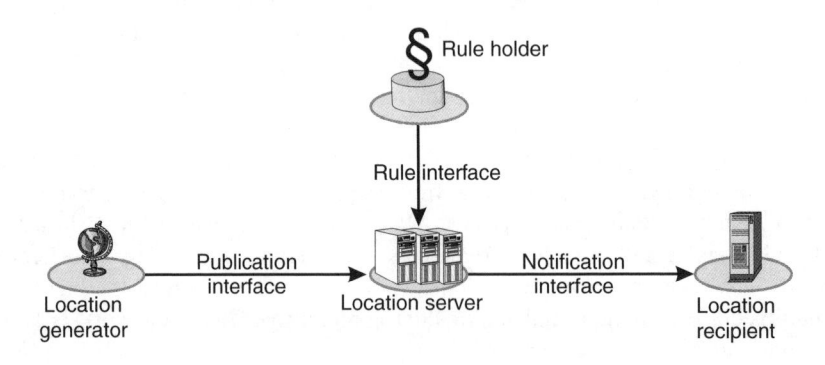

Figure 11.20 Geopriv entities.

The location transport protocol is called *using protocol* in Geopriv, and it is deployed at the publication and notification interfaces. In the version of the Geopriv specifications that was available at the time of writing, this protocol is only described at a very abstract level. Basic requirements are identified instead of a concrete specification, for example, the need for secure communications and mutual authentication of the actors involved in the exchange of location objects and their authorization for accessing them. From the querying and reporting patterns mentioned in Chapter 10, only periodic reporting is explicitly mentioned.

However, Geopriv aims to adopt a general event notification framework, which is specified as an extension for SIP and which allows to specify application-specific events (the so-called *event packages*) and to handle subscriptions and notifications on the basis of these events in the course of a SIP session (see (IETF RFC 3265 2002)). For example, the CPP makes use of this framework in that it defines changes in the presence state of a user as events (IETF RFC 3856 2004). If other participants of an instant messaging application subscribe for these events, they are automatically notified if the presence state changes, for example, from "busy" to "off-line". In the same manner, it would be possible to determine location-related events in order to implement more sophisticated reporting patterns like distance or zone-based reporting for supporting LBSs.

Geopriv follows a two-step approach for specifying privacy rules and for making them available. First, a basic and very simple set of privacy rules can be attached to each location object, which is thus available on the fly for evaluation and consideration by each actor that receives a location object. These rules will be explained in the following section. Second, additional, more sophisticated rules like location, time, and accuracy constraints can be deposited at the rule handler. When a location object passes the location server, the privacy rules inside this object indicate whether additional rules from the rule holder have to be considered when processing or forwarding the location object. In this case, the location server has to receive these rules over the rule interface before proceeding with processing or forwarding the location object. To speed up this process, the location server can alternatively subscribe at the rule holder for receiving up-to-date privacy rules in a push manner. At the time of writing, the underlying mechanisms were still being worked out, and hence they cannot be covered in detail here.

11.6.2 Location Objects

In Geopriv, a location object is basically an extension of presence information. For providing interoperability between different instant messaging systems, the IETF has specified the *Presence Information Data Format* (PIDF) (see (IETF RFC 3863 2004)). According to this RFC, a simple PIDF object may appear as follows:

```
<presence xmlns="urn:ietf:params:xml:ns:pidf"
          xmlns:im="urn:ietf:params:xml:ns:pidf"
          entity="pres:bob@example.de">
    <tuple id="bk45ak">
        <status>
            <basic>open</basic>
            <im:im>busy</im:im>
        </status>
        <contact>mailto:bob@example.de</contact>
        <timestamp>2004-11-06T18:55:23Z</timestamp>
    </tuple>
</presence>
```

The root is given by the `<presence>` element, which contains the namespace declaration the presence information is based on, an `entity` attribute for indicating the publisher of the information in terms of an URL as well as zero, one, or several `<tuple>` elements. The latter carry the actual presence information. Since there might be several of them, it is possible to segment a presence information into several pieces, each of them relating to a particular application, a particular device, or a particular time it has been generated. To differentiate between them, each is labeled by an unambiguous identifier. A <tuple> element, in turn, consists of one optional <basic> element and a number of optional extension elements. The <basic> element adopts the values "open" or "close", which indicate the availability to receive instant messages. By using extension elements, it is possible to deliver additional, application-specific information like the state of the person the presence information is related to, for example, in terms of "busy" or "away" (as demonstrated by the element <im:im>), or location information as is done by Geopriv. Finally, the <contact> element enables to deliver a contact address and the <timestamp> element is used for indicating the time the information was generated. A <tuple> element may contain additional elements, which, however, are not further listed here.

As can be derived from this example, presence information has a similar structure as location data. For example, it contains the identity of a target as well as additional data items that specify the target's current state. Also, presence information is disseminated along a supply chain consisting of suppliers and consumers; that is why, analogus to location data, it is important to safeguard the target's privacy interests with regard to its presence information, too. Owing to these similarities, the Geopriv working group has proposed an extension of PIDF for encapsulating location data, which is referred to as *PIDF-Location Object* (PIDF-LO).

The extension is simply achieved in that a <geopriv> element is included as an optional subelement of <status>. The <geopriv> element consists of two subelements <location-info> and <usage-rules>. The former carries the location information, while the latter includes the privacy rules that are to be applied on this information. A PIDF-LO may then appear as follows:

```
<presence xmlns="urn:ietf:params:xml:ns:pidf"
          xmlns:im="urn:ietf:params:xml:ns:pidf:im"
          xmlns:gp="urn:ietf:params:xml:ns:pidf:geopriv10"
          xmlns:gml="urn:opengis:specification:gml:schema-xsd:feature:v3.0"
          entity="pres:bob@example.de">
    <tuple id="bk45ak">
        <status>
            <basic>open</basic>
            <im:im>busy</im:im>
            <gp:geopriv>
                <gp:location-info>
                    <gml:location>
                        <gml:Point gml:id="point1" srsName="epsg:4326">
                            <gml:coordinates>
                                48:08:57N 11:35:47E
                            </gml:coordinates>
                        </gml:Point>
                    </gml:location>
                </gp:location-info>
                <gp:usage-rules>
                    <gp:retransmission-allowed>
                        no
                    </gp:retransmission-allowed>
```

```
                <gp:retention-expiry>
                    2005-12-31T23:59:59Z
                </gp:retention-expiry>
                <gp:ruleset-reference>
                    http://ruleholder.com/location-objects/
                </gp:ruleset-reference>
            </gp:usage-rules>
            <gp:method>GPS</gp:method>
        </gp:geopriv>
    </status>
    <contact>mailto:bob@example.de</contact>
    <timestamp>2004-11-06T18:55:23Z</timestamp>
  </tuple>
</presence>
```

The existence of <location-info> is mandatory. It carries the actual location information, which may be specified in any XML-based format, which is unambiguously identified by the imported XML schema. Geopriv uses GML 3.0 as a baseline location format, which must be supported by all Geopriv-compliant implementations. While GML also supports the specification of complex features such as topologies and polygons, only support of the "feature.xsd" schema is strictly required for Geopriv implementations. In addition, implementations might support other formats, too, for example, the civil location format.

Each PIDF-LO may contain two basic privacy rules, which indicate whether the recipient of the PIDF-LO is allowed to share its content with other actors and an absolute date at which the recipient is no longer permitted to possess the location information. The former is specified in the <retransmission-allowed> element, while the latter is given by the <retention-expires> element. As these two rules might be too rudimentary for many applications or privacy requirements of targets, it is possible to specify additional rules that are maintained by the rule holder presented earlier. If a PIDF-LO is subject to such external rules, this is indicated by the <ruleset-reference> element. It contains the address of the rule holder, where the rules related to the target given by the entity attribute can be found. If <ruleset-reference> is available, the recipient of the PIDF-LO has to request them at the rule holder.

Finally, it is possible to indicate the positioning method that has been used for locating the target. This is done in the <method> element. Possible values supported so far are "GPS", "A-GPS", "Manual", "DHCP", "Triangulation", "Cell", and "802.11". Strictly speaking, this element offends against the common practice of other LCS approaches to hide details of positioning from the LBS application. However, as elements for specifying the quality of location data are missing in the current Geopriv specifications, the <method> element provides the only means for the receiver of a location object to make conclusions of the expected accuracy of location data on its own.

11.6.3 Geopriv Outlook

In recent times, VoIP has gained increasing momentum for fixed telephony and is expected to completely replace traditional circuit switched fixed telephony within the next one or two decades. Sooner or later, VoIP is expected to enter into the mobile communications market, too. For example, future releases of 3G cellular networks has been designed as "All-IP" solutions, where the traditional CS domain entirely disappears, and the still increasing number of WLAN hotspots will accelerate this process. One enabling technology for VoIP and other session-related applications is certainly SIP, and it is in fact the underlying signaling framework in almost all implementations today. With increasing popularity of Internet-based session applications, there is a strong demand for making them location based, especially for the purposes of emergency services. Geopriv addresses these demands and is designed to integrate into SIP.

At the time of writing, it was too early to discuss the pros and cons of Geopriv and to compare it with the other LCS approaches presented in this chapter, because work on it was still in progress. A particular advantage certainly is that Geopriv explicitly cares for sophisticated privacy mechanisms, which has been insufficiently addressed by other LCS approaches. Unfortunately, most concepts so far have merely been specified on a very abstract level and therefore require much more detailing. The interested reader can find up-to-date information at the Web site of this group (see (IETF Geopriv Working Group)).

11.7 Conclusion

Location services bridge the gap between positioning and the LBS application and thus, besides positioning, represent another core function for realizing LBSs. For offering a location service, one has to deal with two aspects: first, how to realize it in the positioning infrastructure and second, how to make it accessible for external actors. The first aspect concerns the control and coordination of positioning and the gathering and transformation of position fixes. For cellular networks, this is subject to standardization by telecommunication authorities, which is an essential prerequisite for interoperability between equipment (e.g., terminals and network components) originating from different vendors. Unfortunately, similar guidelines for realizing location services in local and indoor positioning infrastructures are still absent, which, as mentioned earlier, also holds for indoor positioning. Therefore, local systems applied today are based on proprietary solutions only.

The second aspect, that is, the interface for requesting and receiving location data, is covered by various protocols or APIs that are presented in this chapter. All approaches offer querying and reporting mechanisms as introduced in Chapter 10, however, with different degrees of flexibility. For example, all of them care for periodic reporting, but zone-based reporting is only explicitly addressed by Parlay/OSA and distance-based reporting is covered by none of them. However, it must be stressed that above all these features must be supported by the positioning infrastructure, and 3GPP envisages advanced reporting approaches ("change-of-area") not before Release 6. This is the worse as reporting is a key function for tracking people and thus for the realization of advanced proactive services. On the other hand, the reader should also keep in mind that reporting and tracking require positioning to be performed permanently, which might be an expensive process resulting in a high signaling overhead, especially at the air interface.

Furthermore, MLP and Parlay/OSA are not suited for arranging the location server in the mobile terminal (compare with Scenario (5) of Chapter 10). They must be offered by a location server in the fixed network, unless the terminal would have been assigned a permanent IP address and would be equipped with server functionality. The WAP Location Framework is the only approach that explicitly envisages to directly gather location data from the terminal over the WAP push mechanism. In this way, it is possible to bypass the GMLC or any fixed location server, which, however, does not necessarily meet the demands of the respective operator, because it loses the control of positioning and consequently cannot charge for it.

All approaches care for basic privacy mechanisms such as secure communication including authentication, but suffer more or less with respect to the checking of privacy constraints. Geopriv introduces a new approach here in that a basic set of privacy constraints is carried together with location data and additional constraints can be requested from a rule holder. Future releases of MLP will also address the latter aspect in that the PCP provides connectivity between a location server and a PCE.

From a market perspective, the de facto standard today is certainly MLP, which most operators prefer to WAP or Parlay. This is due to a number of reasons. First, MLP is meanwhile available in its third release and is therefore subject to several improvements; that is why it can be considered to be more mature than other approaches. Second, it can be deployed as a stand-alone solution, while the WAP Location Framework and the Parlay Mobility SCF are just "service features" of a more complex infrastructure, thereby being more expensive in its installation and operation. Third, MLP is universal in that its realization is not restricted to the GMLC, but can basically be deployed in any location server, including those operated in local, noncellular infrastructures.

Nevertheless, it is useful to give some additional remarks on that. Initially, WAP and MLP have been specified by two independent consortia. Although there were efforts for cooperation between them, both have passed their own specifications. However, as the WAP Forum and LIF have joined the OMA in the meantime, it is strongly desired that future efforts on location protocols are bundled and the existence of two divergent standards will not be continued (Adams et al. 2004). It seems that a decision has been made for MLP, as work on the next release is just in progress at the time of writing and the WAP Location Framework has never achieved a significant impact.

On the other hand, the significance of the Parlay/OSA Mobility SCF is closely related to the success of Parlay/OSA in general. As mentioned earlier, it is a general concept for making internal network functions available to external actors and thus the location service is only one of several features. Parlay/OSA is an integral part of the 3G service architecture, which has barely been installed so far, because operators first concentrated on the establishment of 3G access networks and related modifications in their core networks. However, later stages of 3G network expansion will certainly focus on new service concepts and in this context Parlay/OSA might become an alternative for MLP. However, in spite of the competition between both groups, OMA and Parlay have agreed on cooperation in 2003 for sharing technical information during the specification development phase.

The future success of Geopriv is hard to estimate, as there exist no final specifications and no products at the time of writing. Geopriv focuses on Internet-based applications and integrates well into SIP. However, SIP and related services could also be supported by MLP, and the Parlay/OSA Mobility SCF could explicitly address them, too. The advantage of Geopriv is that it provides more flexibility with regard to location representation; for example, it supports descriptive locations, and provides for better integration of privacy aspects.

12

LBS Middleware

The term *middleware* stems from the distributed computing area and refers to a set of standardized APIs, protocols, as well as infrastructure services for supporting the rapid and convenient development of distributed services and applications based on the client/server paradigm. When the first distributed applications became important in the early 1990s, application developers were increasingly faced with a multitude of heterogeneous programming languages, hardware platforms, operating systems, and communication protocols, which complicated both the programming and deployment of distributed applications. The term middleware refers to a layer that is arranged on top of operating systems and communications stacks and thus hides heterogeneity from the applications through a set of common, well-defined interfaces. In this way, the distributed client and server components an application is made up of can be programmed in the same manner as if they were executed on the same host. The application programmer does not have to care for aspects resulting from distribution, but can concentrate on the actual functions of the application.

The basis for nearly all middleware approaches was founded by the *International Organization for Standardization* (ISO), which fixed the common principles and structures of middleware in a framework known as *Reference Model for Open Distributed Processing* (RM-ODP). The main objective of ODP is to achieve distribution, interworking, and portability in an environment of heterogeneous IT resources and multiple organizational domains of different actors. ODP groups the functions of middleware into different transparency mechanisms, such as *location*, *failure*, *persistence*, and *transaction transparency*. Each of them provides a number of APIs and services to the developer for masking the complexity associated with the respective functions. For example, location transparency refers to the mechanisms for locating the server components of a distributed application in terms of a network location (i.e., it does not refer to a geographic location). It can be realized by naming or trading services, which map the name or the desired type of a server component onto its network address in order to make it accessible for other client components. Failure transparency, on the other hand, refers to recovery procedures that are automatically triggered in a transparent way during execution if one or several components fail. For a detailed description of ODP and its associated transparency mechanisms, see (ISO/IEC 1996–1998).

Location-based Services: Fundamentals and Operation Axel Küpper
© 2005 John Wiley & Sons, Ltd

The common principles of ODP have been adopted by many middleware approaches, such as the different implementations of the *Common Object Request Broker Architecture* (CORBA), Java's *Remote Method Invocation* (RMI), or several approaches for web services. All of them provide several infrastructure services and support different communication patterns, for example, synchronous and asynchronous interactions. However, all of them focus on demands resulting from distribution and communications, while there is a lack of functions for meeting the special demands of LBSs. This chapter identifies these demands and presents a conceptual, but not yet realized LBS middleware. Subsequently, the chapter introduces the *Location API for J2ME* and the *OpenGIS Location Services*, which have first adopted the idea of an LBS middleware and turned it into concrete specifications.

12.1 Conceptual View of an LBS Middleware

In analogy to the idea of traditional middleware as introduced earlier, an LBS middleware should provide a set of APIs, protocols, and infrastructure services that can be used by a broad range of LBS applications. Owing to the interorganizational character of LBS operation, the middleware should spread over the entire LBS supply chain ranging from the LBS target, position originator and location provider to the LBS and content provider as well as the LBS user. This means that it should incorporate the different infrastructures located at these actors and protocols used in between and hide their heterogeneity from the respective LBS applications. An LBS middleware need not be created from scratch. Rather, it can adopt the functions of traditional middleware created for communications and distribution, and extend them by functions needed for LBSs. Thus, the functional range of LBS middleware can be considered from at least two points of view: first, from the point of view of communications middleware incorporating the transparency mechanisms as provided by CORBA or web services, and second, from the point of view of what is referred here as *positioning transparency*. Figure 12.1 shows a generic approach for an LBS middleware that covers the latter point of view.

Positioning transparency describes the mechanisms for hiding all aspects of positioning from the application, including the selection of positioning methods, the distribution of location data, its transformation into other formats of representation, as well as its combination with geographic content or the combination of location data originating from different targets. This also concerns management aspects like the description and evaluation of quality of location data, the saving of the privacy of targets, and accounting. Simply speaking, an LBS middleware provides facilities for putting together all technologies, mechanisms, and protocols that have been presented in the previous chapters of this book.

As can be derived from Figure 12.1, the middleware is organized in different layers situated between the application on the one hand and positioning methods as well as geographic content maintained in GISs on the other. The application usually consists of several server and client-side components that are executed at the LBS provider and the LBS user. On the top, the application makes use of *core services*, which are functional building blocks for processing any kind of location information, for example, (reverse) geocoding, point-of-interest search, the compilation of navigation data, or distance calculations. They represent common functions designed for reuse by a broad range of different applications through a set of well-defined APIs. The core services process location information derived from location services and geographic content and return the result to the application. Some GIS

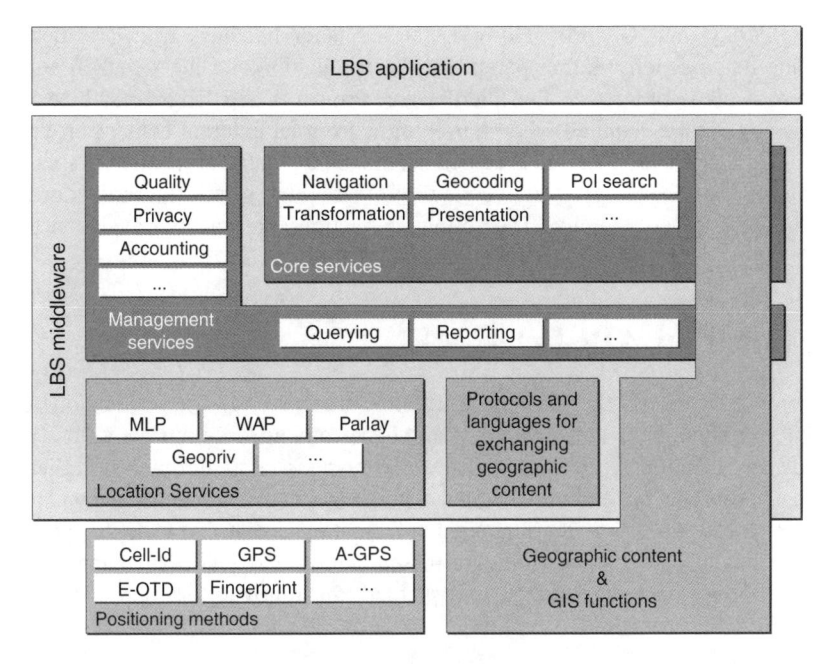

Figure 12.1 LBS middleware.

products available at the market today integrate these core services and thus offer tailored solutions specifically for LBSs. However, the drawback of this approach is that the application strongly depends on the interfaces of a concrete product. For being independent from GIS solutions of a certain vendor, it is desirable to adopt open interfaces for masking their particularities. As will be explained in a subsequent section, this approach is followed by the Open GIS Location Services.

Another important set of functions is represented by *management services*, which are responsible for both *supporting* and *controlling* the execution of the application and related core services. They support them in that they map requirements of the application onto location services or other subservices for retrieving location data and geographic content. Examples are the enforcement of quality demands with respect to location data or the selection of querying and reporting strategies for proactive LBSs. On the other hand, management must be able to control the application and related core services in that, for example, privacy concerns of a target are not violated or financial transactions related to the delivery of location information are restricted. The management services may involve several components and subservices along the supply chain, for example, in order to specify and check privacy options at the HLR, or from release 6 of GSM/UMTS, at the Privacy Profile Register (PPR) of a cellular network.

The lower layer of an LBS middleware comprises the protocols for location services and exchanging geographic content. For location services, the various approaches have been covered in the last chapter. From today's point of view, it is hard to estimate whether there will be a single, universal protocol that can be deployed in conjunction with all positioning infrastructures and methods or whether there will be a coexistence of several protocols like MLP, Parlay, and Geopriv. However, if the latter becomes true it is desired that the resulting heterogeneity is transparent for the application and the selection and control of these protocols is obliged to the middleware. Protocols and languages for exchanging geographic content are required when retrieving it from an external content provider.

Note again that the architecture sketched here is not available as a concrete specification. The intention was rather to give an overview of an area that is still the subject of intensive research in the LBS community. The following sections present two available platforms that partly fulfill the range of functions described here.

12.2 Location API for J2ME

The Location API for J2ME supports positioning transparency for applications executed at a mobile terminal. It is not a middleware shared among different actors and does not incorporate any location services or high-level core services as introduced earlier. It only selects an appropriate positioning method depending on the demands of the application and delivers location data to it. If applied for creating a distributed LBS application, a programmer has to care for location-data dissemination and related management aspects by himself. Before the Location API is explained, the following section at first gives an overview of J2ME.

12.2.1 Overview of J2ME

The Java technology comprises three development suites, which are called *Java 2 Enterprise Edition* (J2EE), *Java 2 Standard Edition* (J2SE), and *Java 2 Micro Edition* (J2ME). They differ from each other in the complexity and the number of features they provide to the developer and thus focus on different environments. J2EE has been designed for building large-scale enterprise applications, which are spread across a distributed network of many computers and have specific demands on scalability, the amount of data to be processed, and other things. J2SE delivers a core set of tools and APIs for constructing desktop applications or web-based applets. This section focuses on J2ME, which aims at applications executed on mobile and embedded devices, which typically are constrained by their limited capabilities in terms of processing power, storage, and user interfaces. Figure 12.2 gives an overview of the architectures of all suites, which will now be described for J2ME.

The major problem when developing J2ME was that it had to be applicable on a diverse range of devices with significant differences in the resources they provided. For example, earlier mobile phones provided RAM and ROM of merely a few kilobytes and were equipped with very small monochrome displays, while the features of the latest generation of PDAs and smartphones can be nearly compared with the capabilities of desktops. In addition, devices of similar capabilities but from different vendors come with a number of proprietary facilities and interfaces, for example, Bluetooth, cameras, and MP3 players. In spite of these difficulties, J2ME provides a set of common interfaces and APIs that allows development of portable applications that, to a large extent, are independent of vendor-specific hardware platforms.

J2ME copes with the problems mentioned earlier in that it provides a modular approach for dynamically tailoring and configuring the runtime environment to the capabilities of the respective devices. For this purpose, J2ME introduces the concepts of *configurations* and

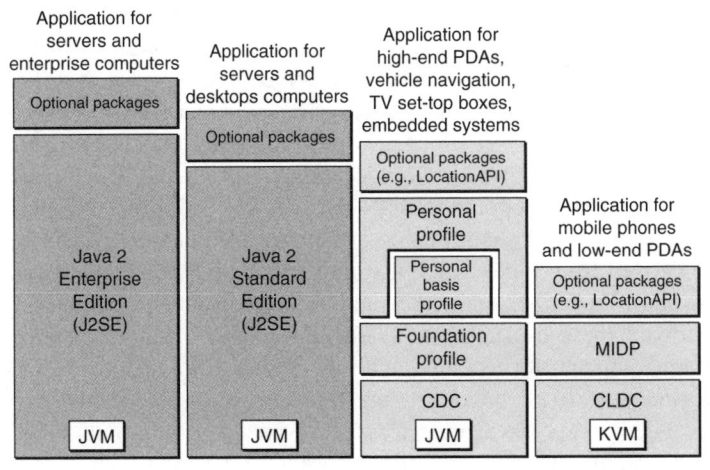

Figure 12.2 Overview of Java development suites.

profiles. Configurations represent a horizontal classification of devices according to their processing power and memory size. Devices with similar resources are covered by a special configuration, which prescribes a minimum platform of virtual machine, language, and library features a J2ME-compliant device of that configuration must support. However, devices with similar resources may still focus on different applications, for example, smartphones, and TV set-top boxes. Therefore, it is necessary to offer more adjusted features additionally that meet the demands of different application areas and in this way allows for a vertical classification of devices. This is realized by profiles that sit on top of a certain configuration and target devices in a specific vertical market. An application is always written for one or several profiles and is thus portable to all devices supporting these profiles.

J2ME currently supports two configurations, which are referred to as *Connected Limited Device Configuration* (CLDC) and *Connected Device Configuration* (CDC). The former focuses on resource-constrained mobile devices like pagers, mobile phones, and low-end PDAs, which have simplified user interfaces, low memory size, and intermittent network connections of low bandwidth. To cope with these limited resources, a slim virtual machine has been developed, which is known as the *Kilobyte Virtual Machine* (KVM) (see Figure 12.2). The CDC, on the other hand, provides a complete *Java Virtual Machine* (JVM) that can be used for J2SE and J2EE. The CDC is intended for devices with at least a few megabytes of memory size, wireless or wired connectivity, and user interfaces with various degrees of sophistication. Examples are set-top boxes, high-end PDAs, and vehicle navigation systems.

The profile that is of utmost relevance for mobile phones and thus for developing LBS applications is the *Mobile Information Device Profile* (MIDP), which runs on top of the CLDC. Applications that make use of this profile are also referred to as *MIDlets*. The MIDP provides a rich set of APIs for styling sophisticated user interfaces supporting the development of mobile games, for establishing secure and nonsecure network connections including HTTP, HTTPS, datagram, sockets, server sockets, and serial port communication, as well as for receiving event-based information in a push fashion. MIDlets are downloadable on demand, like applets.

For the CDC, three profiles are available. The *foundation profile* is a network-capable implementation of the CDC for embedded devices without a user interface. It can be extended with the *personal profile*, which provides support for realizing graphical user interfaces based on the full Java *Abstract Window Toolkit* (AWT) and which hence focuses on devices like high-end PDAs or game consoles. For devices with limited display capabilities, a subset of the personal profile is available, which is called *personal basis profile*.

As can be derived from Figure 12.2, the profiles can be extended with various optional packages that allow to further tailor the runtime environment to the special capabilities of devices. Most of these packages are available for MIDP. Modern mobile phones and smartphones are equipped, for example, with the *Wireless Messaging API* for supporting messaging via SMS and cell broadcast, the *Bluetooth API* for ad hoc communications between several devices, or the *Mobile Media API* for displaying multimedia content like audio and video. Among the diverse optional packages, the most important one for developing LBS applications based on J2ME is the *Location API*, which will be introduced in the next section.

12.2.2 Location API for J2ME

The Location API is an optional package for J2ME for obtaining location information from the implementation of a positioning method. Remember from Chapters 7–9 that there exists a broad range of methods differing from each other in the accuracy and format of derived position fixes and the supporting network infrastructures. Most of them are controlled by network-dependent signaling messages, which are also used to transfer location information generated in the terminal to network components like the GMLC for further processing. However, the related standards do not provide any means for processing location information inside the device and to make it accessible to local applications executed there. It is exactly this deficiency that the Location API targets at. It provides a set of common and generic APIs that can be used by local applications for obtaining location information independent of a particular positioning method. Although the API exclusively resides at the mobile device, it explicitly addresses both terminal-based and network-based positioning. However, the latter must be supported by network-dependent signaling for transferring the position fixes from the network to the terminal. An example of this is the mobile-originated location request in combination with basic self-location described in Section 11.1.3.2. The Location API works on top of either the CLDC (from version 1.1) and MIDP (from version 2.0) or the CDC and the personal profile. Its full specification can be retrieved from (Java Community Process 2003).

The API offers a number of classes for obtaining location information from the implementation of a positioning method, selecting an appropriate positioning method, representing location information, and for defining points of interest, the so-called *landmarks*, and storing them persistently in an internal database of the device. Each implemented positioning method is represented and can be accessed by an object of the class `LocationProvider` (which should not be confused with the role "location provider" of the LBS supply chain as defined in Chapter 6). The class defines methods for receiving the current or the last known location on request, as well as for requesting the internal state of

LBS MIDDLEWARE

the positioning implementation (in terms of "available", "out of service", or "temporarily unavailable"). Optionally, it is possible to make use of the interfaces `LocationListener` and `ProximityListener` in conjunction with a `LocationProvider` object. The former delivers periodic location reports to the application, while the latter triggers a location report as soon as the implementation detects that the current location of the terminal is within a predefined proximity radius of the registered coordinates. Figure 12.3 gives an overview of the possible interactions between `LocationProvider` and application.

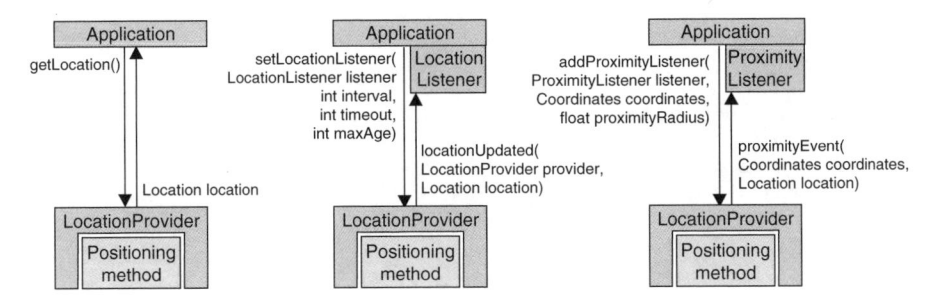

Figure 12.3 Interaction patterns between application and `LocationProvider`.

The selection of a suitable positioning method is dependent on the implementation and is hidden from the application. However, the application can define criteria that are included in the invocation of the factory method of the `LocationProvider` class when instantiating it. The Location API then selects the positioning method that best of all fulfills these criteria and returns it to the application as an instance of `LocationProvider`. The definition of criteria is supported by the `Criteria` class, which offers a rich set of methods for fixing most of the quality parameters identified in Chapter 6, such as the desired horizontal and vertical accuracy, as well as the maximum tolerable TTFF and power consumption (in terms of "low", "high", and "medium"). Furthermore, it can be specified whether positioning should also deliver the altitude, speed, and orientation, and whether requesting location information is allowed to incur financial costs to the user. The latter is a useful option if the device supports both free-of-charge positioning, like pure GPS, as well as cellular positioning, for example, A-GPS and E-OTD in conjunction with ciphered assistance data provided by broadcast signaling, for which the user is charged a fee.

Location information is encapsulated in an object of the `Location` class, which contains a number of other objects representing different types of location data. A single position fix is represented as an object of the `Coordinates` class, which delivers the position in terms of latitude, longitude, and altitude with regard to the WGS-84 datum. In addition, it is possible to obtain the horizontal and vertical accuracies, as well as the speed and orientation of movement and a timestamp of the generation of location information. The application can also request information of the positioning method used, which is indicated as a combination of the underlying basic technology (i.e, Cell-Id, short range, satellite, AoA, ToA, or TDoA) and the site where the position fix has been calculated (i.e., terminal-based or network-based). Optionally, if supported by the positioning method, the `Location` object may also deliver address information as an instance of the `AddressInfo` class. This class contains various fields for defining descriptive locations such as street names, postal codes, cities, countries, building names, floors, rooms, and so on.

The following examples demonstrate the usage of the API for creating a `Location-Provider` instance depending on certain criteria and receiving location information from it:

```
try {
    Criteria cr=new Criteria();
    cr.setHorizontalAccuracy(20);
    cr.setAltitudeRequired(true);
    cr.setPreferredPowerConsumption(medium);
    cr.setCostAllowed(true);

    LocationProvider lp=Location.Provider.getInstance(cr);

    Location l=lp.getLocation(30);
    Coordinates c=l.getQualifiedCoordinates();

    if(c!=null) {
        double lat=c.getLatitude();
        double lon=c.getLongitude();
        float alt=c.getAltitude();
        ...
    }
}
```

The selected positioning method should deliver position fixes of at least 20 m horizontal accuracy, provide altitude information, and have a medium power consumption only. Furthermore, the user may be charged for each location request. A method that would fulfill these criteria is A-GPS. After an instance of `LocationProvider` has been created, the location information is requested with a timeout value of 30 s. Finally, the derived coordinates are extracted and further processed.

The Location API also supports the conversion of a string representation of coordinates into a double representation and vice versa and the calculation of distances. Furthermore, an application can make use of the so-called *landmark stores* for storing, deleting, and retrieving landmarks from a persistent database inside the mobile device. A landmark can be used to represent points of interest and it contains fields for specifying coordinates, address information, a name, and a description.

12.3 OpenGIS Location Services

The OGC has released a specification for an LBS middleware, which is denoted as *OpenGIS Location Services* (OpenLS). The core component is a *GeoMobility server*, which works in conjunction with an application server, both located in the domain of an LBS provider. OpenLS supports many of the functions of the reference LBS middleware presented at the beginning of this chapter. In particular, it hides the complexity of GIS operations and location dissemination from the application developer in that it provides a range of ADTs for representing location data and geographic content as well as several services for interrelating a user's position with geographic content and presenting the results to the user.

LBS MIDDLEWARE

Figure 12.4 gives an overview of how the GeoMobility server integrates in the LBS supply chain. It is located in the domain of an LBS provider and interacts with the application server of this domain on the one hand and with an external GMLC of a cellular operator on the other. The GeoMobility server is equipped with a spatial database for storing and accessing geographic content. Optionally, it can be connected to other spatial databases of external content providers if geographic content needed for an LBS is not available locally. GeoMobility and application server do not necessarily be separated physically, but can also be executed on the same node.

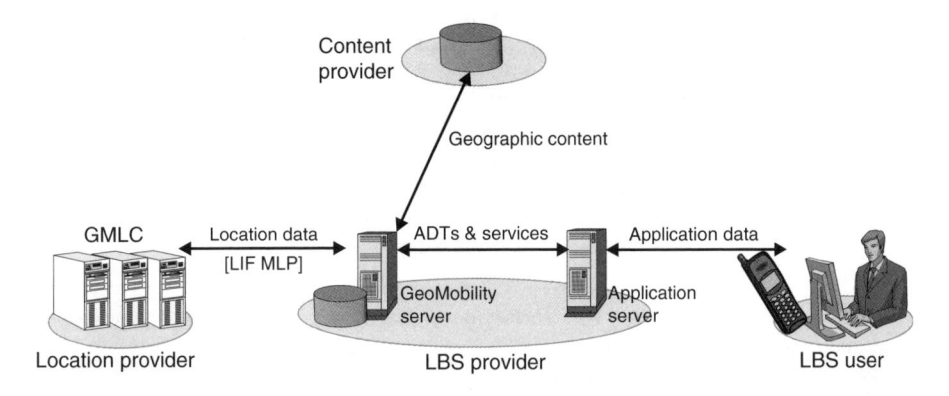

Figure 12.4 The GeoMobility server in the LBS supply chain.

The main task of the GeoMobility server is to host and execute LBS applications, which, when combined with management functions and value-added services of the application server, are offered as LBSs to the users. The range of possible applications is not limited and can basically comprise any application mentioned in this book, for example, restaurant finders, community and navigation services, mobile marketing or gaming. Thus, it is important to note that the GeoMobility server does not define any standard applications, but allows to develop or tailor applications for and to the special needs of certain groups of users. For this purpose, it offers a number of core services that can be used by an LBS application. The resulting layered architecture between application, core services, and spatial content is depicted in Figure 12.5.

The OpenLS Version 1.0 specification envisages the following core services:

- a *directory service* for finding nearby points of interest,
- a *gateway service* for requesting and receiving location data from a remote GMLC via the MLP,
- a *geocoder service* for translating descriptive locations into spatial locations,
- a *reverse geocoder service* for translating spatial locations into descriptive ones,
- a *route service* for determining travel routes and navigation information between two or more points, and

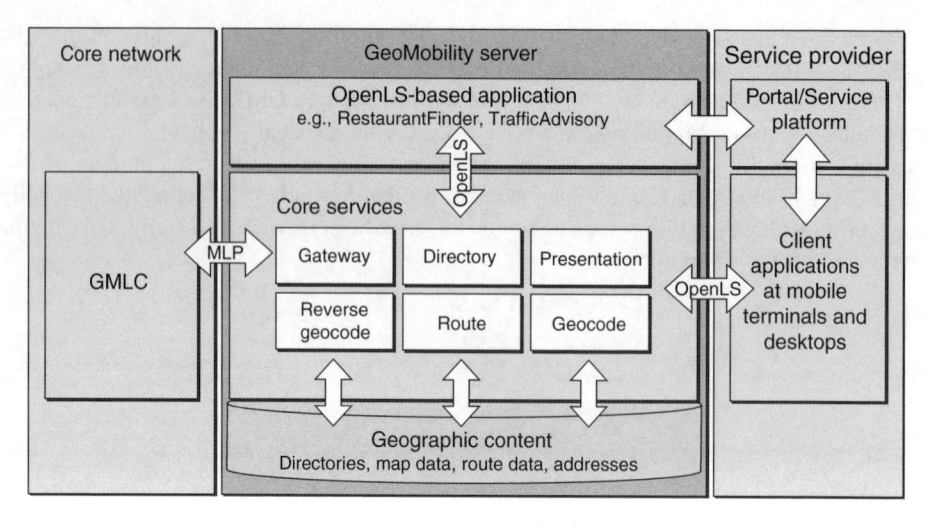

```
                    ┌─────────────────────────────────────────────────────┐
  Core network      │  GeoMobility server          │  Service provider    │
                    │  OpenLS-based application    │  Portal/Service      │
                    │  e.g., RestaurantFinder,     │  platform            │
                    │  TrafficAdvisory             │                      │
  GMLC   ⇔ MLP  ⇔   │  Core services: Gateway,     │  Client applications │
                    │  Directory, Presentation,    │  at mobile terminals │
                    │  Reverse geocode, Route,     │  and desktops        │
                    │  Geocode                     │                      │
                    │  Geographic content:         │                      │
                    │  Directories, map data,      │                      │
                    │  route data, addresses       │                      │
                    └─────────────────────────────────────────────────────┘
```

Figure 12.5 OpenLS architecture.

- a *presentation service* for the portrayal of maps and map overlays like points of interest, addresses, and routes.

While the functions of the core services are clearly identified and it is also comprehensible how to compile an LBS application from them, the role of the application server in Figure 12.4 is less clear. The OpenLS specification characterizes its tasks as follows:

- **Portal.** The application server may act as a portal that offers access to several location and non location-based applications, from which the user can select one to use. As known from the Internet, such a portal provides the user with additional information about services, terms of usage, and facilities for subscription.

- **Session management.** Session management provides functions for managing the life cycle of a user session, including authentication, request handling, and session termination. In particular, it must take care that the user's state is kept between consecutive requests, which is needed if the previous session history is to be taken into account during service provisioning.

- **Personalization.** For many services, it is desired that the user can configure the service's behavior and appearance, which is commonly known as *personalization*. For example, in the context of LBSs, he might specify his preferred points of interest, or he might define personal descriptive locations, like "home", "office", or "gym", and attach coordinates to them.

- **Management.** Each service must be managed in that the user is charged for service usage, the quality of service (QoS) is monitored, and the user's privacy is saved. Usually, these functions are obliged to the application server.

However, the OpenLS specification does not deal with these issues, but concentrates on the specification of the core services.

LBS MIDDLEWARE

The interfaces of core services are specified in XML and consist of a request and response message, which, in turn, include a number of complex information elements known as *Abstract Data Types* (ADTs). In (OGC 2004), the ADTs are specified as *XML schemas*, and for future releases a SOAP version is announced, too. The interaction with an OpenLS core service is basically the same as that with an LCS in MLP (see Chapter 11), that is, a request message carries a number of data elements for configuring the respective service, and a response message contains the result. Upon invocation, a core service decodes the XML information elements of the request and maps them onto operations of the underlying GIS and spatial database respectively. The result is then XML-encoded and returned to the invoking component.

Figure 12.6 illustrates three scenarios of a simple interaction between the LBS user's terminal, the application, and the core services. The scenarios differ in the way the user accesses an LBS, that is, through a WAP browser in conjunction with a WAP pull proxy located in the cellular network in Scenario (A), a dedicated J2ME client application in Scenario (B), or the terminal's SMS function in conjunction with a cellular network's SMS-Center (SMS-C) in Scenario (C). However, the scenarios only represent examples; invocations made from a stationary terminal based on other client applications are of course possible, too.

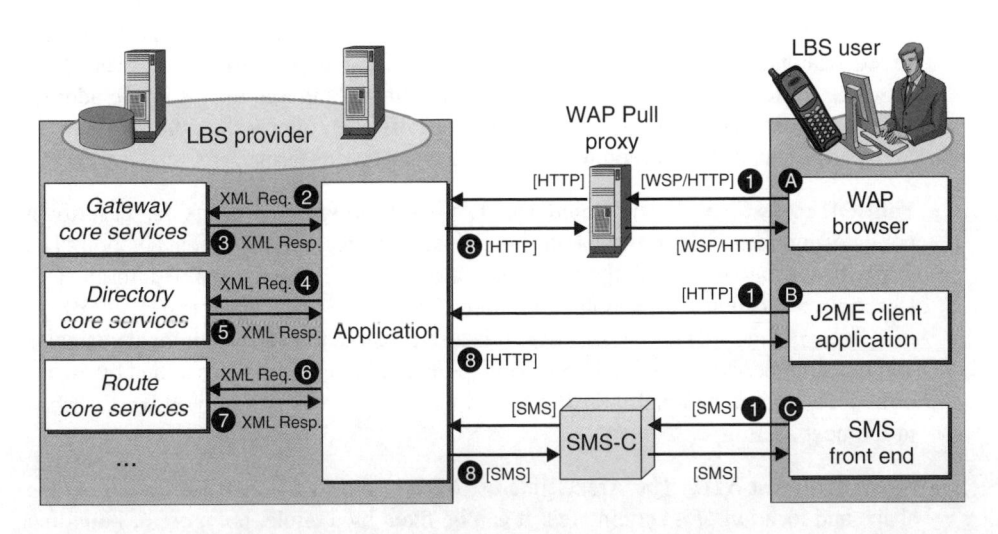

Figure 12.6 Request/Response cycle of OpenLS.

Depending on the respective scenario, the application is invoked through HTTP or WSP messages or by sending an SMS. In Scenarios (A) and (B), the application might be realized by the *Common Gateway Interface* (CGI) of a web server, a servlet, or alternative web technologies. For Scenario (C), the application must be connected to the SMS-C of the respective operator, for which various protocols exist. In the next step, the application encodes the request from the user and identifies either a single or a number of core services needed for fulfilling the request. The types and number of core services as well as the sequence of their invocation depend on the respective application.

12.3.1 Information Model

The core services mentioned earlier are based on the OpenLS information model, which defines ADTs for specifying XML elements that serve as input or output parameters of the core services. The ADTs are available as XML schemas in (OGC 2004). Some of them reuse elements of the GML geometry collection, which has been introduced in Chapter 3.

The following ADTs are used to specify different types of location:

- **Position ADT.** This ADT is used to represent a spatial location as derived from positioning of a target. The position is expressed by the `<gml:Point>` element and may be optionally extended with an uncertainty shape, which can be defined as `<gml:Polygon>`, `<gml:Multipolygon>`, `<gml:CircleByCenterPoint>`, an ellipse, or a circular arc. In addition, it is possible to specify the quality of position (in terms of vertical and horizontal accuracies), the time the position fix is derived, and the target's speed and direction of motion.

- **Address ADT.** The Address ADT is used to represent descriptive locations in terms of street addresses. The user can decide whether to specify an address by an informal string representation (enabled by an element called `<freeformAddress>`) or by using a formal structure of street names, intersection of streets, places, postal codes, street locators, building locators, and supplemental address information. The ADT provides sufficient flexibility in order to cope with different country-specific address representations. Therefore, it is required to specify the country the address refers to by means of well-defined country codes (e.g., "DE" for Germany).

- **PointOfInterest ADT.** The PointOfInterest ADT provides a means for describing points of interest such as restaurants or sightseeing together with their position. The position can be expressed by the `<gml:Point>` element or the Address ADT. Descriptive attributes of a point of interest include `<ID>`, `<POIName>`, `<phoneNumber>` and `<description>`. It is also possible to define additional attributes as a list of name/value pairs. Alternatively, points of interest can be specified by means of national or international industrial classification schemes (which will be described in the next section).

- **AreaOfInterest ADT.** The AreaOfInterest ADT is simply a means for describing the shape and location of a certain area. It can be fixed by a circle, polygon, or bounding box and does not contain any descriptive attributes like the PointOfInterest ADT. It can be used as a search parameter within a request to the GeoMobility server, for example, in order to find out all points of interest located in the area specified by this ADT.

- **Location ADT.** This ADT is an abstract auxiliary ADT, which is used to represent any of the other location-related ADTs introduced earlier. It serves as a placeholder for these ADTs and is contained in the interface specifications of services that are able to commonly process elements of the Position, Address, PointOfInterest, and AreaOfInterest ADTs.

LBS MIDDLEWARE

Besides these location-related ADTs, the information model also maintains ADTs for representing routing or map information:

- **RouteSummary ADT.** The RouteSummary ADT can be used to describe the overall characteristics of a route, including an estimated time needed to travel the complete route, the total distance covered by the route, and a rectangular bounding box describing the geographic area that covers the route.

- **RouteGeometry ADT.** This ADT is used to represent a route as a piecewise linear list of coordinates connected by straight-line segments. The geometry is described by a `<gml:LineString>` element.

- **RouteInstructionList ADT.** Elements of the RouteInstructionList ADT carry turn-by-turn route instructions and advisories, for example, in plain text.

- **Map ADT.** The Map ADT contains elements for carrying rendered maps in binary format embedded in XML or, alternatively, an URL where the data of the map can be found. In addition, it provides elements for describing the properties of a map, such as resolution, bounding box, center point, height, and width.

Figure 12.7 gives an overview of the usage of these ADTs in the different core services. The model depicted in this figure is also referred to as *OpenLS information model*. It describes the interdependencies between the different core services by means of common ADTs. The following section presents an introduction to the core services.

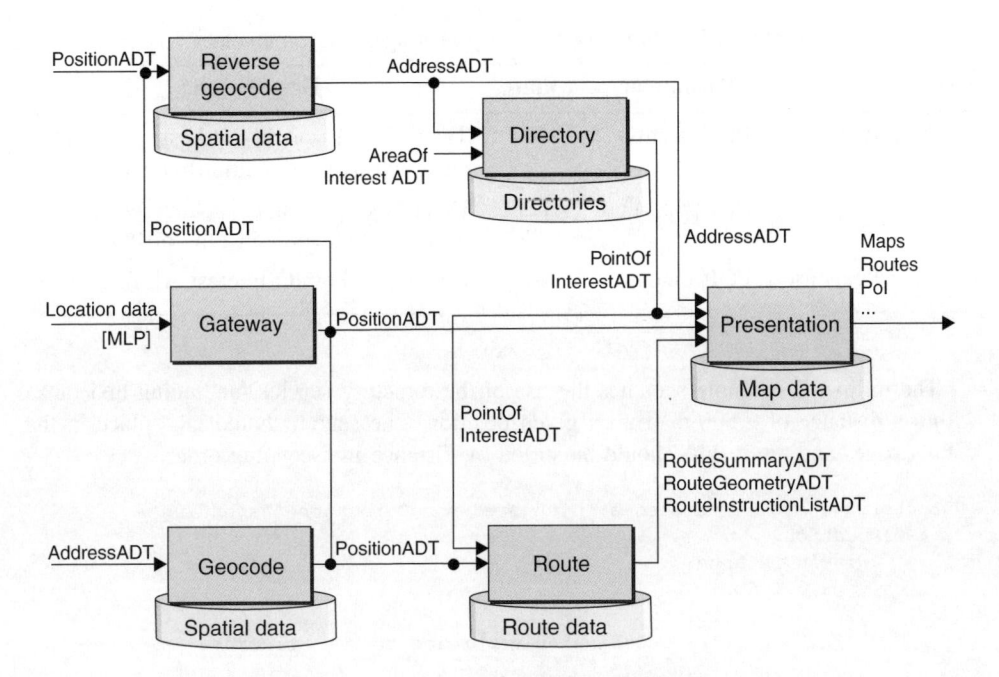

Figure 12.7 OpenLS information model.

12.3.2 Core Services

In (OGC 2004), the functions and interfaces for core services are specified. Each implementation of the GeoMobility server based on this specification has to support at least these interfaces and provide the corresponding functions. However, it is left to the vendors of OpenLS implementations to enhance the core service by additional functions and also to develop additional services.

The interfaces of core services provide one or several operations, each of it consisting of a pair of request and response messages. A message consists of a header and a body part that are based on XML schema definitions. The structure of the header is the same for all core services and operations. It carries authentication data like the name of the client and a password, a session identifier, the type of device, and other information. The body part, on the other hand, carries the service-specific operations and their parameters. They are introduced in the following sections.

12.3.2.1 Directory Service

The directory service enables the search for points of interest at a certain address, within a specified distance or nearest to a specified location, or within a given area. It accepts various search parameters like the type of point of interest, the type of product searched for as well as any keywords. The parameters of the Directory Request and Directory Response messages are listed in Table 12.1.

Table 12.1 Directory service – messages and parameters

Message	Parameter	Mand.	Description
DirectoryRequest	POILocation	No	Type of POI search: Address, WithinDistance, Nearest, or WithinBoundary.
	POIProperties	Yes	Determines search parameters, for example, keywords or type of POI.
DirectoryResponse	POIContext	No	Contains a PointOfInterest ADT element.

The following example sketches the use of the directory service for finding taxi ranks within a distance of 2,000 m from a given position. The search should take place in the yellow pages and the results should be sorted by distance in ascending order.

```
<DirectoryRequest sortCriteria="Distance" sortDirection="Ascending">
    <POILocation>
        <WithinDistance>
            <Position>
                <gml:Point>
                    <!-- Coordinates according to GML structure -->
                <gml:Point>
            </Position>
            <MaximumDistance value="2000"/>
        </WithinDistance>
```

```
        </POILocation>
        <POIProperties directoryType="YellowPages">
            <POIProperty name="NACE_division" value="I">
            <POIProperty name="NACE_group" value="60">
            <POIProperty name="NACE_class" value="22">
        </POIProperties>
<DirectoryRequest>
```

Of particular importance is the `<POIProperties>` element, which, in this example, specifies the type of the point of interest according to industrial classification schemes. In general, such schemes enable the classification of activities, products, expenditures, trade, health care, education, employment, and environment. They provide a hierarchical structure for tagging these matters with unambiguous identifiers. Initially, the motivation behind such schemes was to make national statistics comparable between different countries, but they also serve very well for referencing points of interest in the context of LBSs. The PointOfInterest ADT supports three classification schemes, which are named *National American Industry Classification Scheme* (NAICS), *Standard Industrial Classification* (SIC), and *Nomenclature Générale des Activités Économiques dans les Communautés Européennes* (NACE). An overview of these schemes can be derived from the web site of the statistical division of the United Nations (see (UN Statistical Division Web site)).

For processing the request above, it is assumed that the GeoMobility server maintains taxi ranks according to the NACE scheme, which comprises a hierarchical classification of *divisions*, *groups*, and *classes*. Taxi ranks belong to division "I" (Transport, storage, and communication), and within this division to the group "60" (Land Transport). This group comprises services such as "Transport via railways", "Scheduled passenger transport", and also "Taxi operation", which has been assigned the identifier "60.22". To give another example, restaurants are included in division "H" (Hotels and restaurants), which also contains hostels, camping, and bars, and are represented by the identifier "55.30". However, it must be stressed that points of interest are not necessarily organized according to such a scheme. It is also possible to use any other proprietary scheme and to describe them as simple attribute/value pairs.

As a response, the directory service may deliver the following:

```
<DirectoryResponse>
    <POIContext>
        <POI ID="1" POIName="IsarFunk" phoneNumber="498912345678">
            <POIAttributeList>
                <ReferenceSystem>
                    <NACE division="I" group="60" class="22"/>
                </ReferenceSystem>
            </POIAttributeList>
            <Address countryCode="DE">
                <freeFormAddress>
                    Rosenheimerstr. 16, 81678 Munich
                </freeFormAddress>
            </Address>
        </POI>
        <Distance value="500"/>
    </POIContext>
</DirectoryResponse>
```

A response can contain zero, one, or several <POIContext> elements, each containing a particular point of interest and, if desired, its distance to the reference location that has been defined in the request. If several elements are returned, they are sorted according to the criteria and order specified in the request.

12.3.2.2 Gateway Service

Basically, the gateway service acts as a relay or proxy of the MLP and thus implements all services (apart from emergency services) described in Section 11.3. Likewise, messages and parameters of the gateway service are very similar to those used in the MLP and hence they are not further described here.

12.3.2.3 Location Utility Service

The location utility service consists of two subservices that are called geocoder and reverse geocoder service. The former deals with the transformation of an address (in terms of place name, street address, postal code, or building name) into a geographic position, while the latter provides the inverse function. The messages and elements of the location utility service are provided in Table 12.2.

The following example demonstrates the usage of the geocoder service. The request contains an Address ADT element, which encapsulates a street address in free form that should be translated into a geographic position.

```
<GeocodeRequest>
    <Address countryCode="DE">
        <freeFormAddress>Rosenheimerstr. 16, 81678 Munich</freeFormAddress>
    </Address>
</GeocodeRequest>
```

The Geocode Response message then delivers a list of pairs, each pair consisting of the coordinates as <gml:Point> element and the associated street address as Address ADT element.

```
<GeocodeResponse>
    <GeocodeResponseList numberOfGeocodedAddresses="1">
        <GeocodedAddress>
            <gml:Point>
                <!-- Coordinates according to GML structure -->
            </gml:Point>
            <Address countryCode="DE">
                <freeFormAddress>
                    Rosenheimerstr. 16, 81678 Munich
                </freeFormAddress>
            </Address>
        </GeocodedAddress>
    </GeocodeResponseList>
</GeocodeResponse>
```

Table 12.2 Location utility service – messages and parameters

Message	Parameter	Mand.	Description
GeocodeRequest	Address	Yes	Carries one or several Address ADT elements to to be transformed.
GeocodeResponse	GeocodeResponseList		Delivers a list of elements, each embedding an Address ADT element and the associated geographic position as `<gml:Point>` element.
ReverseGeocode-Request	Position	Yes	Position ADT element to be mapped onto one or several addresses.
	ReverseGeocode-Preference	No	Describes the type of desired return parameters in terms of street address, postal code, building, etc.
	SearchArea	No	Geographic area according to AreaOfInterest ADT in order to narrow down or widen the search area specified by position.
ReverseGeocode-Response	Address	Yes	Reverse geocoded address as Address ADT element.
	Point	Yes	Exact geographic position of returned address.
	SearchCentreDistance	No	Distance of point from position specified in the request.

The reverse geocoder service works in opposite direction, that is, it takes a Position ADT element and returns a single or a list of corresponding Address ADT elements.

12.3.2.4 Presentation Service

The presentation service renders maps for display on a mobile terminal. Each map can be combined with several layers. It is distinguished between *basemap layers* for representing the common real-world infrastructure, such as road networks and buildings, and *overlays* for depicting a subscriber's individual locations and routes on the basemap layers. The service takes the elements of the Location and RouteGeometry ADTs as input and creates a graphical map that is either directly delivered to the subscriber or that can be downloaded from the application server. The quality of presentation can be dynamically adapted to the display capabilities of the used mobile terminal. Table 12.3 gives an overview of the messages and related parameters the presentation service consists of.

The first time an application makes use of an implementation of the presentation service, it has to request its basic capabilities by invoking the `GetPortrayMapCapabilities-Request` message. This message does not include any parameters, but simply causes the service to return lists of supported reference systems, basemap layers, image formats,

Table 12.3 Presentation service – messages and parameters

Message	Parameter	Mand.	Description
GetPortrayMap-CapabilitiesRequest			Used to request the features supported by the presentation service.
GetPortrayMap-CapabilitiesResponse	AvailableSRS	Yes	List of source reference systems supported by the implementation, e.g. UTM.
	AvailableLayers	Yes	List of supported basemap layers, e.g. road networks, public transportation, etc.
	AvailableFormats	Yes	List of supported mime-types for encoding maps, e.g., image/png, image/gif, etc.
	AvailableStyles	Yes	List of available styles for map creation, e.g., Victorian, postmodernistic, etc.
PortrayMapRequest	Output	Yes	Carries the desired map attributes and determines the geographic area to be covered.
	Basemap	No	Specifies the desired basemap layers.
	Overlay	No	Specifies one or several desired overlays by means of the Location ADT or RouteGeometry ADT.
PortrayMapResponse	Map	Yes	Describes attributes of a generated map and provides an URL the map can be downloaded from.

and styles in a `GetPortrayMapCapabilityResponse` message. From the available options declared in these lists, the application can then select those desired for portraying a map. The following example demonstrates the use of the `PortrayMapRequest` message.

```
<PortrayMapRequest>
    <Output width="132" height="176" format="image/jpg">
        <BBoxContext>
                <gml:Pos><!-- upper right corner of the map --></gml:Pos>
                <gml:Pos><!-- lower left corner of map --></gml:Pos>
        </BBoxContext>
    </Output>
    <Basemap filter="Include">
        <Layer name="RoadNetwork">
            <Style><Name>postmodernistic</Name></Style>
```

LBS MIDDLEWARE

```
        </Layer>
        <Layer name="PublicTransportation">
            <Style><Name>postmodernistic</Name></Style>
        </Layer>
    </Basemap>
    <Overlay zorder="0">
        <RouteGeometry><!-- Route subscriber-->POI --></RouteGeometry>
    </Overlay>
    <Overlay zorder="1">
        <Position><!-- Subscriber's position --></Position>
    </Overlay>
    <Overlay zorder="1">
        <POI ID="5"><!-- POI --></POI>
    </Overlay>
</PortrayMapRequest>
```

It instructs the service to compose a map of a certain geographical area specified by the BBoxContext element, and of a certain resolution and format given by the attributes of the Output element. The map should contain basemap layers for showing the road and public transportation networks of the selected area and to display the subscriber's position, a predetermined point of interest, and a precalculated route in between. The response to this request may then appear like this:

```
<PortrayMapResponse>
    <Map>
        <Content width="132" height="176" format="image/jpg">
            <URL>http://www.location-based-services.org/example_map.jpg</URL>
        </Content>
        <BBoxContext>
            <gml:Pos><!-- upper right corner of the map --></gml:Pos>
            <gml:Pos><!-- lower left corner of map --></gml:Pos>
        </BBoxContext>
    </Map>
</PortrayMapResponse>
```

This PortrayMapResponse message delivers the basic parameters of the generated map (resolution, covered geographic area, and image format) and provides an URL that the subscriber can access to download the map. Alternatively, the binary data of the map can be delivered directly to the subscriber.

12.3.2.5 Route Service

The route service is used to realize navigation applications. It calculates a route from a starting point to an ending point, maybe taking into consideration several waypoints in between that the LBS user wants to pass. Optionally, the application can request to return turn-by-turn instructions in plain text as well as in a map that depicts the route as an overlay. Furthermore, it is possible to store routes, once they have been derived, at the server and to fetch them later when needed. Table 12.4 gives an overview of the messages and parameters of the route service.

Table 12.4 Route service – messages and parameters

Message	Parameter	Mand.	Description
DetermineRoute-Request	RoutePlan	Yes	Defines criteria for the creation of routes. Contains elements for defining waypoints and to determine whether to calculate the shortest, fastest, and the least-traffic route and the preferred mode of transport.
	RouteHandle	Yes	Contains a reference to a previously determined route.
	RouteInstruction-Request	No	Used to request the delivery of turn-by-turn instructions.
	RouteGeometry-Request	No	Used to request the return of the route geometry and related parameters.
	RouteMapRequest	No	Used to request the delivery of a route map and its desired parameters.
DetermineRoute-Response	RouteHandle	No	Contains a reference handle to the route at the application server for referring to it in subsequent requests.
	RouteSummary	Yes	Summarizes the main route characteristics in terms of expected traveling time, total route distance, and other parameters.
	RouteGeometry	No	Contains the route geometry as a polyline as `<gml:LineString>` element.
	RouteInstruction-List	No	Provides turn-by-turn instructions.
	RouteMap	No	Delivers references to route maps stored at the server as well as map parameters.

The following example demonstrates how to invoke the route service:

```
<DetermineRouteRequest provideRouteHandle="true">
    <RoutePlan>
        <RoutePreference>Fastest</RoutePreference>
        <WayPointList>
            <StartPoint>
                <Address countryCode="DE">
                    <freeFormAddress>
                        Oettingenstr. 67, 80533 Munich
                    </freeFormAddress>
                </Address>
```

```
            </StartPoint>
            <ViaPoint><!-- Intermediate points --></ViaPoint>
            <StopPoint>
                <Position>
                    <gml:Point><!-- Stop position --></gml:Point>
                </Position>
            </StopPoint>
        </WayPointList>
    </RoutePlan>
    <RouteGeometryRequest value="true"/>
</DetermineRouteRequest>
```

This request carries the route plan that instructs the service to calculate the fastest route along the points stored in the waypoint list. The waypoints are of type Location ADT and can thus be specified as addresses, positions, and points or areas of interest. In addition, the request configures the service to return the route geometry, which describes it as polyline of coordinates, and to return a route handle for later referencing the route. The response to this request may then appear like this:

```
<DetermineRouteResponse>
    <RouteHandle routeID="12">
    <RouteSummary>
        <TotalTime>P0Y0M0DT0H15M</TotalTime>
        <TotalDistance value="800"/>
        <BoundingBox>
            <gml:Pos><!-- upper right corner --></gml:Pos>
            <gml:Pos><!-- lower left corner --></gml:Pos>
        </BoundingBox>
    </RouteSummary>
    <RouteGeometry>
        <gml:LineString><!-- Route specification --></gml:LineString>
    </RouteGeometry>
</DetermineRouteResponse>
```

It contains a route handle and a route summary, which specifies the expected time needed to travel the route, the total distance, and a bounding box that reflects the geographic area covering the route. Furthermore, the route itself is specified as a `<gml:LineString>` element. The route handle is useful for requesting turn-by-turn instructions or a route map in subsequent requests to the route service.

12.4 Conclusion

This chapter has identified the main functions of an LBS middleware by means of a generic approach and subsequently introduced the Location API for J2ME and the OpenLS. The Location API can be seen as a *device-centric approach*. It aims at realizing location-based applications executed on a mobile device rather than on complex, distributed LBSs. It selects an appropriate positioning method under consideration of the application's quality requirements and delivers location data generated by this method to the local application. Thus, it can be seen as an important contribution to developing and deploying portable applications that work on a broad range of devices originating from different vendors.

Support for location services for transferring location data to external actors is so far missing and would have to be implemented manually by the programmer. OpenLS, on the other hand, can be regarded as a *network-centric approach*. It focuses on supporting server applications at an LBS provider through a number of common core services with open, vendor-neutral interfaces. For gathering location data, it makes use of the MLP; the support of other location protocols like Parlay is so far missing. Especially, there is a lack of protocols for fetching location data directly from the mobile devices and thus for bypassing the GMLC. Another deficiency stems from the missing support of management-related tasks like the specification and enforcement of privacy policies.

13

LBS – The Next Generation

The first generation of LBSs went into operation at the beginning of the new millennium. At that time, market analysts saw them as killer applications for the emerging 3G networks and predicted tremendous growth rates and revenues for network operators and service providers offering these services. However, very soon it turned out that LBSs were not that "next big thing" after the success story of GSM and the unexpected hype of the rudimentary SMS. Rather, LBSs were and still are often understood as a niche market for "techies" who use them to find nearby restaurants or ATMs.

So, what are the reasons for the moderate success of the first generation LBS? The following list presents a number of reasons compiled for this book from literature as well as from talks with operators and LBS providers (i.e., this list is not found in official market analyses or representative surveys).

- **Inaccurate positioning technology.** Most of today's LBSs depend on Cell-Id positioning, which, as explained in Chapter 8, is easy and cheap to implement, but delivers only very inaccurate position fixes with large variances. More accurate positioning technologies like E-OTD or A-GPS are not deployed by most operators, unless they are required by law to do so, for example, to offer enhanced emergency services like E-911. The reason for the insufficient implementation of advanced positioning results from the fact that in recent years most operators have tied up capital and manpower in the establishment of 3G networks and related multimedia services, and hence have neglected positioning and LBSs.

- **Less demand for data services in general.** Another reason for the moderate success of LBSs is the lack of demand for mobile data services in general. As indicated earlier, today's LBSs are almost always combined with data services such as a WAP browser or a client application, both executed on top of GPRS. These services, when compared to the wired Internet, are often considered to be deficient because of insufficient user interfaces, unsatisfactory bandwidth, and overpriced fees. Thus, these negative attributes are also associated with LBSs. On the other hand, the realization of voice or multimedia services as LBSs has not yet taken place.

Location-based Services: Fundamentals and Operation Axel Küpper
© 2005 John Wiley & Sons, Ltd

- **Domination of reactive data services.** The market still focuses on reactive data services such as WAP and thus follows the traditional request/response interaction pattern known from the wired WWW. However, the mobile user does not sit in front of a PC to handle his business or to entertain. Rather, he is on the move and wants to be supplied with sophisticated services that fulfill his current needs and adapt to his current situation in a proactive manner. For LBSs, this means that the LBS user wants to be automatically informed about nearby points of interest or about location events that are related to friends, colleagues, or relatives, for example, if these persons come in close vicinity to him. This does not mean that reactive mobile services are generally considered to be obsolete, but that there is a need to develop and offer more proactive ones in addition.

- **No competition.** Standardization authorities have created location services in a manner whereby the operator keeps full control of location data. For example, this becomes manifest in the GMLC (see Chapter 11), which represents the only standardized interface in a cellular network for accessing location data by third party LBS providers, regardless of whether positioning is done by a network- or terminal-based method. On the other hand, standardized protocols for receiving location data generated by terminal-based positioning directly from the mobile terminal, and thus for bypassing the GMLC, have barely been in operation so far. This is worse as most operators either do not offer any access to the GMLC for external actors at all or offer access only via proprietary, unreliable, or overpriced interfaces. As a result, the emergence of competitive markets for LBSs were and still are successfully prevented, which, by the way, is in full contrast to the so-called "multiservice" and "multiprovider" environments that have been proclaimed many times.

- **Absence of research community.** As has been explained several times in this book, the development of LBSs was primarily due to the initiative for enhanced emergency services in the late 1990s. Until then, the research community predominantly covered location-related services only in conjunction with ubiquitous computing, which focused more on local and indoor environments than on large-scale cellular networks that imposed much harder constraints and requirements on the operation of LBSs than local networks. As a consequence, there was lack of research and experience when the first LBSs went into operation. This became apparent, for example, with the absence of appropriate mechanisms for saving privacy, the interorganizational dissemination of location data, and the realization of proactive LBSs, to name only a few.

However, in recent years, the LBS community is undergoing a number of evolutionary changes that give rise to the hope that the circumstances are changing and the next generation of LBSs will become a real success. The most essential evolution is certainly the emergence of small and cheap GPS receivers that can be attached via Bluetooth to mobile devices like smartphones and PDAs. While this configuration is still somewhat inconvenient, because the user has to carry two devices while on the move, the fusion of GPS receivers and smartphones or PDAs will become a standard configuration in the near future. Basically, GPS is just another radio technology besides multiband GSM/GPRS, UMTS, Bluetooth, WLAN, and infrared, all of which the vendors successfully managed to

LBS – THE NEXT GENERATION

integrate in a single mobile device. At the time of writing, the first GPS phones have been created and are now being offered for purchase, predominantly in countries prescribing enhanced emergency services. It is only a question of time when these devices will reach a sufficient market penetration and enter the markets in other countries also. When this happens, it is obvious to use GPS functions not only in emergency situations, but also as a basis for new LBSs.

This is also fostered by the emergence of vendor-neutral operating systems and service platforms such as Symbian or J2ME. These platforms enable application developers to create vendor- and device-independent client applications for mobile devices that interact with servers in the Internet in order to offer and provide new sophisticated services to the user. Furthermore, they enable downloading of program code on demand over the air and thus to dynamically supply the user with applications and services that are tailored to his current situation or his current location. In this way, it is possible to arrange service operation independent of network operators, a trend from which LBSs will certainly benefit, too. However, in contrast to the conventional mobile services, LBSs to be offered in this way assume the availability of open APIs of these platforms for accessing positioning technology inside the mobile device, for example, the GPS unit mentioned earlier, and also other technologies like WLAN fingerprinting. The Location API for J2ME, which has been introduced in Chapter 12, is a first step in the right direction.

In this scenario, the future of cellular positioning methods remains unclear. It is up to the operators to decide whether to implement them in their networks, although there is an increasing tendency in many countries besides the United States, for example, in Europe as well as in Asia, to regulate operators to enable advanced positioning for emergency callers. The main problem is that there is no single positioning method that delivers sufficient and constant accuracy in all environments (indoor vs. outdoor, rural vs. urban) and that highly accurate position fixes must also be attainable for users with legacy terminals, for example, without a GPS unit. An obvious solution would be to establish at least one terminal-based and one network-based approach. For the former, A-GPS would represent a good compromise between implementation costs on the one hand and accuracy on the other, while for the latter U-TDoA might be appropriate.

Thus, it can be expected that the combination of GPS-capable mobile devices and open, programmable mobile operating systems and service platforms will significantly contribute to the second generation of LBSs and determine its features. However, there are a number of unsolved problems. Some of the major challenges are:

- **Privacy.** The main challenge is certainly privacy. This book has covered several technologies for preventing the misuse of an LBS target's location data, especially privacy policies and anonymization. Privacy policies might be an appropriate means for controlling the processing of location data at the different actors of an LBS value chain. However, they are generally burdened by complex user interfaces for configuring them as well as from the fact that they do not protect against an actor's misbehavior or attacks from hackers or insiders. The latter problem is the subject of intensive research in the area of anonymization. Unfortunately, the dilemma here is that current technologies for anonymization are generally weighed down by a tremendous overhead and in the particular case of LBSs are contradictory to the LBSs functions or at least constrict them. Another problem arises from the social and cultural impacts LBSs may have on society. As soon as LBSs establish on a broad basis, many people may feel that they are exposed to a social pressure imposed by superiors or relatives for unintentionally sharing their location data with them. Actually, after the pervasiveness of mobile communications has doomed people to be reachable "anywhere anytime", LBSs provide even a more sophisticated way to control people. Therefore, it might be necessary to think about mechanisms for pretending a false location, or, in other words, for enabling an LBS target person to lie to the LBS users about his location, at least on a temporary basis.

- **Location protocols.** With MLP, the WAP Location Framework, Parlay, and Geopriv there are various alternatives for exchanging location data between the actors of the LBS supply chain. However, these approaches suffer from lack of sufficient support of triggered location reports, which would be essential for tracking LBS targets in the network and thus for building proactive LBSs. Most of them have so far restricted themselves to periodic location reports only, and do not care for distance- and zone-based reporting or for approaches based on dead reckoning. Also, there is a missing flexibility in dynamically configuring the parameters of these reporting strategies depending on the demands of the respective LBS. Furthermore, as mentioned earlier, it would be desirable to obtain position fixes and refined location data directly from the mobile terminal, for example, via GPRS or WLAN, instead of from a central GMLC.

- **Seamless switching between positioning technologies.** As indicated earlier, there is no positioning technology that delivers optimal results irrespective of the environment the target moves in. Some methods, like GPS, work best in rural areas, while others, like E-OTD, are more suited for urban areas. In particular, there are large differences between outdoor and indoor positioning and related applications. For example, typical indoor applications at least require locating targets with the granularity of rooms and floors inside a building, which has so far been impossible to determine via satellite or cellular positioning. In most cases, GPS does not work at all inside buildings, although, in the meantime, there are some initiatives, such as Indoor GPS, for coping with this issue. As a consequence, it would be necessary to develop mechanisms that automatically select the best positioning method from all those that are available on the spot. This includes that the terminal or the network also dynamically switches between different methods for keeping the required level of accuracy, for example, from GPS to WLAN fingerprinting, if the target enters a building.

- **Mobile GIS.** Spatial databases and GIS are important for relating the collected location data of mobile LBS targets with geographic content. When the first GISs appeared in the late 1970s, they focused on applications in the areas of surveying, mapping, and navigation, but nobody thought about their role in the operation of LBSs. As a consequence, spatial databases and GIS provide efficient means for reflecting and relating the location, shape, and topologies of stationary geographic content, but not of mobile objects, which impose entirely new requirements on algorithms of computational geometry, data representation models, and query languages. This deficiency has been addressed by intensive research in this area since the late 1990s, but has so far not been covered sufficiently by commercial products. Another challenge is the development of GIS solutions tailored to the execution in mobile devices taking into consideration their very limited processing and storage capabilities.

- **Advanced LBS middleware.** The development of middleware technologies tailored to the special needs of LBSs is essential for their rapid creation and deployment in competitive markets. The traditional middleware focusing on distributed computing does not provide suitable APIs and infrastructure services. Rather, it is required to offer convenient tools that hide the complexity of positioning and related tasks, for example, quality aspects, privacy, and billing, from the application developer. For this purpose, the middleware has to span the entire LBS supply chain ranging from the LBS target to the LBS user.

In recent years, the LBS community is experiencing an increasing number of initiatives that are addressing these challenges. They are taking place in the form of intensive cooperation between research, industry, and standardization, which are working very hard on making the next and future generations of LBSs a success.

Bibliography

3GPP2 C.S0022-0-1. (2001). *Position Determination Service Standard for Dual Mode Spectrum Systems — Addendum.*

3GPP2 N.S0030. *Enhanced Wireless 9–1–1 Phase 2.*

3GPP TS 03.71. *Location Services (LCS); Functional Description — Stage 2.*

3GPP TS 22.071. *Location Services (LCS); Service description; Stage 1.*

3GPP TS 22.228. *Service Requirements for the IP Multimedia Core Network Subsystem (Stage 1).*

3GPP TS 23.002. *Network Architecture.*

3GPP TS 23.041. *Technical Realization of Cell Broadcast Service (CBS).*

3GPP TS 23.060. *General Packet Radio Service (GPRS); Service Description; Stage 2.*

3GPP TS 23.271. *Functional Stage 2 Description of LCS.*

3GPP TS 25.215. *Physical Layer — Measurements (FDD).*

3GPP TS 25.225. *Physical Layer — Measurements (TDD).*

3GPP TS 25.305. *Stage 2 Functional Specification of User Equipment (UE) Positioning in UTRAN.*

3GPP TS 25.331. *Radio Resource Control (RRC) Protocol Specification.*

3GPP TS 25.413. *UTRAN Iu Interface RANAP Signalling.*

3GPP TS 25.423. *UTRAN Iur Interface RNSAP Signalling.*

3GPP TS 25.430. *UTRAN Iub Interface; General Aspects and Principles.*

3GPP TS 25.453. *UTRAN Iupc Interface PCAP Signaling.*

3GPP TR 25.850. *UE Positioning in UTRAN Iub/Iur Protocol Aspects.*

3GPP TS 29.002. *Mobile Application Part (MAP) Specification.*

3GPP TS 29.060. *General Packet Radio Service (GPRS; GPRS Tunnelling Protocol (GTP) Across the Gn and Gp Interface.*

3GPP TS 43.059. *Functional Stage 2 Description of Location Services (LCS) in GERAN.*

3GPP TS 44.031. *Mobile Station (MS) – Serving Mobile Location Centre (SMLC) Radio Resource LCS Protocol (RRLP).*

3GPP TS 44.035. *Broadcast Network Assistance for Enhanced Observed Time Difference (E-OTD) and Global Positioning System (GPS) Positioning Methods.*

3GPP TS 44.060. *Mobile Station (MS) – Base Station System (BSS) Interface; Radio Link Control/Medium Access Control (RLC/MAC) protocol.*

3GPP TS 44.065. *Mobile Station (MS) — Serving GPRS Support Node (SGSN); Subnetwork Dependent Convergence Protocol (SNDCP).*

3GPP TS 44.071. *Mobile Radio Interface Layer 3 Location Services (LCS) Specification.*

3GPP TR 45.811. *Feasibility Study on Uplink TDoA in GSM and GPRS.*

Location-based Services: Fundamentals and Operation Axel Küpper
© 2005 John Wiley & Sons, Ltd

3GPP TS 48.008. *Mobile Switching Centre — Base Station System (MSC–BSS) Interface; Layer 3 Specification.*

3GPP TS 48.018. *General Packet Radio Service (GPRS); Base Station System (BSS) – Serving GPRS Support Node (SGSN); BSS GPRS Protocol (BSSGP).*

3GPP TS 48.031. *Serving Mobile Location Centre — Serving Mobile Location Centre (SMLC — SMLC); SMLCPP Specification.*

3GPP TS 49.031. *Location Services (LCS); Base Station System Application Part LCS Extension (BSSAP-LE).*

3GPP Web site. http://www.3gpp.org/ access year: 2005.

3GPP2 Web site. http://www.3gpp2.org/ access year: 2005.

Aatique, M. (1997). *Evaluation of TDoA Techniques for Position Location in CDMA Systems.* Master Thesis, Faculty of the Virginia Polytechnic Institute and State University.

Adam, N. R., J. C. Worthmann. (1989). Security-Control Methods in Statistical Databases: A Comparative Study. *ACM Computing Surveys*, Vol. 21, No. 4, 515–556.

Adams, P. M., G. W. B. Ashwell, R. Baxter. (2004). Location-Based Services — An Overview of the Standards. D. Ralph, S. Searby (Eds.), *Location and Personalisation: Delivering Online and Mobility Services.* BT Communications Technology Series 8, Institution of Electrical Engineers (IEE), 43–58.

Agrawal, S. C., S. Agrawal. (2003). *Location–based Services.* Tata Consultancy Services, White Paper, http://www.tcs.com/

Akerberg, D. (1988). Properties of a TDMA Picocellular Office Communication System. *Proceedings of IEEE Globecom 1988*, Hollywood, FL, USA, 1343–1349.

Allen, B. D., G. Bishop, G. Welch. (2001). *Tracking: Beyond 15 Minutes of Thought.* Materials of the SIGGRAPH 2001 course. http://www.cs.unc.edu/ tracker/ref/s2001/tracker/

Amir, A., A. Efrat, J. Myllymaki, L. Palaniappan, K. Wampler. (2004). Buddy Tracking — Efficient Proxy Detection Among Mobile Friends. *Proceedings of IEEE Infocom 2004*, Hong Kong, China.

Bahl, P., V. N. Padmanabhan. (2000). RADAR: An In-Building RF-based User Location and Tracking System. *Proceedings of IEEE Infocom 2000*, Tel Aviv, Israel, 775–784.

Benford, S., R. Anastasi, M. Flintham, A. Drozd, A. Crabtree, C. Greenhalgh, N. Tandavanitj, M. Adams, J. Row–Farr. (2003). Coping with Uncertainty in a Location-Based Game. *IEEE Pervasive Computing*, September (2003), 34–41.

Beresford, A. R., F. Stajano. (2003). Location Privacy in Pervasive Computing. *IEEE Pervasive Computing*, Vol. 2, No. 1, 46–55.

Brigham, E. O. (1998). *Fast Fourier Transform and Its Applications.* Prentice Hall.

Bureau International des Poids et Mesures. (1998). *The International System of Units (SI).* Seventh Edition. http://www.bpim.org/

BWCS Consulting. (2002). *Why are LB$ a long time coming?* White paper. http://www.bwcs.com/

BWCS (2003). *The Last Known Location of E-OTD.* White Paper. http://www.bwcs.com/

Caffery, J. (2000). *Wireless Location in CDMA Cellular Radio Systems.* Kluwer Academic Publishers.

De Caluwe, R., G. De Tre, G. Bordogna. (2004). *Spatio-Temporal Databases.* Springer-Verlag.

Castro, P., P. Chiu, T. Kremenek, R. Muntz. (2001). A Probabilistic Room Location Service for Wireless Networked Environments. *Proceedings of the 3rd International Conference on Ubiquitous Computing*, Atlanta, Georgia. Springer-Verlag, 18–34.

Chan, Y. T., K. C. Ho. (1994). A Simple and Efficient Estimator for Hyperbolic Location. *IEEE Transactions on Signal Processing*, Vol. 42, No. 8, 1905–1915.

Chaum, D. (1981). Untraceable Electronic Mail, Return Addresses and Digital Pseudonyms. *Communications of the ACM*, Vol. 24, No. 2, 84–88.

Coordination Group on Access to Location Information for Emergency Services (CGALIES). (2002). *Report on implementation issues related to access to location information by emergency services (E112) in the European Union.* http://europa.eu.int/

Cuellar, J. R. (2002). Location Information Privacy. B. Srikaya (Ed.), *Geographic Location in the Internet.* Kluwer Academic Publishers, 179–208.

Da, R., G. Dedes. (1995). Nonlinear Smoothing of Dead Reckoning Data with GPS Measurements. *Proceedings of Mobile Mapping Symposium.* ASPRS, 173–182.

Dey, A. K., G. D. Abowd. (1999). *Towards a better Understanding of Context and Context–Awareness.* Technical Report GIT–GVU–99–22, Georgia Institute of Technology.

van Diggelen, F., C. Abraham. (2001). *Indoor GPS Technology.* http://www.globallocate.com/

Dinan, E. H., B. Jabbari. (1998). Spreading Codes for Direct Sequence CDMA and Wideband CDMA Cellular Networks. *IEEE Communications Magazine,* Vol. 36, No. 9, 48–54.

Eberspächer, J., H.–J. Vögel, C. Bettstetter. (2001). *GSM — Switching, Services and Protocols,* Second Edition. John Wiley & Sons.

Egenhofer, M. (1989). A Formal Definition of Binary Topological Relationships. *Proceedings of the Third International Conference on Foundations of Data Organization and Algorithms (FODO),* Paris, France, 457–472.

Egenhofer, M., R. D. Franzosa. (1991). Point-Set Topological Relations. *International Journal of Geographical Information Systems,* Vol. 5, No. 2, 161–174.

Enkelmann, W. (2003). FleetNet - Applications for Inter-Vehicle Communication. *Proceedings of the IEEE Intelligent Vehicles Symposium 2003,* Columbus, Ohio, 162–167.

EPSG Web site. http://www.epsg.org/ access year: 2004.

Erwig, M., R. H. Güting, M. Schneider, M. Vazirgiannis. (1999). Spatio-Temporal Data Types: An Approach to Modeling and Querying Moving Objects in Databases. *GeoInformatica,* Vol. 3, No. 3, 269–296.

ETSI ES 202 915-6. (2005). *ETSI Standard Open Service Access (OSA); Application Programming Interface (API); Part 6: Mobility SCF (Parlay 4).*

Eugster, P. T., P. A. Felber, R. Guerraoui, A.-M. Kermarrec. (2003). The Many Faces of Publish Subscribe. *ACM Computing Surveys,* Vol. 35, No. 2, 114–131.

European Communities. *Official Web Site of the European Navigation Satellite System Galileo.* http://europa.eu.int/comm/dgs/energy_transport/galileo/ access year: 2005.

Fang, B. T. (1990). Simple Solution for Hyperbolic and Related Position Fixes. *IEEE Transactions on Aerospace and Electronic Systems,* Vol. 26, No. 5, 748–753.

Farrell, J., T. Givargis. (2000). Differential GPS Reference Station Algorithm — Design and Analysis. *IEEE Transactions On Control Systems Technology,* Vol. 8, No. 3, 519–531.

Federal Geographic Data Committee. (1998). *Geospatial Positioning Accuracy Standards; Part 3: National Standard for Spatial Accuracy.* http://www.fgdc.gov/

Fischmeister, S., G. Menkhaus. (2002). *The Dilemma of Cell-Based Proactive Location–Aware Services.* Technical Report TR-C042, Software Research Lab, University of Constance.

Foy, W. H. (1976). Position Location Solutions by Taylor-Series Estimation. *IEEE Transactions on Aerospace and Electronic Systems,* Vol. 12, No. 2, 187–194.

Fremuth, N., A. Tasch, M. Fränkle. (2003). Mobile Communities – New Business Opportunities for Mobile Network Operators? *Proceedings of the 8th International Workshop on Mobile Multimedia Communications, (MoMuc'03),* Munich, Germany, 341–346.

Friedlander, B. (1987). A Passive Localization Algorithm and its Accuracy Analysis. *IEEE Journal of Oceanic Engineering,* Vol. 12, No. 1, 234–244.

Friis, H. T. (1946). A Note on a Simple Transmission Formula. *Proceedings of IRE Waves Electrons,* Vol. 34, 254–256.

Gamma, E., R. Helm, R. E. Johnson. (1997). *Design Patterns*. Addison-Wesley Professional.

GPS. (1995). *Global Positioning System Standard Positioning Service Signal Specification*. GPS web site at the US Coast Guard, http://www.navcven.uscg.gov/

GPS. (2000). *Navstar GPS Space Segment/Navigation User Interface*. GPS web site at the US Coast Guard, http://www.navcven.uscg.gov/

GPS. (2001). *Global Positioning System Standard Positioning Service Performance Standard*. GPS web site at the US Coast Guard, http://www.navcven.uscg.gov/

Grejner-Brzezinska, D. (2004). Positioning and Tracking Approaches and Technologies. H. A. Karimi, A. Hammad (Eds.), *Telegeoinformatics — Location–Based Computing and Services*. CRC Press, 69–106.

Gruteser, M., D. Grunwald. (2003). Anonymous Usage of Location-Based Services Through Spatial and Temporal Cloaking. *Proceedings of the ACM/USENIX International Conference on Mobile Systems, Applications and Services (MobiSys '03)*, San Francisco, California, 31–42.

GSM Association. (2003). *Location–Based Services. Permanent Reference Document SE.23*. http://www.gsmworld.com/

Hata, M. (1980). Empirical Formula for Propagation Loss in Land Mobile Services. *IEEE Transactions on Vehicular Technology*, Vol. 29, No. 3, 317–325.

Hata, M., T. Nagatsu. (1980). Mobile Location Using Signal Strength Measurements in a Cellular System. *IEEE Transactions on Vehicular Technology*, Vol. VT–29, 245–251.

Hegering, H.-G., S. Abeck, B. Neumair. (1999). *Integrated Management — Concepts, Architectures, and their Operational Application*. Morgan Kaufmann Pubishers.

Hein, G. W., T. Pany. (2002). Architecture and Signal Design of the European Satellite Navigation System Galileo — Status December 2002. *Journal of Global Positioning Systems*, Vol. 1, No. 2, 73–84.

Hein, G. W., J. Godet, J. Issler, J. Martin, P. Erhard, R. Lucas-Rodriguez, T. Pratt. (2002). *Status of Galileo Frequency and Signal Design*. Available from (European Communities).

Hightower, J., G. Borriello. (2001a). Location Systems for Ubiquitous Computing. *IEEE Computer*, Vol. 34, No. 8. August (2001), 57–66.

Hightower, J., G. Borriello. (2001b). Location Sensing Techniques. *Companion Report to the IEEE Computer article on Location Systems for Ubiquitous Computing*. University of Washington, Computer Science and Engineering.

Huber, M., T. Dietl, J. Kammerl, P. Dornbusch. (2003). Collecting and Providing Location Information: The Location Trader. *Proceedings of the 8th International Workshop on Mobile Multimedia Communications, (MoMuc'03)*, Munich, Germany, 305–310.

IEEE 802.11 Working Group Web site. http://grouper.ieee.org/groups/802/11/ access year: 2005.

IETF Geopriv Working Group. http://www.ietf.org/html.charters/geopriv-charter.html access year: 2004.

IETF RFC 3261, J. Rosenberg, H. Schulzrinne, G. Camarillo, A. Johnston, J. Peterson, R. Sparks, M. Handley, E. Schooler. (2002). *SIP — Session Initiation Protocol*.

IETF RFC 3265, A. B. Roach. (2002). *Session Initiation Protocol (SIP)-Specific Event Notification*.

IETF RFC 3693, J. Cuellar, J. Morris, D. Mulligan, J. Peterson, J. Polk. (2004) *Geopriv Requirements*.

IETF RFC 3856, J. Rosenberg. (2004). *A Presence Event Package for the Session Initiation Protocol (SIP)*.

IETF RFC 3859, J. Peterson (2004). *A Model for Presence and Instant Messaging*.

IETF RFC 3863, H. Sugano, S. Fujimoto, G. Klyne, A. Bateman, W. Carr, J. Peterson. (2004). *Presence Information Data Format*.

IETF Web site. http://www.ietf.org/ access year: 2004.

BIBLIOGRAPHY

ISO/IEC 10746-1, 2, 3, 4 — ITU-T Recommendation X.901, X.902, X.903, X.904. (1996–1998). *Open Distributed Processing - Reference Model.*

IST Emily Project. (2002). *Deliverable 18: Business Models Report.*

IST Locus Project. (2001). *Deliverable 3: Implementation options for Enhanced 112 Emergency Services.*

ITU–T Recommendation E.164. *Numbering Plan for the ISDN Era.* Access year: 2004.

ITU–T Recommendation E.214. *Structure of the Land Mobile Global Title for the Signalling Connection Control Part (SCCP).* Access year: 2004.

ITU–T Recommendation Q.700. *Introduction to CCITT Signalling System No. 7.* Access year: 2004.

Jain, R., Y. Lin, S. Mohan. (1995). A Forwarding Strategy to Reduce Network Impacts of PCS. *Proceedings of IEEE Infocom 1995*, Los Alamitos, California, 481–489.

Jain, R., Y. Lin, C. Lo, S. Mohan. (1994). A Chaining Strategy to Reduce Network Impacts of PCS. *IEEE Journal on Selected Areas in Communications*, Vol. 12, No. 8, 1434–1444.

Jakobs, K., C. Pils, M. Wallbaum. (2001). Using the Internet in Transport Logistics — The Example of a Track & Trace System. *Proceedings of the 1st International Conference on Networking, (ICN 2001)*, Colmar, France. Springer-Verlag, 194–203.

Java Community Process — JSR-179 Expert Group. (2003). *Location API for Java 2 Micro Edition.*

Kaaranen, H., A. Ahtiainen, L. Laitinen, S. Naghian, V. Niemi. (2001). *UMTS Networks — Architecture, Mobility and Services.* John Wiley & Sons.

Kaemarungsi, K., P. Krishnamurthy. (2004). Modeling of Indoor Positioning Systems Based on Location Fingerprinting. *Proceedings of IEEE Infocom 2004*, Hong Kong, China, http://www.ieee-infocom.org/2004/

Karlich, S., T. Zahariadis, N. Zervos, N. Nikolaou, B. Jennings, V. Kollias, T. Magedanz. (2004). A Self-Adaptive Service Provisioning Framework for 3G+/4G Mobile Applications. *IEEE Wireless Communications*, Vol. 11, No. 5 (October 2004), 48–55.

Klukas, R. W. (1997). *A Superresolution Based Cellular Positioning System Using GPS Time Synchronization.* Ph. D. Thesis, UCGE Report 20114, Department of Geomatics Engineering, University of Calgary, Calgary, Canada.

Kosch, T., C. Schwingenschlögl, L. Ai. (2002). Information Dissemination in Multihop Inter–Vehicle Networks — Adapting the Ad–hoc On–Demand Distance Vector Routing Protocol. *Proceedings of IEEE International Conference on Intelligent Transportation Systems, (ITSC)*, Singapore.

Krakiwsky, E. J. (1990). *The Method of Least Squares: A Synthesis of Advances.* UCGE Report 10003, Department of Geomatics Engineering, University of Calgary, Calgary, Canada.

Küpper, A. (2001). *Nomadic Communication in Converging Networks.* Ph. D. Thesis, Aachen University of Technology, VDE Verlag.

Laitinen, H., J. Lähteenmäki, T. Nordström. (2001). Database Correlation Method for GSM Location. *Proceedings of IEEE VTC 2001 Spring Conference*, Rhodes, Greece, 2504–2508.

Langheinrich, M. (2002). A Privacy Awareness System for Ubiquitous Computing Environments. *Proceedings of the 4th International Conference on Ubiquitous Computing.* Springer-Verlag, 237–245.

Lehne, P. H., M. Pettersen. (1999). An Overview of Smart Antenna Technology for Mobile Communications Systems. *IEEE Communications Surveys*, Vol. 2, No. 4, 2–13.

Leick, A. (2004). *GPS Satellite Surveying*, Third Edition. John Wiley & Sons.

Leonhardi, A., K. Rothermel. (2002). Protocols for Updating Highly Accurate Location Information. A. Behcet (Ed.), *Geographic Location in the Internet.* Kluwer Academic Publishers, 111–141.

Lin, Y.–B. (1997). Reducing Location Update Cost in a PCS Network. *IEEE/ACM Transactions on Networking*, Vol. 5, No. 1, 25–33.

Lin, Y.-B., Y.-R. Haung, Y.-K. Chen, I. Chlamtac. (2001). Mobility Management: From GPRS to UMTS. *Wireless Communication and Mobile Computing*, Vol. 1, No. 4, John Wiley & Sons, 339–359.

Lochert, C., H. Hartenstein, J. Tian, H. Füßler, D. Hermann, M. Mauve. (2003). A Routing Strategy for Vehicular Ad Hoc Networks in City Environments. *Proceedings of the IEEE Intelligent Vehicles Symposium 2003*, Columbus, Ohio, 156–161.

Longley, A. G., P. L. Rice. (1968). *Prediction of Tropospheric Radio Transmission Loss Over Irregular Terrain — A Computer Method*. ESSA Technical Report, ERL 79–ITS 67, U.S. Government Printing Office, Washington, District of Columbia.

Magedanz, T., R. Popescu–Zeletin. (1996). *Intelligent Networks — Basic Technology, Standards and Evolution*. International Thomson Computer Press.

McDonnell, R., K. Kemp. (1996). *International GIS Dictionary*. John Wiley & Sons.

mGain. (2003). *Web site of the EU's Information Society Technologies (IST) project mGain*. http://www.mgain.org/

Mizusawa, G. A. (1996). *Performance of Hyperbolic Position Location Techniques for Code Division Multiple Access*. Master Thesis, Faculty of the Virginia Polytechnic Institute and State University.

Mukherjee, A., S. Bandyopadhyay, D. Saha. (2003). *Location Management and Routing in Mobile Wireless Networks*. Artech House.

Myles, G., A. Friday, N. Davies. (2003). Preserving Privacy in Environments with Location-Based Applications. *IEEE Pervasive Computing*, Vol. 2, No. 1, 56–64.

National Research Council, Committee on the Future of The Global Positioning System. (1995). *The Global Positioning System: A Shared National Asset*, Washington, District of Columbia. National Academy Press, 1995.

Nehmzow, U., B. McGonigle. (1993). Robot Navigation by Light. *Self-Organisation and Life: From Simple Rules to Global Complexity. Proceedings of the Second European Conference on Artificial Life*, Brussels. ULB, 835–844.

NENA Web site. http://www.nena9-1-1.org/ access year: 2005.

NGS Web site. http://www.ngs.noaa.gov/ access year: 2005.

NIMA Web site. http://www.nima.mil/ access year: 2005.

Nyquist, H. (1928). Certain Topics in Telegraph Transmission. *Transactions of the AIEE*, Vol. 47, 617–644.

OGC. (1998). *OpenGIS Simple Features Specification for CORBA/SQL/OLE-COM*. http://www.opengis.org/

OGC. (1999). *OpenGIS Abstract Specification — Topic 0: Abstract Specification Overview*. http://www.opengis.org/

OGC. (2001). *OpenGIS Implementation Specification: Coordinate Transformation Services*. http.//www.opengis.org/

OGC. (2003). *OpenGIS Geography Markup Language (GML) Implementation Specification*. http://www.opengis.org/

OGC. (2004). *OpenGIS Location Services (OpenLS): Core Services*. http://www.opengis.org/

OGC Web site. http://www.opengeospatial.org/ access year: 2005.

Okasaka, S., S. Onoe, S. Yasuda, A. Maebara. (1991). A New Location Updating Method for Digital Cellular Systems. *Proceedings of the IEEE 41st Vehicular Technology Conference*, St. Louis, Mississippi, 345–350.

Okumara, T., E. Ohmori, K. Fukuda. (1968). Field Strength and its Variability in VHF and UHF Land Mobile Service. *Review Electrical Communication Laboratory*, Vol. 16, No. 9–10, 825–873.

OMA Location Working Group, http://www.openmobilealliance.org/tech/wg_committees/loc.html access year: 2005.

OMA LOC MLP. *Mobile Location Protocol* (2004).

OMA Web Site. http://www.openmobilealliance.org/ access year: 2005.

Ott, G. (1977). Vehicle Location in Cellular Mobile Radio System. *IEEE Transactions on Vehicular Technology*, Vol. VT–26, 43–46.

Pahlavan, K., P. Krishnamurthy. (2002). *Principles of Wireless Networks – A Unified Approach*. Prentice Hall.

Parlay Group Web site. http://www.parlay.org/ access year: 2005.

Perkins, C. E. (1997). *Mobile IP — Design Principles and Practices*. Prentice Hall.

Plassmann, D. (1994). Location Management for MBS. *Proceedings of the IEEE Vehicular Technology Conference*, Stockholm, Sweden, 649–653.

Proietti, M. (2002). Carrier Choices in Location — The System Integrator's View. *GPS World*, Vol. March 2002, 23–28.

Rappaport, T. S. (2002). *Wireless Communications — Principles and Practice*, Second Edition. Prentice Hall.

Rigaux, P., M. Scholl, A. Voisard. (2002). *Spatial Databases with Application to GIS*. Morgan Kaufmann Publishers.

Rizos, C. (1999). *Introduction to GPS*. University of New South Wales.

Robinson, A. H., J. L. Morrison, P. C. Muehrcke, A. J. Kimerling, S. C. Guptill. (1995). *Elements of Cartography*, Sixth Edition. John Wiley & Sons.

Roos, T., P. Myllymäki, H. Tirri, P. Misikangas, J. Sievänen. (2002). A Probabilistic Approach to WLAN User Location Estimation. *International Journal of Wireless Information Networks*, Vol. 9, No. 3, 155–163.

Roth, J. (2002). *Mobile Computing — Grundlagen, Technik, Konzepte*. d-punkt-Verlag.

Roth, J. (2004). Data Collection. J. Schiller, A. Voisard (Eds.), *Location-Based Services*. Morgan Kaufmann Publishers, 175–205.

Royal Institute of Technology, Sweden. (2000). *WIPS Technical Documentation*.

Salmon, P. H. (2004). Location Calls to the Emergency Services. D. Ralph, S. Searby (Eds.), *Location and Personalization: Delivering Online and Mobility Services*. BT Communications Technology Series 8, Institution of Electrical Engineers (IEE), 31–42.

Sanders, G., L. Thorens, M. Reisky, O. Rulik, S. Deylitz. (2003). *GPRS Networks*. John Wiley & Sons.

Schäfer, G. (2004). *Security in Fixed and Wireless Networks: An Introduction to Securing Data Communications*. John Wiley & Sons.

Scharf, D., R. Bayer. (2002). CoPark — EIN MOBILER DATENBANKBASIERTER DIENST und Ein Neuartiges Konzept zur Parkraumbewirtschaftung in Ballungszentren. *Proceedings of the Workshop "Mobile Datenbanken und Informationssysteme 2002"*, Magdeburg, Germany.

Schelkunoff, S. A., H. T. Friis. (1952). *Antennas: Theory and Practice*. John Wiley & Sons.

Schiller, J. (2000). *Mobile Communications*. Addison–Wesley.

Schmidt, A., K. van Laerhoven. (2001). How to Build Smart Appliances? *IEEE Personal Communications*, Vol. 8, No. 4 (August 2001), 66–71.

Schmidt, A., M. Beigl, H.–W. Gellersen. (1999). There is More Context than Location. *Computer & Graphics Journal*, Vol. 23, No. 6, 893–902.

Schüler, T. (2001). *On Ground–Based GPS Tropospheric Delay Estimation*. Schriftenreihe, Vol. 73, Studiengang Geodäsie und Geoinformation, Universität der Bundeswehr München.

Shankar, P. M. (2002). *Introduction to Wireless Systems*. John Wiley & Sons.

Shannon, C. E. (1949). Communication in the Presence of Noise. *Proceedings Institute of Radio Engineers*, Vol. 37, No.1, 10–21.

Sinnreich, H., A. B. Johnston. (2001). *Internet Communications Using SIP: Delivering VoIP and Multimedia Services with Session Initiation Protocol*. John Wiley & Sons.

Small, J., A. Smailagic, D. P. Siewiorek. (2000). *Determining User Location for Context Aware Computing through the Use of a Wireless LAN Infrastructure*. Institute for Complex Engineered Systems, Carnegie Mellon University, Pittsburgh, Pennsylvania. http://www-2.cs.cmu.edu/aura/publications.html access year: 2005.

SnapTrack. (2003). *Location Technologies for GSM, GPRS and UMTS Networks*. White Paper. http://www.snaptrack.com/

Snyder, J. P. (1987). *Map Projections — A Working Manual*. US Government Printing Office.

Song, H.-L. (1994). Automatic Vehicle Location in Cellular Communications Systems. *IEEE Transactions on Vehicular Technology*, Vol. 43, 902–908.

Sourceforge.net Web site. http://www.sourceforge.net/ access year: 2005.

Stallings, W. (2002a). *Wireless Communications and Networks*. Prentice Hall.

Stallings, W. (2002b). *Cryptography and Network Security — Principles and Practice*, Third Edition. Prentice Hall.

Sweeney, L. (2002). k-anonymity :A Model for Protecting Privacy. *International Journal on Uncertainty, Fuzziness and Knowledge-based Systems*, Vol. 10, No. 5, 557–570.

Symmetricom. (2002). *Location of Mobile Handsets — The Role of Synchronization and Location Monitoring Units*. White Paper. http://www.symmetricom.com/

Tabbane, S. (1997). Location Management for Third-Generation Mobile Systems. *IEEE Communications Magazine*, Vol. 35, No. 8 (August 1997), 72–84.

Thurston, J., T. K. Poiker, J. P. Moore. (2003). *Integrated Geospatial Technologies — A Guide to GPS, GIS, and Data Logging*. John Wiley & Sons.

TIA/EIA IS-817. (2000). *Position Determination Service Standard for Analog Systems*.

TIA/EIA-136-740. (2001). *TDMA Third Generation Wireless - System Assisted Mobile Positioning Through Satellite (SAMPS) Teleservices*.

TIA/EIA IS-801-1. (2001a). *Position Determination Service Standards for Dual Mode Spread Spectrum Systems*.

TIA/EIA IS-801-1. (2001b). *Position Determination Service Standards for Dual Mode Spread Spectrum Systems — Addendum 1*.

Torrieri, D. J. (1984). Statistical Theory of Passive Location Systems. *IEEE Transactions on Aerospace and Electronic Systems*, Vol. 20, No. 2, 183–198.

TruePosition. (2003). *An Examination of U-TDOA and Other Wireless Location Technologies: Their Evolution and Their Impact on Today's Wireless Market*. White paper. http://www.trueposition.com/

UN Statistical Division Web site. http://unstats.un.org/unsd/class access year: 2004.

Ververidis, C., G. C. Polyzos. (2002). Mobile Marketing Using Location Based Services. *Proceedings of the 1st International Conference on Mobile-Business*, Athens, Greece. http://www.mobiforum.org/mBusiness-2002/

Viterbi, A. J. (1995). *CDMA — Principles of Spread Spectrum Communication*. Prentice Hall PTR.

W3C, L. Cranor, M. Langheinrich, M. Marchiori, M. Presler-Marshall, J. Reagle. (2002). *The Platform for Privacy Preferences 1.0 (P3P1.0) Specification*. http://www.w3.org/TR/P3P/

W3C Web site. http://www.w3.org/ access year: 2004.

Wallbaum, M., S. Diepolder. (2005). Benchmarking for Wireless LAN Location Systems. *Proceedings of the 2nd IEEE Workshop on Mobile Commerce and Services*, (WMCS05), Munich, Germany. IEEE Computer Society.

Wallbaum, M., T. Wasch. (2004). Markov Localization of Wireless Local Area Network Clients. R. Batti, M. Conti, R. Cigno (Eds.), *Proceedings of the First IFIP TC6 Working Conference on Wireless On-Demand Network Systems*, WONS 2004. Springer-Verlag, 1–15.

Walther, U., S. Fischer. (2002). Metropolitan Area Mobile Services to Support Virtual Groups. *IEEE Transactions on Mobile Computing*, Vol. 1, No. 2, 96–110.

Wang, J. (1993). A Fully Distributed Location Registration Strategy for Universal Personal Communication Systems. *IEEE Journal in Selected Areas of Communications*, Vol. 12, No. 6, 850–860.

Want, R., A. Hopper, V. Falcao, J. Gibbson. (1992). The Active Badge Location System. *ACM Transactions on Information Systems*, Vol. 10, No. 1, 91–102.

WAP Forum. (2001a). *Wireless Application Protocol Architecture Specification.*

WAP Forum. (2001b). *Location XML Document Formats.*

WAP Forum. (2001c). *Location Framework Overview.*

WAP Forum. (2001d). *Location Protocols.*

WAP Forum. (2001e). *WAP Push Architectural Overview.*

Ward, A., A. Jones, A. Hopper. (1997). A New Location-Technique for the Active Office. *IEEE Personal Communication*, Vol. 4, No. 5, 42–47.

Westin, A. (1970). *Privacy and Freedom*, New York. Atheneum.

Wilton, P., J. Colby. (2005). *Beginning SQL*. John Wiley & Sons.

Youssef, M., A. Agrawala, A. Udaya Shankar. (2003). WLAN Location Determination via Clustering and Probability Distributions. *Proceedings of the First IEEE International Conference on Pervasive Computing and Communications (PerCom)*. IEEE Computer Society, 155–164.

Zhang, J. (2002). Location Management in Cellular Networks. I. Stojmenović (Ed.), *Handbook of Wireless Networks and Mobile Computing*. John Wiley & Sons, 27–47.

Index

3GPP, 1, 129
4-intersection model, 49
9-intersection model, 49

A-FLT, 230
A-GPS, 179, 186, 192, 217, 225–230, 252, 269, 321, 337, 339
    acquisition time, 186, 225
    assistance data, 186, 226
    terminal-assisted, 226
    terminal-based, 226
accelerometer, 141
access network, 93, 117
accounting, 272
accuracy, 125, 129, 130, 141, 186
acquisition time, 184, 241
Active Badge, 130
active set, 214, 219
ActiveBadge, 242
ActiveBat, 243
actor, 249
adaptive frame synchronization, *see* timing advance
ADT, 48, 325
ALI, 289–291
almanac, 169, 171, 228, 229
altitude, 21, 25
amplitude, 62
AMPS, 90, 185, 229
angle velocity, 169
angulation, *see* positioning
ANI, 8, 289, 290
anonymity, 260
anonymization, 261, 263, 265–269, 278, 300

antenna, 62, 70–72, 75
    array, 72, 80, 138, 143
    dipole, 71
    directional, 70
    far-field distance, 75
    gain, 71
    half-wave dipole, 71, 75
    isotropic, 70
    omnidirectional, 70, 130
    sectorized, 71, 130, 143
    vertical quarter-wave antenna, 71
AoA, 211, 220
API, 59
apogee, 157, 160
application area, 267
application data, 249, 251
ArcIMS, 36
ArcInfo, 36
ArcView, 36
ASK, *see* modulation
aspect, 28
assistance data, 125
AT, 199
ATD, 199, 204
ATD/RTD change, 200
atmosphere, 73
attenuation, 75, 91, 148, 235
    Friis free space equation, 75, 148
    inverse square law, 75
AuC, 94
augmented reality, 36
authentication, 99, 102, 261
autocorrelation, *see* spread spectrum
azimuth, 29

Location-based Services: Fundamentals and Operation   Axel Küpper
© 2005 John Wiley & Sons, Ltd

base station, 91–93, 96, 186, 188
    neighbor, 202, 203, 205, 206, 222
    reference, 203, 205, 219
Bayesian modeling, 142
BCCH, 190, 198, 199, 217
BCH, 217
Bluetooth, 5, 240, 338
BOC, 180
British National Grid, 33
BSA, 233
BSC, 93, 96, 106
BSS, 93, 95, 96, 234, 259
BTS, 93
building block, 276
burst, 188

caching, 254
CAD/CAM, 36
calibration, 236
carrier, 63, 66
carrier frequency, 63
carrier phase ranging, 144, 145, 173
CBC, 191, 227
CC, 105
CDM, see multiplexing
CDMA, see multiple access
Cdma2000, 90, 96, 185, 225, 229
CdmaOne, 90, 96, 185, 225, 229
celestial equator, 162
cell allocation, 91, 186
cell update, 115
cell-based method, 211
Cell-Id, 119, 186, 192–194, 218–220, 231, 253, 337
cellular networks, 91–97
central meridian, 31
centripetal force, 158, 159
CGALIES, 9
CGI, 106, 325
change-of-area event, 287, 300, 312
channel, 79
    logical, 190, 216–217
    physical, 187, 190, 216–217
    transport, 216–217
channel reuse distance, 92
channelization code, 213

chip, 82
chipping sequence, 82
CI, 106, 190
CIA, 261
circuit-switched, 93, 95, 97, 112
clock, 146–148, 152
    accuracy, 146
    atomic, 147, 173
    correction, 228
    drift, 147, 173, 200
    drift rate, 147
    OCXO, 147
    quartz crystal, 147, 173
    stability, 146
    synchronization, 127, 133, 136, 146, 173
    TCXO, 147
    time offset, 146, 173
code, 82
code phase ranging, 145, 173
Columbus, Christopher, 140
computational geometry, 36, 53–55
conceptual schema, 47, 50
confidentiality, 261
constraint data model, 59
consumer, 251, 254
content provider, 251
context information, 2
context-aware service, 2
coordinate system, 19–23
    Cartesian, 20–21, 132, 141
    Ellipsoidal, 20–23
    orientation, 20, 23
    origin, 20, 21, 23
    scale, 20
Coordinate Transformation Services, 34
coordinates, 19
CORBA, 59, 306, 316
core network, 93, 110
core service, 316, 323
covariance matrix, 138, 176
CPICH, 217, 221, 222
Cricket, 244
cross correlation, see spread spectrum

INDEX

D-AMPS, 90, 225, 229
D-GPS, 177–179, 186, 228
    acquisition data, 226
DAMPS, 90
data
    analog, 61
    digital, 61
data burst, 95, 197
datum, 19, 23–27
    global, 25
    horizontal, 23–25
    local, 25
    vertical, 23, 25–27
datum origin, 25
DBMS, 36, 43–45
    object-oriented, 43, 51
    object-oriented relational, 44
    relational, 43, 50
dead reckoning, *see* positioning
deduced reckoning, *see* dead reckoning
demodulation, 63
description, 37
descriptive attribute, 37
design matrix, 135, 137
deviation limit, 206
diffraction, *see* multipath propagation
directivity, 70
dispersion, 73
dissemination, 272
distortion, 27, 29, 64
    angular, 29
    areal, 29, 30
    direction, 29
    distance, 29
    scale, 29, 30
DoD, 171
Doppler effect, 78–79
Doppler shift, 78, 145, 173, 228
Doppler, Christian, 78
Doppler spread, 79
drift, *see* clock
DRNC, 214
DTDs, 294
dual-frequency ranging, 167, 172, 175
duplexing, 81, 82, 212

E-112, 9, 288
E.164, 284
E-911, 8, 89, 229, 230, 273, 288–294, 337
    Phase I, 290–291
    Phase II, 291–294
e-cash, 268
E-FLT, 230
E-OTD, 186, 187, 191, 194–208, 221, 226, 227, 230, 231, 252, 269, 321, 337
    assistance data, 202
    change limit, 200
    circular, 197
    deviation limit, 200
    hyperbolic, 194–196
    reporting period, 200
    terminal-assisted, 202, 207
    terminal-based, 202, 207
E-OTD assistance data, 203
easting, 32, 151
eccentricity, 160
ECEF, 20, 27, 132, 174, 176
ecliptic, 162
EDGE, 90, 96
EHF, 69
EIR, 94
Ekahau, 239
electromagnetic wave, *see* signal
ELF, 69
ELIS, 307
ellipsoid height, *see* geodetic height
ELRS, 307
emergency response agency, 289, 290, 292
emergency service, 273, 277
    wired, 289–290
    wireless, 290–294
emergency service zone, 289, 290
emergency services, 288–294
enquiry and information services, 257
ephemeris, 168, 169, 174, 228, 229
EPSG, 34, 299
equatorial plane, 21, 157, 161
escape velocity, 161
ESME, 292–294

ESNE, 292
ESRD, 290, 292
ESRK, 292
Euclidean distance, 135, 142
European datum, 25
Explorer I, 155

far-field distance, *see* antenna
FCC, 8
FDD, 81, 186
FDM, *see* multiplexing
FDMA, *see* multiple access, 186
feature, 36, 45–46
FGDC, 149
fingerprinting, *see* positioning
first harmonic, 65
flattening, 23, 24, 151
fleet management, 254
floating car data, 5, 141
FLT, 230
forwarding pointer, 103
Fourier, Jean-Baptiste, 65
Fourier series, 65
Fourier transformation, 66
frame, 81
frequency, 62
frequency hopping, 81
frequency reuse distance, 80
FSK, *see* modulation, 187
functional entity, 276
fundamental frequency, 65

Galileo, 128, 156, 179–183
  availability, 181
  E2-L1-E1 carrier, 180
  E5a-E5b carrier, 180
  E6 carrier, 180
  integrity, 179, 181
  services, 182–183
Gauß-Krüger, 33
geocenter, 20, 21, 25
geocoding, 35, 330
geodesy, 23
geodetic datum, *see* datum
geodetic height, 21, 26
geographic content, 36, 251
geographic data model, 36

geoid, 23, 26
geoid height, 27
geometry, 153
GeoMobility Server, 322
Geopriv, 307–313, 340
  CPP, 307
  location generator, 308
  location object, 309–311
  location recipient, 308
  location server, 308
  PIDF, 307
  PIDF-LO, 310
  rule, 308
  using protocol, 308
GERAN, 96, 275
GGSN, 96, 110
  address, 110
GIS, 35, 59, 89, 125, 316, 322, 325, 340
GLONASS, 128, 156
GML, 326
GMLC, 253, 261, 275, 278, 280, 284, 287, 293, 300, 306, 338, 340
GMSC, 94, 96, 106
GMSK, 187
GMT, 23
Gold code, *see* spread spectrum
Gold sequence, 213
GPRS, 1, 89, 90, 95–96, 110, 114, 187, 257, 301, 338
GPS, 20, 25, 35, 127, 128, 133, 136, 145, 155, 162–177, 180, 197, 199, 254, 258, 338
  acquisition time, 172, 240
  assistance data, 166, 167
  C/A-code, 167, 168, 171, 173
  control segment, 163
  DoP, 172, 174–177
  error budget, 174
  HOW, 168
  indoor, 240–241
  L1 carrier, 166, 171, 181
  L2 carrier, 166, 167
  L5 carrier, 167, 175, 181
  M-signal, 168
  navigation message, 165, 167–170

INDEX

357

P-code, 167, 168, 173
pilot signal, 165–168
PPS, 170
PRN code, 167
PRN-number, 167
ranging codes, 165
SA, 170, 174
satellite constellation, 164–165
spreading code, 165–168
SPS, 170
SV, 165
SV health, 168
SVN, 165
TLM, 168
TOW, 168
UERE, 174
gravitational force, 159
Greenwich Meridian, 21, 23
GSM, 89, 91, 93–95, 103, 229, 338
    air interface, 186–190
    positioning components, 190–192
    topology, 103
GSM Association, 1
GSN, 96
GTD, 195
GTP, 110
guard band, 81
guard time, 81, 188
gyroscope, 141

half-wave dipole, *see* antenna
handover, 90, 93, 99, 107, 210, 215
hearability problem, 152, 211, 215, 222
HLR, 94, 96, 103, 105, 106, 273, 278, 287
hopping sequence, 82
Horus, 239
HTTP, 302
hyperframe, 187
Hz, 63

identifier abstraction, 260, 266–268, 278
IDL, 59, 306
IEEE 802.11, 233
IMEI, 272
IMSI, 104, 108, 259, 260, 272, 280, 287, 298

IMT-2000, 90
inclination angle, 157, 161, 164, 169
indexing, 59
Indoor GPS, 173, 340
inertial navigation
    see dead reckoning, 140
information content abstraction, 260, 268–269
infrared, 338
instant messaging, 307
integer ambiguity, 144, 145, 173
integrity, 261
IntelliWhere, 36
interarrival time, 112
interference
    cochannel, 71, 80, 86, 92
    frequency-selective, 78
    neighbor channel, 80, 81, 86
International Meridian Conference, 23
intimacy, 260
inverse square law, *see* attenuation
ionization, 74
ionosphere, 73, 152, 169, 174
ionospheric delay, 175
IP address, 110
IPDL, 222
    burst mode, 222
    continuous mode, 222
Iridium, 128
ITU, 68

J2EE, 318
J2ME, 318–320, 325, 339
    CDC, 319
    CLDC, 319
    configuration, 318
    foundation profile, 320
    Java AWT, 320
    JVM, 319
    KVM, 319
    MIDP, 319
    personal basis profile, 320
    personal profile, 320
    profile, 319
J2SE, 318

k-anonymity, 268
Kepler, Johannes, 160
Kepler's laws of planetary motion, 160
Keplerian Elements, 160, 162, 169, 174

LAC, 106
LAI, 106, 107, 109, 190
Lambert, Johann Heinrich, 31
landmark, 320
latency, 126
lateration, *see* positioning
latitude, 21–22, 162
layer, 45
LBS
    client/server, 252
    community service, 4, 163, 257, 268, 285, 323
    enhanced emergency service, 8
    enquiry and information service, 4
    fleet management and logistics, 5
    health care, 163
    Instant Messaging, 4
    mobile gaming, 6, 323
    mobile marketing, 6, 163, 323
    navigation, 163
    navigation service, 323
    peer-to-peer, 252, 253, 258, 268
    proactive, 251, 252, 257
    reactive, 251, 252
    restaurant finder, 323
    toll systems, 9
    traffic telematics, 5
    value-added service, 7
LBS middleware, 315–336, 341
LBS provider, 250
LBS user, 251, 258, 273
LCS, 271–313
    client, 273, 274, 294, 295
    GSM/UMTS, 273–287
    server, 273
least square fit, 133, 135, 137
LEO, 79
LER, 300
LFSR, 213
LIF, 294
line, 300

line of sight, 21, 35, 75, 76, 128, 139, 152, 197
LMU, 188, 190, 196–198, 200, 203, 206, 208, 231
    associated, 217
    stand-alone, 217, 221
localization, 99
location, 17–34
    current, 271
    descriptive, 18, 89, 99
    initial, 271
    last known, 271
    network, 18, 89
    physical, 17
    spatial, 18–33, 35, 89, 99
    virtual, 17
Location API for J2ME, 316, 318–322, 335, 339
location area, 100–103, 106, 107, 259
location calculation and release procedure, 279, 281
location data, 249, 251, 256, 271
location dissemination, 254, 258, 308
location management, 89, 96, 99–290
    CS domain, 103–109
    PS domain, 109–119
    state model, 107, 115, 117
location preparation procedure, 279–282
location protocols, 340
location provider, 250, 251, 271
location register, 96
location request
    mobile originating, 273, 282–283
        autonomous self-location, 274, 282
        basic self-location, 274, 282, 320
        transfer to third party, 274, 282
    mobile terminating, 273, 280–282
        deferred, 273, 281
        immediate, 273, 280
location sensing, 129
location service, 1
location update, 99–102, 107–109, 113
    on cell crossing, 115
    on LA crossing, 108, 115
    on location-area crossing, 107

INDEX

on routing-area crossing, 115
on URA crossing, 115
periodic, 107, 115
location-aware service, 1
location-based service, 1
location-related service, 1
longitude, 21–22, 162

m-sequence, *see* spread spectrum
MAP, 95, 109
map, 27–33, 36
map projections, 27–33
mapping, 163
Maxwell, James Clerk, 72
MCC, 105
mean anomaly, 162
mean sea level, 25
medium, 62
    density, 72
    guided, 68
    refraction index, 72
    unguided, 68
medium access, *see* multiple access, 152
MEO, 79
Mercator, 30
Mercator projection, *see* projection
meridian, *see* longitude
meter, 21, 72
miAware, 36
middleware, *see* LBS middleware
misclosure vector, 135, 140
mix network, 267
mix zone, 267
MLP, 294–301, 306, 313, 330, 340
    ELIS, 296
    ELRS, 296
    PCE, 300
    SLIR, 295, 303, 306
    SLRS, 296
    SUPL, 300
    TLRS, 296, 297, 300, 303, 307
MNC, 105
mobile gaming, 257
Mobile IP, 110
mobile-originated call, 99
mobile-originated packet, 110
mobile-originated traffic, 99

mobile-terminated call, 99, 106
mobile-terminated data, 102
mobile-terminated packet, 99, 110
mobility management, 94, 97–99, 127
modulation, 63–64, 211
    analog, 63
    ASK, 63, 65
    digital, 63
    FSK, 63
    PSK, 63, 84, 167
    QAM, 64
    QPSK, 63
monitor period, 199
MS, 93
MSC, 94, 103, 273, 275, 278, 281, 291, 292
MSC area, 103, 106, 108
MSISDN, 105, 106, 109, 259, 260, 272, 280, 287, 298, 302
MSN, 105
MSRN, 105, 106, 108, 109
multiframe, 187
multiframe offset, 204
multipath propagation, 64, 76–196
    diffraction, 76, 77
    reflection, 76
    scattering, 76, 77
    shadowing, 77, 128
multiple access, 79–86, 143, 211
    CDMA, 82, 86, 90, 92, 148, 152, 166, 167, 181
    FDMA, 80, 81, 90, 92, 152
    SDMA, 80
    TDMA, 81, 82, 90, 92, 144
multiplexing, 79–86
    CDM, 82–86
    FDM, 80–81
    SDM, 80
    TDM, 81–82
multislot operation, 188, 190

NACE, 329
NAD-25, 27, 34
NAICS, 329
NAVSTAR, 156
NDC, 105
near–far effect, 215

neighbor cell, 197
neighbor channel interference, 187
NENA, 289
network model, 41
neural network, 142
Newton's laws of motion, 158
Newton, Sir Isaac, 23, 158
NGS, 34
Nibble, 239
NIMA, 27, 33, 34
NMT, 90
Node B, 96
noise, 64
normal burst, 198
northing, 32, 151
Nyquist, Harry, 67
Nyquist theorem, 67

observable, 123, 138, 142
observable pattern, 255
OBU, 5, 9, 10, 128, 141, 254
OCXO, see clock, 188
odometer, 141
OGC, 59, 294
    simple features, 59
OMA, 294
OMC, 94
OMS, 94
OpenLS, 316, 317, 322–336
    core service, 328–335
    directory service, 323, 328–330
    gateway service, 323, 330
    geocoder service, 323
    information model, 326–327
    presentation service, 324, 331–333
    reverse geocoder service, 323
    route service, 323, 333–335
    utility service, 330–331
operating costs, 126, 127
orbit, 157, 169
    altitude, 157, 164
    circular, 157, 159
    elliptical, 157, 160
    GEO, 157
    HEO, 160
    inclined, 157
    LEO, 157

    MEO, 157, 164
    plane, 157
    polar, 157
    shape, 157
orthogonality, see spread spectrum, 213
orthometric height, 26
oscillator, 146
OSI, 301
OTA, 302
OTD, 196, 202, 204
OTD measurement, 197
OTD measurements, 202–205, 221
OTDoA-IPDL, 211, 216, 217, 220–227, 230, 231, 278
oven, 147
overhead
    signaling, 125
OVSF, see spread spectrum, 213

P-CCPCH, 217
P-TMSI, 111, 116
P3P, 264
packet encapsulation, 110
packet polling procedure, 211
packet session, 112
packet-switched, 95, 97, 112
paging, 99–102, 106, 111, 113
paging area, 101
PAP, 302, 305
parallel, see latitude
Parlay/OSA, 305–307, 340
    interactive emergency location request, 307
    interactive request, 306
    Mobility SCF, 313
    network-induced emergency location report, 307
    periodic request, 307
    SCF, 306
    triggered request, 307
    user location service, 306
path loss, 75, 148
path-loss gradient, 76, see attenuation
pattern matching
    see positioning, 142
PCE, 313

INDEX

PCH, 190
PCP, 300, 313
PDC, 90
PDE, 229
PDN, 95
PDP
    address, 111
    context, 115, 116
    type, 111
PDP address, 110
PDTCH, 190
PDU, 111
perigee, 157, 160, 162
    argument of, 161
permittivity, 74
personal mobility, 98–99, 104
personalization, 324
perturbing forces, 174
phase, 62
piggybacking, 256, 257, 304
pilot, *see* signal
PLMN, 284
PMD, 287
POI, 274
point, 300
point of interest, 4, 35, 46, 47, 52, 316
point-in-polygon check, 40, 54
policy holder, 265
policy repository, 264
polygon, 300
portal, 324
position, 18
    initial, 294
    last known, 294
    updated, 294
position estimate, 140
position fix, 125, 131, 133, 134, 148, 251, 256
position measurement establishment procedure, 279
position originator, 250, 251
Positioning, 99, 123–154, 271
    accuracy, 148–151
    angulation, 123, 138–140, 143
    AoA, 138, 152

Cell-ID, 131
cellular, 185–232
CGI, 131
CoO, 131
dead reckoning, 140–141, 256
DoA, 138
error potential, 133, 137, 139
error sources, 151–153
fingerprinting, 142, 152, 235
hybrid, 142–143
indoor, 233–245
infrastructure, 124
    cellular, 125, 129
    indoor, 125, 129
    integrated, 125, 127
    satellite, 125, 128
    stand-alone, 125, 127
lateration, 131–138, 143, 187, 188, 211, 235
    circular, 123, 132–136, 157, 165, 173, 194
    hyperbolic, 123, 136–138, 143, 186, 194, 208, 230
network-assisted, 128
network-based, 128, 131, 256, 278, 292
non radiolocation, 123
pattern matching, 142
precision, 148–151
proximity sensing, 130–131, 143, 186, 211, 235
radiolocation, 123
satellite, 155–184
TDoA, 145
terminal-assisted, 128, 203
terminal-based, 128, 131, 165, 203, 256, 278, 292
ToA, 136, 144, 157
positioning measurement establishment procedure, 281
POTS, 93
power consumption, 126
power control, 215
power spectral density, 84
PPR, 287, 300, 317
precision, 125

predicate, 49
    direction, 49
    metric, 49
    topological, 49–50
presence, 4
presence information, 307
Prime Meridian, 21, 23
privacy, 257–269, 272, 339
    accuracy constraints, 263
    actor constraints, 262, 274, 283
    class
        call/session-related, 284
        call/session-unrelated, 284
        PLMN operator, 284
        universal, 284
    constraint, 262, 274
    identity constraints, 263
    location constraints, 263
    notification constraints, 263, 274, 283
    option, 274, 281
    options, 283–287
    policy, 260–262
    service constraints, 263, 286
    time constraints, 263
projected surface, 27
projection, 19
    azimuthal, 29
    conformal, 29–31
    conical, 28
    cylindrical, 28, 30
    equal-area, 29
    equidistant, 29
    Mercator, 30–31
    planar, 28
    Transversal Mercator, 31
    Transverse Mercator, 31
    UTM, 31–33
projection surface, 27
proximity sensing, *see* positioning
PSAP, 8, 289, 290, 293
pseudo angle, 139
pseudo range, 135
pseudonym, 266
pseudorange, 133
PSK, *see* modulation, 180

public-private partnership, 7
pull proxy, 301
pulse ranging, 144
push proxy, 305
Pythagoras, 23
Pythagoras theorem, 132

QAM, *see* modulation
QoS, 111, 274, 277, 281
QPSK, *see* modulation, 212
query, 50–53
querying, 254–255, 308

RA update, 116–117
RADAR, 142, 239
radio pattern, 70, 75
radius
    equatorial, 24
    polar, 24
RAI, 116
range measurements, 143–148, 172–173
raster attribute, 39
raster mode, 37–39
reference AT change, 199
reference cell, 197
reference ellipsoid, 21, 23
reference point, 249
reference system, 19, 124, 272
reflection, 74, *see* multipath propagation
refraction, 72–74, 152, 172, 174, 177
refraction index, *see* medium
Remote Procedure Call, 254
reporting, 254–257, 308
    distance-based, 256, 257
    immediate, 255
    periodic, 256, 257
    zone-based, 256, 257, 287
requestor, 286, 297
reserve, 260
reverse geocoding, 35, 330
RFID, 131, 233, 239–240, 244
right ascension of the ascending node, 161
right ascension system, 162
RIT measurements, 197–201, 203, 206, 221
RLP, 300

INDEX

RM-ODP, 315
RMI, 316
RMS, 149
RNC, 96, 111, 217, 226
RNS, 96
roaming, 90, 93, 99, 108, 287
role, 249
roll-out costs, 126, 127
round trip time, 143
routing area, 114, 259
routing-area update, 115, 116
RRC protocol, 118
RSS, 143, 148, 234
RTD, 196, 199, 202, 204
RTT, 186, 189, 217, 218, 220, 229
Rx timing deviation, 211, 217

SAS, 217
scattering, *see* multipath propagation
scene analysis, *see* pattern matching
scrambling code, 213
SDH, 94
SDM, 72, *see* multiplexing
SDMA, *see* multiple access
selective routing, 7
Selective Routing Database, 290, 291
semimajor axis, 24, 160
semiminor axis, 24
service area, 103
session management, 324
SFN, 212, 225
SFN–SFN observed time difference, 217, 221
SGSN, 96, 109, 110, 116, 273, 275, 278, 283
    address, 110
shadowing, *see* multipath propagation
Shannon, Claude Elwood, 67
shift keying, *see* modulation
SIC, 329
side lobe, 81
sidereal day, 164
signal, 61–63
    analog, 62
    bandwidth, 66–68
    digital, 62
    effective bandwidth, 67
    filter, 67
    frequency domain, 65
    infrared, 69, 123
    microwave, 69
    noise, 67
    pilot, 123, 130, 144, 165, 186
    propagation, 68
    radio, 69, 123
    side lobe, 67
    spectrum, 66–70
    speed, 63, 72–75
    time domain, 64, 65
    ultrasound, 123
signal branch, 214
signaling, 95
    broadcast, 202, 203, 207, 208, 226
    CAS, 289, 291
    computational, 125
    NCAS, 290, 292
    point-to-point, 202, 203, 206, 208, 226
SIM, 93, 99, 105
SIP, 307, 312
SLPP, 283
Smallworld, 36
SMLC, 190, 194, 196, 198, 200, 202, 203, 206, 209, 217, 226, 229, 261, 275
SMS, 1, 301, 337
SMS-C, 325
SMSS, 93, 95
SN, 105
SNDCP, 111
SNR, 67, 234
SOAP, 295, 325
solar day, 164
solitude, 260
SONET, 94
spaghetti model, 41
spatial component, 36
spatial data model, 37
spatial database, 35, 59, 89, 323, 325, 340
spatial object, 37, 39
    arc, 41
    line, 39

spatial object (*continued*)
    node, 41
    point, 39
    polygon, 41
    polyline, 39, 41
spatial–temporal database, 59
speed of light, 72
spread spectrum,
    autocorrelation, 84–86, 145
    cross-correlation, 84–86, 167
    Gold code, 86, 167
    m-sequence, 86
    orthogonality, 86
    OVSF, 86
    spreading code, 84, 92, 145, 165
    spreading factor, 84
    Walsh code, 86
spreading code, *see* spread spectrum
spreading factor, *see* spread spectrum, 212
Sputnik, 155
SQL, 43, 52, 59
SRNC, 214, 219
SS7, 95, 109, 257
SSL, 262, 295
standard deviation, 149
stratosphere, 73
superframe, 187
supplementary service, 1, 285
supplier, 251, 254
supply chain, 249–254, 258, 302, 304, 308
surveying, 163, 180
Symbian, 339
synchronization burst, 198
System Information List, 199

T3124 time, 210
tail bits, 188
target, 250, 251, 258
Taylor series, 133
Taylor series expansion, 137, 140
TCH, 190
TCXO, *see* clock
TDD, 82
TDM, *see* multiplexing

TDMA, *see* multiple access, 186
TDMA frame, 187, 188, 190, 199, 204
TDoA, 138
terminal mobility, 97–99, 104
tessellation, 38
theme, 45–46
Tigris, 36
time offset, *see* clock
time slot, 81, 187, 188
timing advance, 186, 188–190, 192–194
timing measurement, 136, 194
TLS, 262, 295
TMSI, 105–107, 109, 259, 260, 266
ToA, *see* positioning
topological model, 41
topological relationship, 41–43, 45
tourist guide, 257
training, 236
Transit, 155
transmission medium, *see* medium
transparency, 315
    positioning, 316
Transverse Mercator projection, *see* projection
troposphere, 73, 152
tropospheric delay, 175
true anomaly, 162
TTFF, 126, 225, 231, 321
tunneling, 110–111
two-location algorithm, 102

U-TDoA, 186, 187, 191, 208–211, 230, 231, 253, 269, 287, 339
ubiquitous computing, 129
UE, 96
UMTS, 89, 90, 92, 96–97, 117, 229, 338
    air interface, 211–217
    positioning components, 217
universal gravitational constant, 159
URA, 114, 220
user segment, 163
UTC, 23, 169
UTM, *see* projection, 272
UTRAN, 96, 211, 275, 278
    FDD, 212, 218, 220, 221, 223
    TDD, 212, 218, 220, 221, 223

vector mode, 37, 39–41
Vernal Equinox, 162
VLR, 94, 96, 103, 105, 107, 273, 283
VoIP, 311
VPN, 110

W3C, 264
Walsh code, *see* spread spectrum
WAP, 1, 301, 313, 338
    attachment service, 304
    deferred query service, 303
    immediate query service, 303
    Location Attachment Functionality, 303
    Location Framework, 301–306, 340
    Location Query Functionality, 303
    pull proxy, 325
wavelength, 63, 75, 173
WCDMA, 211
WGS-84, 25, 27, 34, 174, 272, 299, 307

WhereMops, 239
WIPS, 131, 242
WLAN, 5, 130, 233–234, 338
    active scanning, 235
    fingerprinting, 236–239
    passive scanning, 234
WLAN fingerprinting, 233, 244, 340
    deterministic, 238
    empirical approach, 238
    modeling approach, 238
    probabilistic, 238
WML, 301
WSDL, 306
WSP, 295, 301, 302
WTP, 301

X.25, 96
XML schema, 325

yield, 125